AF411327

The Amazing World of IDPs in Human Diseases

The Amazing World of IDPs in Human Diseases

Editors

Simona Maria Monti
Giuseppina De Simone
Emma Langella

MDPI • Basel • Beijing • Wuhan • Barcelona • Belgrade • Manchester • Tokyo • Cluj • Tianjin

Editors
Simona Maria Monti
Biostructures and Bioimaging
Institute-CNR
Italy

Giuseppina De Simone
Biostructures and Bioimaging
Institute-CNR
Italy

Emma Langella
Biostructures and Bioimaging
Institute-CNR
Italy

Editorial Office
MDPI
St. Alban-Anlage 66
4052 Basel, Switzerland

This is a reprint of articles from the Special Issue published online in the open access journal *Biomolecules* (ISSN 2218-273X) (available at: https://www.mdpi.com/journal/water/special_issues/hydraulics_numerical_methods).

For citation purposes, cite each article independently as indicated on the article page online and as indicated below:

LastName, A.A.; LastName, B.B.; LastName, C.C. Article Title. *Journal Name* **Year**, *Volume Number*, Page Range.

ISBN 978-3-0365-1028-6 (Hbk)
ISBN 978-3-0365-1029-3 (PDF)

Contents

About the Editors

Simona Maria Monti PhD, Senior Researcher at CNR. The scientific activity of Dr SM Monti is focused on chemical biology investigations concerning structural and functional characterization of proteins and bioactive peptides involved in biomedical pathways and human disease. An important part of Dr Monti's research is focused on the study of Carbonic Anhydrases, which are ubiquitous metallo-enzymes that catalyze the inter conversion of the carbon dioxide into bicarbonate ion. In humans, Carbonic Anhydrases are widely distributed and play important roles in a variety of physio/pathological processes. Indeed, abnormal levels/activities of Carbonic Anhydrases are associated to different pathologies such as glaucoma, neurological disorders and cancer. Recently Dr Monti extended her topics to intrinsically disordered proteins and their biomedical involvement. Dr Monti has two international patents and two more have been recently deposited. Dr. Monti is responsible for many national and international research projects, collaborating with a wide range of scientists and institutions; she is author of more than 110 indexed publications among which are Science, PNAS, and JACS and has a h-index of 37 (Google Scholar).

Giuseppina De Simone PhD, Research Director at CNR. The research activities of Dr. Giuseppina De Simone are focused on several aspects of structure/function relationships in macromolecules with a biological interest. These studies have been dedicated to a variety of macromolecular systems with different structural complexities and have been carried out by combining X-ray crystallographic methods with several other techniques, including enzyme expression and purification, enzyme functional characterization, molecular modelling, kinetic and spectroscopic analysis. Dr. De Simone is author of more than 150 scientific publications on international journals among which PNAS, Nat Cell Biol, JACS, Chem. Rev and has a h-index of 46 (Google Scholar). She has been the Principal Investigator of several research grants from national and international institutions. Currently, Dr. De Simone is associate editor of the Journal of Enzyme inhibition and Medicinal Chemistry.

Emma Langella PhD, Researcher at CNR. Dr. Emma Langella graduated in Chemistry (summa cum laude) at the University of Naples "Federico II" in 2000 and received her Ph.D. in Chemical Sciences from the same University in 2003. In 2002 she was visiting scientist at University of Rome "La Sapienza". Since 2011, she is researcher at the Institute of Biostructures and Bioimaging in Naples. The research activity of Dr. Langella is mainly focused on the in silico study of biological systems involved into physiological and pathological processes. Her main research areas include: (i) application of molecular dynamics simulation techniques to proteins, protein complexes, macromolecules and oligonucleotides to gain insights into macromolecular conformational preferences in atomic detail, (ii) molecular docking and theoretical binding free energy calculations for computer-aided drug design. Recent scientific activities concern the study of macromolecular targets involved in several human diseases (Carbonic Anhydrases, beta-Amyloid peptide, Prion Protein). Dr. Langella is involved into many national and international research projects.

Editorial

The Amazing World of IDPs in Human Diseases

Simona Maria Monti *, Giuseppina De Simone and Emma Langella

Institute of Biostructures and Bioimaging, CNR, via Mezzocannone 16, I-80134 Naples, Italy;
gdesimon@unina.it (G.D.S.); emma.langella@cnr.it (E.L.)
* Correspondence: marmonti@unina.it

Citation: Monti, S.M.; De Simone, G.;
Langella, E. The Amazing World of
IDPs in Human Diseases. *Biomolecules*
2021, *11*, 333. https://doi.org/
10.3390/biom11020333

Received: 1 February 2021
Accepted: 4 February 2021
Published: 23 February 2021

Publisher's Note: MDPI stays neutral
with regard to jurisdictional claims in
published maps and institutional affiliations.

It has been clearly established that some proteins or protein regions are devoid of any stable secondary and/or tertiary structure under physiological conditions, but still possess fundamental biological functions [1]. These intrinsically disordered proteins (IDPs) or regions (IDRs) have peculiar features due to their plasticity, such as the capacity for binding their biological targets with high specificity and low affinity, and the possibility of interaction with numerous partners [2,3]. IDPs and IDRs are especially prevalent in eukaryotes, suggesting that disorder in nucleated cells is associated with many key functions, such as signaling and regulation. However, a correlation between intrinsic disorder and various human diseases such as cancer, diabetes, amyloidosis, and neurodegenerative diseases is also evident, highlighting the importance of this topic [4,5]. For the present Special Issue, we have invited researchers to contribute with original research articles, as well as reviews, on the amazing world of the IDPs or IDRs involved with human diseases. We have brought together an internationally recognized team of researchers who work in this field. The contributing authors have presented important and novel aspects of disorder, either correlated with specific medical diseases, or aimed to increase the basic understanding of intrinsic disorder.

The work presented by Tsvetkov and colleagues investigated the degradation of IDPs via a unique class of 26S proteasome that is free of ATP [6]. The authors found that NADH-stabilized 26S proteasome complexes promote the efficient degradation of many IDPs that might not require ATP-dependent unfolding, such as p27, Tau, c-Fos, and more. This interesting finding exemplifies a new principle of how mitochondria, with a key role in NADH production, might be involved in IDP/IDR homeostasis.

Ortega-Alarcon and co-workers presented an investigation on MeCP2 [7], which is an intrinsically disordered multi-domain protein and a potential pharmacological target associated with Rett syndrome (RTT). The authors report an in-depth biophysical study of two mutant variants of MeCP2 associated with RTT using different protein constructions in order to evaluate the effect of the protein-disordered regions on structural stability, conformation, and DNA binding ability. The results obtained lead to a more general reflection on the molecular context-dependent effects induced by mutations in proteins.

The interaction between the intrinsically disordered NUPR1 and the human importinα3 (Impα3) was investigated in depth by Neira et al. [8] using several spectroscopic and biophysical techniques, including NMR and molecular docking. The authors focused on the affinity of the Nuclear Localization Sequence (NLS) of NUPR1 towards Impα3, taking into account several mutants of the NLS region, and demonstrated that the phosphorylation of Thr68 induces a conformational switch in the NLS region of NUPR1 which hampers binding to Impα3. From a more general point of view, the study allows the detection of key residues able to modulate NUPR1 interactions with its partners.

Pajkos and colleagues carried out an interesting in silico analysis concerning the evolutionary origin of intrinsically disordered regions that are specifically targeted in cancer [9]. The authors focused on a subset of cancer genes belonging to the class of IDPs and used a novel conservation and phylogenetic-based strategy. By means of several case studies, the authors conclude that the disordered cancer risk regions showed remarkable

1

conservation with ancient evolutionary origin, highlighting their importance in biological processes.

The paper by Wong et al. analyzes the enrichment patterns of missense mutation-causing single nucleotide variants (SNVs) that are associated with disease and cancer, as well as those present in the healthy population, in protein–protein complexes showing IDR-mediated interactions [10]. Notably, data analysis indicates a strong enrichment at the interface core of interacting IDRs in disease mutations and its depletion in neutral ones, thus supporting the disruption of IDR interactions as a common mechanism for many diseases. In conclusion, the authors highlight the importance of understanding and predicting the effect of missense mutations on disease susceptibility.

The investigation of *Helicobacter pylori* UreG structural dynamics in solution was carried out by Pierro and colleagues [11], who reported the effects of physiological cofactors Ni(II) and GTP on protein mobility by using techniques such as isothermal titration calorimetry and site-directed spin labeling coupled to electron paramagnetic spectroscopy. The obtained results, which showed that the concomitant addition of both Nickel(II) and GTP induces a modification of the structure and mobility in two regions of the protein, may provide perspectives for future research on molecules with anti-bacterial activities to overcome anti-microbial resistance (AMR).

An interesting study was conducted by Gadhave and colleagues on the disorder content of the ubiquitin proteasome system (UPS) [12], which plays a key role in the pathogenesis of various types of cancers and neurodegenerative diseases. By means of five different IDP prediction tools, authors classified the disease-associated UPS proteins in highly ordered, moderately disordered, and highly disordered proteins. Concurrently, multiple post-translational modification sites were identified, mainly located in the disordered regions of proteins. Since these proteins interact with their biological partners for the normal functioning of protein homeostasis, a complete elucidation of the roles of the identified IDPRs and disorder-based binding regions in the pathogenesis of diseases is of great importance for biomedical research.

Finally, two reviews enriched our Special Issue with a wide and extensive discussion on certain topics [13,14]. The review by Sokolik and co-workers gave a structural overview on the fascinating WASp-interacting protein (WIP), which is a regulator of actin cytoskeleton assembly and remodeling, a cellular multi-tasker, and a key member of a network of protein–protein interactions [13]. The authors provided a deeper understanding of the mechanisms by which WIP mediates its biological functions, which have an impact on health and disease, paving the way for a better understanding of key biological processes with potential therapeutic implications. Last, but not least, Kim and coworkers overviewed the features revealed by NMR spectroscopy of monomeric alpha-synuclein, for which the aggregation is strongly correlated with Parkinson's disease [14].

In conclusion, in this Special Issue, examples of recent progress in the IDPs/IDRs have been reported with the aim to contribute to advancing the field of the intrinsic disorder issue in human disease and encourage further research on the topic.

Conflicts of Interest: The authors declare no conflict of interest.

References

1. Uversky, V.N.; Oldfield, C.J.; Midic, U.; Xie, H.M.; Xue, B.; Vucetic, S.; Iakoucheva, L.M.; Obradovic, Z.; Dunker, A.K. Unfoldomics of human diseases: Linking protein intrinsic disorder with diseases. *BMC Genom.* **2009**, *10*, S7–S17. [CrossRef] [PubMed]
2. Oldfield, C.J.; Cheng, Y.; Cortese, M.S.; Brown, C.J.; Uversky, V.N.; Dunker, A.K. Comparing and combining predictors of mostly disordered proteins. *Biochemistry* **2005**, *44*, 1989–2000. [CrossRef]
3. Langella, E.; Buonanno, M.; Vullo, D.; Dathan, N.; Leone, M.; Supuran, C.; De Simone, G.; Monti, S.M. Biochemical, biophysical and molecular dynamics studies on the proteoglycan-like domain of carbonic anhydrase IX. *Cell. Mol. Life Sci.* **2018**, *75*, 3283–3296. [CrossRef] [PubMed]
4. Langella, E.; Buonanno, M.; De Simone, G.; Monti, S.M. Intrinsically disordered features of carbonic anhydrase IX proteoglycan-like domain. *Cell. Mol. Life Sci.* **2020**. [CrossRef]

5. Uversky, V.N.; Oldfield, C.J.; Dunker, A.K. Intrinsically disordered proteins in human diseases: Introducing the D2 concept. *Annu. Rev. Biophys.* **2008**, *37*, 215–246. [CrossRef]
6. Tsvetkov, P.; Myers, N.; Adler, J.; Shaul, Y. Degradation of intrinsically disordered proteins by the NADH 26S proteasome. *Biomolecules* **2020**, *10*, 1642. [CrossRef] [PubMed]
7. Ortega-Alarcon, D.; Claveria-Gimeno, R.; Vega, S.; Jorge-Torres, O.C.; Esteller, M.; Abian, O.; Velázquez-Campoy, A. Molecular context-dependent effects induced by rett syndrome-associated mutations in MeCP2. *Biomolecules* **2020**, *10*, 1533. [CrossRef] [PubMed]
8. Neira, J.L.; Rizzuti, B.; Jiménez-Alesanco, A.; Palomino-Schätzlein, M.; Abian, O.; Velázquez-Campoy, A.; Iovanna, J. A phosphorylation-induced switch in the nuclear localization sequence of the intrinsically disordered NUPR1 hampers binding to importin. *Biomolecules* **2020**, *10*, 1313. [CrossRef] [PubMed]
9. Pajkos, M.; Zeke, A.; Dosztányi, Z. Ancient evolutionary origin of intrinsically disordered cancer risk regions. *Biomolecules* **2020**, *10*, 1115. [CrossRef]
10. Wong, E.T.C.; So, V.; Guron, M.; Kuechler, E.R.; Malhis, N.; Bui, J.M.; Gsponer, J. Protein–protein interactions mediated by intrinsically disordered protein regions are enriched in missense mutations. *Biomolecules* **2020**, *10*, 1097. [CrossRef]
11. Pierro, A.; Etienne, E.; Gerbaud, G.; Guigliarelli, B.; Ciurli, S.; Belle, V.; Zambelli, B.; Mileo, E. Nickel and GTP modulate *Helicobacter pylori* UreG structural flexibility. *Biomolecules* **2020**, *10*, 1062. [CrossRef]
12. Gadhave, K.; Kumar, P.; Kapuganti, S.K.; Uversky, V.N.; Giri, R. Unstructured biology of proteins from ubiquitin-proteasome system: Roles in cancer and neurodegenerative diseases. *Biomolecules* **2020**, *10*, 796. [CrossRef]
13. Sokolik, C.G.; Qassem, N.; Chill, J.H. The disordered cellular multi-tasker WIP and its protein–protein interactions: A structural view. *Biomolecules* **2020**, *10*, 1084. [CrossRef] [PubMed]
14. Kim, D.-H.; Lee, J.; Mok, K.H.; Lee, J.H.; Han, K.-H. Salient features of monomeric alpha-synuclein revealed by NMR spectroscopy. *Biomolecules* **2020**, *10*, 428. [CrossRef] [PubMed]

 *biomolecules*

Article

Unstructured Biology of Proteins from Ubiquitin-Proteasome System: Roles in Cancer and Neurodegenerative Diseases

Kundlik Gadhave [1], Prateek Kumar [1], Shivani K. Kapuganti [1], Vladimir N. Uversky [2,3] and Rajanish Giri [1,*]

[1] School of Basic Sciences, Indian Institute of Technology Mandi, VPO Kamand, Himachal Pradesh 175005, India; kundlikgadhave264@gmail.com (K.G.); kumar.prateek3@yahoo.com (P.K.); k.shivanikrishna@gmail.com (S.K.K.)

[2] Department of Molecular Medicine and Byrd Alzheimer's Research Institute, Morsani College of Medicine, University of South Florida, Tampa, FL 33620, USA; vuversky@usf.edu

[3] Institute for Biological Instrumentation of the Russian Academy of Sciences, Federal Research Center "Pushchino Cientific Center for Biological Research of the Russian Academy of Sciences", Pushchino, 142290 Moscow, Russia

* Correspondence: rajanishgiri@iitmandi.ac.in; Tel.: +1-905-267134; Fax: +1-905-267138

Received: 16 April 2020; Accepted: 19 May 2020; Published: 21 May 2020

Abstract: The 26S proteasome is a large (~2.5 MDa) protein complex consisting of at least 33 different subunits and many other components, which form the ubiquitin proteasomal system (UPS), an ATP-dependent protein degradation system in the cell. UPS serves as an essential component of the cellular protein surveillance machinery, and its dysfunction leads to cancer, neurodegenerative and immunological disorders. Importantly, the functions and regulations of proteins are governed by the combination of ordered regions, intrinsically disordered protein regions (IDPRs) and molecular recognition features (MoRFs). The structure–function relationships of UPS components have not been identified completely; therefore, in this study, we have carried out the functional intrinsic disorder and MoRF analysis for potential neurodegenerative disease and anti-cancer targets of this pathway. Our report represents the presence of significant intrinsic disorder and disorder-based binding regions in several UPS proteins, such as extraproteasomal polyubiquitin receptors (UBQLN1 and UBQLN2), proteasome-associated polyubiquitin receptors (ADRM1 and PSMD4), deubiquitinating enzymes (DUBs) (ATXN3 and USP14), and ubiquitinating enzymes (E2 (UBE2R2) and E3 (STUB1) enzyme). We believe this study will have implications for the conformation-specific roles of different regions of these proteins. This will lead to a better understanding of the molecular basis of UPS-associated diseases.

Keywords: ubiquitin-proteasome system; intrinsically disordered proteins; protein misfolding; molecular recognition features; cancer; neurodegenerative diseases; protein degradation

---

1. Introduction

Before 1970, lysosomes were thought of as exclusive cellular machinery to execute misfolded protein degradation. However, in 1977, work by Etlinger et al. reported the presence of a second intracellular ATP-dependent mechanism for degradation of proteins [1]. Later, in 1979 and the early 1980s, ATP and the ubiquitin-dependent protein degradation system was discovered by Avram Hershko, Aaron Ciechanover, and Irwin Rose [2–4]. This system is currently known as ubiquitin proteasomal system (UPS), and this work earned them the Nobel Prize in Chemistry (2004) [2–4]. A protein quality control (PQC) system present in eukaryotic cells is believed to be active in lysosome, UPS, autophagy, and endoplasmic reticulum (ER). In the ER, heat shock proteins (HSPs) bind to misfolded proteins and mediate their refolding. However, misfolded proteins that failed refolding are transported into the cytosol, where they are targeted to UPS or lysosome for their degradation [5,6]. Importantly, UPS plays a vital role in DNA repair, cell growth, immune function, cell-cycle regulation, and numerous non-proteolytic functions, including regulation of histone modification and involvement in vesicular trafficking pathways, and the deregulation in any component of UPS has been associated with several diseases [7–9].

In UPS, the misfolded proteins get ubiquitinated by a cascade of enzymes, including ubiquitin-activating E1s as well as ubiquitin-conjugating E2s and E3 ubiquitin ligases (Figure 1). The ubiquitinated substrates are then transferred to the 26S proteasome for degradation [10,11]. 26S proteasome (~2.5 MDa) is the major protease inside the cell, and has two sub-complexes: the 20S core particle (CP), which is responsible for the main proteolytic activity of the proteasome, and the 19S regulatory particle (RP), which helps in the unfolding of ubiquitinated proteins, their subsequent de-ubiquitination, and their translocation into the CP cavity [12]. The specific lysine (K) residues are required for the conjugating of ubiquitin (Ub) to the substrate proteins, including K6, K11, K27, K29, K33, K48, K63, or Met1 [10,13]. This process results in the generation of either monoubiquitinated, multi-monoubiquitinated, or polyubiquitinated proteins, with these different modes of ubiquitylation determining the fate of the target protein—i.e., degradation of a protein via the 26S proteasome, targeting it to a lysosome or prompting it for participation in other cellular processes [13,14]. Ubiquitin (Ub), a 76-amino-acid ubiquitously distributed polypeptide, is required to tag target proteins for proteasome-mediated degradation [2,15]. Mutations in the *UBB* gene encoding ubiquitin-B and molecular misreading of this gene that introduces dinucleotide deletions (e.g., ΔGA, ΔGU) can generate mutated Ub forms, which are associated with human diseases. For example, the UBB+1 (Ubiquitin-B+1) form of Ub generated as a result of molecular misreading is linked to Alzheimer's disease (AD), other tauopathies, and polyglutamine (PolyQ) diseases (e.g., Huntington's disease (HD)) [16–18], with the resulting UBB+1 form being shown to inhibit proteasomal proteolysis [19]. These UBB+1 mutants were found with Aβ accumulations in Alzheimer's and Down syndrome patients [18]. Ub-activating (E1) enzyme catalyzes the first step of ubiquitin activation in the ubiquitination process, where it binds to Ub and transfers it to E2 enzyme [20]. Missense mutations in the *UBA1* gene lead to X-linked spinal muscular atrophy (SMAX2), and reduced UBA1 levels affect UPS-mediated degradation of misfolded proteins leading to neurodegenerative diseases, such as AD [21]. Ubiquitin-conjugating (E2) enzyme catalyzes the second step of ubiquitination, where it accepts Ub from E1 enzyme and transfer it to substrate protein via E3 enzyme [22]. Studies have shown that the impairments of the E2 enzymes or mutations in these proteins are associated with many diseases, such as cancer and neurodegenerative diseases [23]. Ubiquitin-protein ligase (E3) enzyme catalyzes the last step of ubiquitination. E3 binds to a target protein and transfers Ub from the E2 to the target protein. Deregulation of this enzyme is linked to numerous neurodegenerative diseases (AD, Parkinson's disease (PD), Huntingtons disease (HD) and various cancers [24]. Ubiquilins are functionally linked to UPS, where they act as ubiquitin receptors [25]. The human genome encodes four ubiquilin genes, *UBQLN1*, *UBQLN2*, *UBQLN3*, and *UBQLN4*, which encode structurally related and conserved proteins. A fifth ubiquilin gene, UBQLNL, was later identified in humans. Although all ubiquilin family members are present in cytosol, they exert different tissue expression patterns [25]. Ubiquilin 1 is expressed ubiquitously and binds numerous cytosolic or transmembrane proteins, and its dysfunction is linked to neurodegenerative diseases such as AD, PolyQ diseases (e.g., HD), and cancer [6]. Ubiquilin 2 is associated with the regulation of pathways involved in protein degradation, such as UPS, the endoplasmic-reticulum-associated protein degradation (ERAD) pathway, and autophagy. Interestingly, the mutation in *UBQLN2* was recently described in familial amyotrophic lateral sclerosis (ALS) [26].

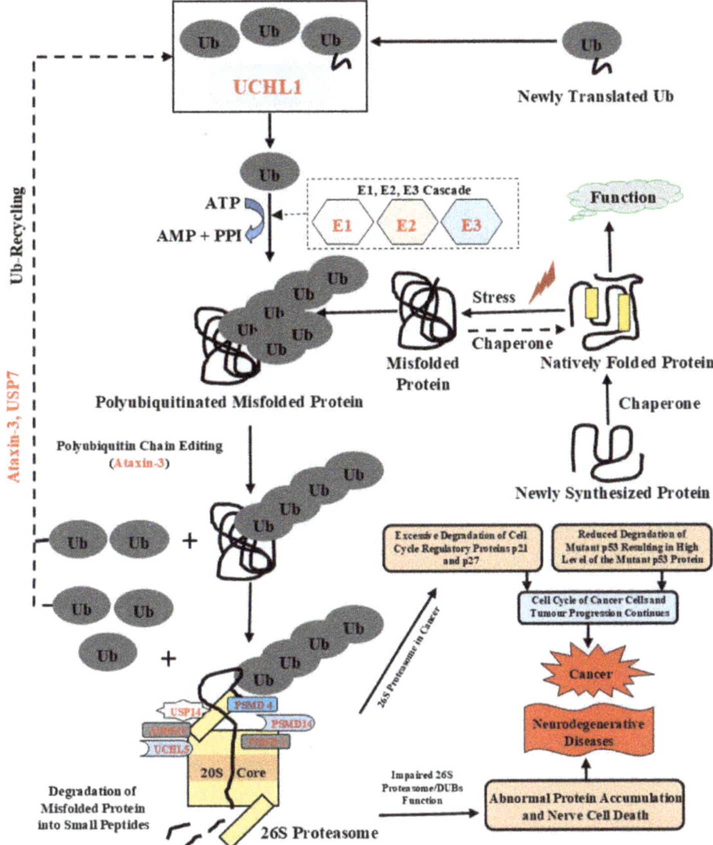

Figure 1. Schematic representation of the ubiquitin proteasomal system. Ubiquitination is an ATP-dependent process performed by three enzymes: E1 (Ub-activating) enzyme, E2 (Ub-conjugating) enzyme, and E3 (Ub-ligase) enzyme. The DUBs, such as ataxin-3, modify the polyubiquitinated chain, to confirm accurate recognition of the misfolded proteins by the 26S proteasome. This covalent modification of misfolded protein targets them to multicatalytic protease complex, the 26S proteasome. Ubiquitination is reversed by DUBs and disassembles polyubiquitin chains. DUBs such as USP7, UCHL1, and ataxin-3 also control and maintain free Ub molecules in the cell. UCHL1 modifies newly translated protein and maintains a pool of mono-Ub. The polyubiquitinated misfolded protein can bind either to the Ub receptor of the 19S regulatory complex or to an adaptor protein that consists of both poly-Ub binding and proteasome binding domain [27]. Once misfolded protein binds to proteasome, the unfolding of the misfolded protein occurs by ATPases followed by removal of the poly-Ub chain by proteasome-associated DUBs and further translocation and degradation of unfolded protein in central proteolytic chamber occurs. Excessive degradation of cell-cycle-regulatory proteins such as p21 and p27 and reduced degradation of mutant p53 leads to a continuous cell cycle of cancer cells and tumor progression leads to the development of cancer [28]. Additionally, impairment in function of 26S proteasome, ubiquitinating enzymes, and DUBs can lead to nerve cell death and the progression of neurodegenerative diseases. Ub: Ubiquitin, E1: Ub-activating enzyme, E2: Ub-conjugating enzyme, E3: Ub-ligase enzyme.

Ubiquitination and deubiquitination are dynamic processes that involve transient protein–protein interactions. There are ~100 deubiquitinating enzymes (DUBs) in the human genome that control several cellular processes in a very dynamic and specific manner. Among these DUB-controlled processes are the progression of the cell cycle, degradation of proteins, apoptosis, activation of kinases, chromosome segregation, gene expression, protein localization, and DNA repair [29]. In the UPS, DUBs are involved in several processes, including de-novo ubiquitin synthesis; ubiquitin precursor processing; cleavage and trimming of polyubiquitin chains; and ubiquitin recycling [30]. Ubiquitin carboxyl-terminal hydrolase isozyme L1 (UCHL1) is a small 223-amino-acid protein, which maintains the pool of mono-Ub required for ubiquitination and is also involved both in the processing of ubiquitin precursors and ubiquitinated protein [31]. The mutation I93M in UCHL1 was reported in PD patients. Furthermore, studies in animal models showed that this mutation led to the inhibition of α-synuclein degradation via the 26S proteasome [32]. Ubiquitin carboxyl-terminal hydrolase isozyme L5 (UCHL5) is a crucial DUB enzyme associated with the 19S regulatory subunit of the 26S proteasome that cleaves the Poly-Ub chain of the target protein [33].

The ubiquitin-specific protease (USP) family of DUBs, with more than 50 members, is the largest family among all DUBs [29]. Ubiquitin carboxyl-terminal hydrolase 7 (USP7) is a 135-kDa DUB enzyme that cleaves ubiquitin from the polyubiquitin chains of target proteins [10,34]. It is associated with many cellular processes, and its dysfunction leads to various pathological conditions, such as cancer, metabolic, and neurological pathologies [10]. Ubiquitin carboxyl-terminal hydrolase 14 (USP14) is a proteasome-associated DUB enzyme, which is activated after the specific association with 26S proteasome and catalyzes the cleavage of ubiquitin subunits from the target protein before its degradation by the proteasome [35,36]. Interestingly, USP14 reduces the degradation of pathogenic/toxic proteins (tau protein, TDP-43, α-synuclein) by the 26S proteasome and is therefore associated with many neurodegenerative diseases, such as AD, ALS, PD, and HD, etc. [36]. 26S proteasome non-ATPase regulatory subunit 14 (PSMD14) is a 310-residue DUB enzyme important for Ub recycling from the proteasome substrate proteins, and its upregulation has been reported in cancer, where it promotes proliferation and migration of cancer cells [37,38]. PolyQ expansion in the C-terminus of ataxin-3 results in conformational changes that further lead to altered subcellular localization, loss of function and binding properties, changed proteolytic cleavage, and aggregation of ataxin-3 [39]. Three receptors of the 26S proteasome in the RP, including Proteasomal ubiquitin receptor ADRM1 (ADRM1), 26S proteasome non-ATPase regulatory subunit 2 (PSMD2), and 26S proteasome non-ATPase regulatory subunit 4 (PSMD4) capture the target protein by binding to Ub and target protein shuttle factors [40,41]. The poly-Ub chains of the target protein are cleaved by the 19S-associated DUBs enzymes (PSMD14, USP14 and UCHL5), and then target protein is unfolded and translocated into the CP for its degradation [41,42]. The pathological conditions associated with the UPS occur due to either gain of function leading to abnormal or enhanced degradation of the target protein or due to loss of function mutations in the enzymes of UPS or in the recognition motif in the target substrates that stabilizes certain proteins [43].

Intrinsically disordered proteins (IDPs) and IDP regions (IDPRs) are the proteins or regions of proteins that lack well-defined, three-dimensional unique structures and show structural transition upon binding with their biological partners [44–49]. IDPs/IDPRs are commonly found in all organisms, being more abundant in eukaryote proteomes [50–52]. Studies have shown that IDPs play a crucial role in protein–protein interaction [53–55]. Our previous study on intrinsic disorder analysis of amyloid cascade signaling of AD reports the presence of abundant intrinsic disorder in most of the proteins [56]. Using bioinformatics analysis, previous studies have reported the presence of intrinsic disorder in ubiquitinating enzymes [57–59]. This paper covers extensive intrinsic disorder and MoRF analysis of 15 proteins, which are decisive for the functioning of UPS, uniquely associated with the clearance of target proteins, and which are important therapeutic targets for cancer as well as neurodegenerative diseases. Numerous mutations in genes that encode proteins involved in the UPS are linked to such diseases. Proteasome inhibitors, such as bortezomib, carfilzomib, and ixazomib, have been approved in 2003, 2012, and 2015, respectively, for the treatment of certain hematological cancers, and some inhibitors are in clinical trials [60]. Several neurodegeneration-related proteins, such as Aβ, Tau, and α-synuclein, are intrinsically disordered, and distorted protein–protein interactions of those proteins are the main process in the disease pathology [61]. Deciphering the roles of intrinsic disorder in UPS proteins in cellular processes, such as protein–protein interaction, protein recognition, protein degradation, and various signaling pathways could provide important knowledge needed for further advances in the development of new drug targets for cancer and neurodegenerative diseases. Therefore, this work may help for the future establishment of new therapeutic routes for cancer and protein misfolding diseases.

2. Materials and Methods

2.1. Retrieval of Sequences and Structures

The 15 proteins of UPS that play a crucial role in this pathway and have been identified in the pathogenesis of human diseases were selected for disorder analysis, which is summarized in Table 1. For sequence-based disorder analysis, the reviewed protein sequences of all proteins have been retrieved in the FASTA format from the UniProt [62] database (Table 1), and their associated crystal structures were fetched from the RCSB PDB database. The crystal structures of some proteins are available in truncated form. The IDPs and MoRF regions in the available structures have been represented in different colors.

Table 1. Proteins involved in the ubiquitin proteasomal system (UPS).

Sr. No.	Protein/Gene Name	Length (Amino Acids)	Function in UPS	Altered Function of Protein in UPS	Involvement in Diseases	UniProt ID	References
1	Ubiquitin-like modifier-activating enzyme 1 (UBA1)	1058	Catalyzes first step in ubiquitination. Binds with Ub, activates it and transfers it to E2 enzyme	Reduced level of UBA1 affects UPS-mediated protein degradation, missense mutations in UBA1 gene lead to SMAX2	Neurological disorders such as AD, SMAX2, and HD	P22314	[21,63,64]
2	Ubiquitin-conjugating enzyme E2 R2 (UBE2R2)	238	Catalyzes second step in ubiquitination. Accepts Ub from E1 enzyme and binds with E3 enzyme	Mutation and dysregulation in UBE2R2 affect UPS function	Dysregulation leads to cancer or neurodegenerative diseases	Q712K3	[23,65]
3	E3 ubiquitin-protein ligase enzyme CHIP (STUB1)	303	Catalyzes final step of ubiquitination. Binds with target protein and transfers Ub from E2 enzyme to target protein.	Deregulation of E3 enzyme affects UPS-mediated degradation process.	Deregulation leads to cancer and neurodegenerative diseases such as AD, PD, and HD	Q9UNE7	[23,24,66]
4	Polyubiquitin-B (UBB)	229	Tags target proteins for proteasomal degradation	Frameshift mutation in ubiquitin-B forms UBB+1, which disturbs UPS-mediated protein degradation	UBB+1 accumulation with Aβ in AD and Down's syndrome	P0CG47	[18,19]
5	Ubiquilin-1 (UBQLN1)	589	Regulates protein degradation through UPS, autophagy, and ERAD	Defects in UBQLN1 lead to perturbed protein degradation via UPS, UBQLN1 downregulation affects APP processing in AD	Cancer, reduced UBQLN1 level found in AD and other neurodegenerative diseases, PolyQ diseases (HD)	Q9UMX0	[6,67]

Table 1. *Cont.*

Sr. No.	Protein/Gene Name	Length (Amino Acids)	Function in UPS	Altered Function of Protein in UPS	Involvement in Diseases	UniProt ID	References
6	Ubiquilin-2 (UBQLN2)	624	Regulates protein degradation via UPS, autophagy, and ERAD	Defects in UBQLN2 lead to perturbed protein degradation, which leads to neurodegenerative diseases	Mutation in UBQLN2 leads to familial amyotrophic lateral sclerosis (ALS)	Q9UHD9	[26,68,69]
7	Ubiquitin carboxyl-terminal hydrolase isozyme L1 (UCHL1)	223	Processing of ubiquitin precursors and ubiquitinated protein. Maintains pool of mono-Ub	Mutation, dysfunction, and downregulation of UCHL1 affects normal UPS function	Cancer and neurodegenerative diseases such as AD and PD	P09936	[11,31,70]
8	Ubiquitin carboxyl-terminal hydrolase isozyme L5 (UCHL5)	329	Proteasome-associated DUB that cleaves 'Lys-48'-linked polyubiquitin chains	Upregulation or downregulation of UCHL5	Oncogenesis	Q9Y5K5	[71]
9	Ubiquitin carboxyl-terminal hydrolase 7 (USP7)	1102	Cleaves Ub from polyubiquitin chains of target protein substrate	Poly-Q repeats, mutation, variation in expression level, and dysfunction in USP7	Dysfunction leads to cancer, metabolic and neurological pathologies	Q93009	[10]
10	Ubiquitin carboxyl-terminal hydrolase 14 (USP14)	494	Proteasome-associated DUB that cleaves Ub from Poly-Ub protein before degradation by the proteasome. Negatively regulates proteasome activity.	USP14 activation inhibits degradation of pathogenic, neurotoxic proteins	Neurodegenerative diseases such as AD, ALS, PD, and HD.	P54578	[36,72]
11	Ataxin-3 (ATXN3)	361	DUB that is involved in polyubiquitin chain trimming	PolyQ expansion in ataxin-3 at its C-terminus	Spinocerebellar Ataxia Type 3 (SCA3)	P54252	[73,74]
12	Proteasomal ubiquitin receptor (ADRM1)	407	Receptor for Ub in RP of 26S proteasome that captures target protein by binding to Ub. It also binds and activates DUB enzyme UCHL5	-	-	Q16186	[75]
13	26S proteasome non-ATPase regulatory subunit 2 (PSMD2)	908	Receptor for Ub in RP of 26S proteasome.	-	-	Q13200	[41]
14	26S proteasome non-ATPase regulatory subunit 4 (PSMD4)	377	Receptor for Ub in RP of 26S proteasome, captures target protein by binding to Ub.	-	-	P55036	[41]

Table 1. *Cont.*

Sr. No.	Protein/Gene Name	Length (Amino Acids)	Function in UPS	Altered Function of Protein in UPS	Involvement in Diseases	UniProt ID	References
15	26S proteasome non-ATPase regulatory subunit 14 (PSMD14)	310	PSMD14 is a 19S-proteasome-associated DUB enzyme that deubiquitinates substrate protein during proteasomal degradation.	Upregulation of PSMD14 leads to dysfunction of UPS	Increase in level of PSMD14 leads to carcinogenesis	O00487	[38,76]

2.2. Identification of Intrinsically Disordered Protein Regions (IDPRs)

Several commonly used disorder predictors, such as PONDR® VSL2 [77], PONDR® VL3 [78], PONDR® VLXT [79], PONDR® FIT [80], and IUPred [81] were utilized for the intrinsic disorder analysis. The mean PPID was calculated for each protein by considering the outputs of five predictors. These predictors use artificial neural networks (ANN) and machine-learning-based algorithms to predict specific disorder regions. The detailed description of the functioning of these predictors has been explained in our previous studies [82–86].

2.3. Molecular Recognition Features (MoRFs) Prediction

The MoRFs are disorder binding motifs that are found in specific disordered regions. These regions were predicted using four different web servers, MoRFCHiBi_Web [87], ANCHOR [88], MoRFpred [89], and DISOPRED3 [90]. Each predictor uses a different data sets and ANN-based models for prediction, which are described in our MoRF-based studies on Zika virus, Chikungunya virus, Rotavirus, SARS-CoV-2 proteomes, and Alzheimer's-disease-associated amyloid cascade signaling proteins [56,83,84,91,92]. Along with these, another web-based predictor, D2P2 [93] has also been used, which predicts disordered regions as well as motifs in proteins.

2.4. Protein–Protein Interaction Using STRING

The functioning of disordered regions from important proteins of UPS are explained in the current study. Furthermore, to explore the interaction among these proteins and with other proteins, they have been analyzed using the STRING v11 database [94]. The database contains experimentally determined information on protein–protein interaction as well as computationally predicted possible interactions, and these are constructed as a map.

2.5. Representation of IDPs and MoRFs

The predicted IDP regions based on mean PPID value and MoRFCHiBi_Web-predicted MoRF regions were shown on available crystal or NMR structures using the Maestro GUI (v11, Schrödinger Inc., Menlo Park, NY, USA). All the color schemes for representing these regions are given in respective figure legends.

3. Results and Discussion

3.1. Intrinsic Disorder in the Proteins of Ubiquitin Proteasomal System

UPS is employed by the eukaryotic cell to get rid of excess, unnecessary, or misfolded short-lived regulatory proteins [95]. Figure 1 represents the complete process of ubiquitination and proteasome-mediated degradation of the proteins. The 26S proteasome complex is assembled symmetrically with the 20S CP at the center flanked by two 19S RPs on either side [12]. The 20S particle has a cylindrical structure and is formed by four stacked heptameric rings comprised of the following subunits: $\alpha_{1–7}$,

$\beta_{1\text{-}7}$, $\beta_{1\text{-}7}$ and $\alpha_{1\text{-}7}$. This is 28 subunits in total. β_1 has caspase-like, β_2 has trypsin-like, and β_5 has chymotrypsin-like peptidase activities [96]. These sites are hidden inside the CP to protect cellular proteins from nonspecific degradation. N-termini of the α subunits form the gates to the cylinder. Six ATPases are present in the RP (700 kDa). RP has 19 subunits that are divided into the base and lid complexes. The RPN11/PSMD14 subunit of the lid has deubiquitinating activity, whereas the AAA-ATPases form a hexameric (trimer of dimers) ring in the base. This ring is formed by RPT1–6 proteins. Apart from these, RPN1/PSMD2, RPN2, RPN10/ PSMD4, and RPN13/ADRM1 also form the base complex. RPN10/PSMD4 and RPN13/ADRM1 act as receptors for ubiquitin, whereas RPN1/PSMD2 and RPN2 form a toroid structure and act as a scaffold by binding to different subunits of the proteasome [97–99].

Here, we looked at the intrinsic disorder predisposition of 15 UPS proteins (Table 2), which play a crucial role in this pathway and human diseases (see Table 1). These proteins include Ub, ubiquitinating enzymes (E1, E2, and E3), DUBs (USP7, Ataxin-3, UCHL5, UCHL1, USP14, and PSMD14), a receptor for Ub in 26S proteasome (ADRM1, PSMD2, and PSMD4), and ubiquilins (UBQLN1 and UBQLN2), which are involved in the pathogenesis of diseases and are listed in Table 1 with their role in the normal functioning of UPS as well as altered function of an individual protein in UPS and further effects.

Table 2. Predicted percentage of intrinsic disorder (PPID) in the proteins of UPS.

Protein	PPID_VSL2	PPID_VL3	PPID_VLXT	PPID_FIT	PPID_IUPRED	PPID_MEAN
UBA1 (E1)	17.86	9.45	22.59	6.14	8.13	9.74
UBE2R2 (E2)	48.74	42.44	42.44	28.15	24.79	37.39
STUB1 (E3)	41.91	37.62	46.86	34.32	19.80	37.95
UBB	30.57	17.03	36.24	7.86	11.35	10.04
UBQLN1	88.96	82.17	62.99	85.06	77.76	87.10
UBQLN2	86.86	73.40	70.99	81.41	70.83	80.93
UCHL1	30.94	11.66	34.98	10.31	3.14	7.62
UCHL5	28.27	22.49	25.53	17.02	10.33	16.11
USP7	23.96	10.89	19.06	11.62	11.25	11.62
USP14	37.25	31.17	26.32	20.85	8.91	23.48
ATXN3	58.17	63.16	55.12	47.65	42.38	53.74
ADRM1	70.52	60.93	62.16	52.83	51.35	61.92
PSMD2	20.37	15.53	27.97	15.86	12.22	13.77
PSMD4	63.93	59.15	64.46	47.75	47.21	55.17
PSMD14	29.68	20.97	22.26	16.45	13.23	18.71

Proteins and their mean PPIDs are colored to reflect their disorder status (ordered—green, moderately disordered—blue, and highly disordered—red).

To obtain a global overview, we looked at the predicted percentage of intrinsic disorder (PPIDs) in these proteins evaluated by PONDR® VLXT and PONDR® VSL2. The results of these analyses are summarized in Figure 2 in the form of the 2D-disorder plot presenting the $PPID_{PONDR\ VLXT}$ vs. $PPID_{PONDR\ VSL2}$ plot. According to the overall levels of intrinsic disorder, proteins can be classified as highly ordered (PPID < 10%), moderately disordered (10% ≤ PPID < 30%) and highly disordered (PPID ≥ 30%) [100]. From this broadly accepted PPID-based classification of proteins and mean PPID obtained from five different IDP prediction tools, UBA1 and UCHL1 are highly ordered; USP14, PSMD14, UCHL5, PSMD2, USP7, and UBB are moderately disordered; and UBQLN1, UBQLN2, ADRM1, PSMD4, ATXN3, STUB1, and UBE2R2 are highly disordered proteins. However, from the 2D disorder plot shown in Figure 2, it is clear that only a few UPS proteins analyzed in this study are moderately disordered proteins, with the remaining members of this set being highly disordered. These results indicate the possible role of intrinsic disorder in the pathogenesis of various proteasome-associated diseases.

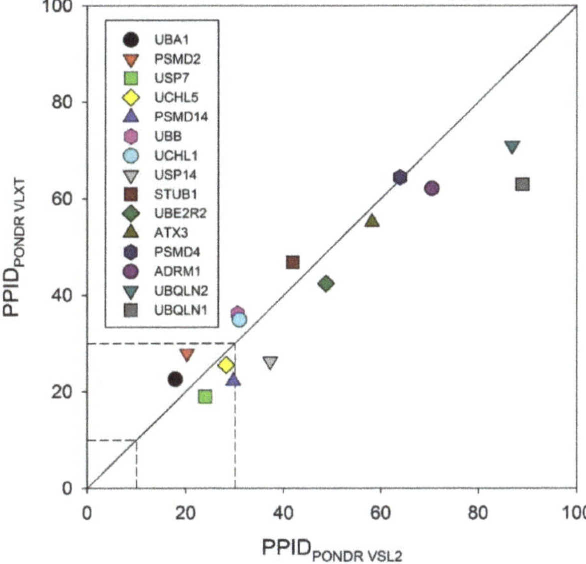

Figure 2. Evaluation of the overall disorder status of 15 UPS proteins associated with human diseases. Here, a 2D disorder plot presents the PPID$_{PONDR\ VLXT}$ vs. PPID$_{PONDR\ VSL2}$ dependence.

We also utilized D^2P^2 database (http://d2p2.pro/) [93] to retrieve additional information on the intrinsic disorder predisposition together with important disorder-related functional information for the members of the set of 15 UPS proteins associated with human diseases. D^2P^2 is a database of predicted disorders for a large library of proteins from completely sequenced genomes. D^2P^2 uses outputs of IUPred [81], PONDR® VLXT [79], PrDOS [101], PONDR® VSL2 [77], PV2 [93], and ESpritz [102]. The output of this database is further enhanced by data related to the location of various posttranslational modifications (PTMs) and predicted disorder-based protein binding sites, known as MoRFs. The D^2P^2-generated functional disorder profiles of 15 UPS proteins are discussed below for individual proteins. Overall, this shows that all these proteins contain noticeable levels of intrinsic disorder, are heavily decorated with various PTMs, and many of them contain multiple MoRFs, suggesting that these proteins are expected to be characterized by high binding promiscuity. Additionally, the disorder-based protein-binding regions/MoRFs for individual proteins, identified by MoRFCHiBi_Web, MoRFpred, DISOPRED3, and ANCHOR, are listed in Table 3. The MoRFs identified by MoRFCHiBi_Web are represented in the available crystal structure of all proteins (see figures of individual proteins).

Next, we analyzed the inter-set intractability of 15 UPS proteins associated with human diseases using a publicly available computational platform, STRING, which integrates extensive information on protein–protein interactions (PPIs), complements it with computational predictions, and returns a PPI network showing all possible PPIs of a query protein(s) [94]. The results of this analysis are represented in Figure 3A, which shows that these proteins are engaged in the formation of a highly interconnected PPI network with 36 edges. In this network, the average node degree is 4.8, and the average local clustering coefficient (which defines how close its neighbors are to being a complete clique; the local clustering coefficient is equal to 1 if every neighbor connected to a given node N_i is also connected to every other node within the neighborhood, and is equal to 0 if no node that is connected to a given node N_i connects to any other node that is connected to N_i) is 0.895. Furthermore, since the expected number of interactions among proteins in a similar size set of proteins randomly selected from human proteome is equal to 5, the inter-set PPI network has significantly more interactions than expected, being characterized by a PPI enrichment p-value of $<10^{-16}$.

Table 3. Identification of MoRF regions for the proteins of UPS.

Protein	MoRFCHiBi_Web	MoRFpred	DISOPRED3	ANCHOR
UBA1 (E1)	1–13, 1048–1057	5–12, 54–60, 423–427, 1051–1058	802–817	1–16, 23–39
UBE2R2 (E2)	166–170, 203–217, 219–226	11–19, 205–213	1–6, 212–238	199–238
STUB1 (E3)	-	-	-	163–169, 198–204, 206–214, 230–239
UBB	40–50, 116–122, 192–202	221–228	-	-
UBQLN1	18–39	34–45, 318–327	1–5, 14–21, 456–474	1–44, 49–54, 72–88, 91–113, 142–168, 192–307, 311–350, 355–496, 507–543
UBQLN2	10–38	30–41, 560–565, 588–592	1–19	1–38, 87–107, 136–161, 193–208, 218–329, 353–377, 398–456, 498–598
UCHL1	-	215–220	-	-
UCHL5	324–328	168–173	1–6, 252–256, 320–329	-
USP7	2–26, 1077–1082, 1090–1102	262–267, 511–516, 1094–1099	1084–1093, 1056–1061, 495–505	1–64
USP14	-	477–482	66–75, 226–232, 489–494	-
ATXN3	56–65, 246–255, 285–290, 312–357	250–254, 282–292, 342–350	1–21, 353–361	215–291, 307–355
ADRM1	21–30	24–28, 399–407	1–19, 385–407	140–202, 208–318, 347–383, 399–407
PSMD2	1–13, 96–102	51–62, 614–618	-	1–30, 35–79
PSMD4	320–345, 365–377	201–205, 329–340, 372–377	196–203, 359–377	201–226, 237–365
PSMD14	1–7	1–9, 249–255	1–12, 16–24	-

We also used STRING to study the engagement of the subunits of 15 UPS proteins in interactions with 500 proteins forming the first shell of the resulting interactome (note that the number of interactors in STRING is limited to 500). In this analysis, the highest confidence level of 0.9 was used. Figure 3B represents the resulting interactome, which includes 515 nodes connected by 21,801 edges. Therefore, this interactome is characterized by an average node degree of 84.7 and shows an average local clustering coefficient of 0.782. The expected number of interactions for the set of human proteins of this size is 7780, indicating that this PPI network, centered at 15 UPS proteins associated with human diseases, has significantly more interactions than expected (PPI enrichment p-value is $<10^{-16}$).

Most of the proteins associated with diseases, such as cancer, AD, PD, diabetes, and cardiovascular disease are either IDPs or contain long IDPRs, and misbehavior or mutations in the IDPs/IDPRs have broad involvement in the pathogenesis of these diseases [44–49]. From overall analysis, we found that most of the UPS proteins are intrinsically disordered and are closely linked with the pathophysiology of many diseases, such as cancer, AD, PD, ALS, HD, etc. (see Tables 1–3). The proteasomal ubiquitin receptors, such as ADRM1 and PSMD4, and extraproteasomal ubiquitin receptors, such as UBQLN1 and UBQLN12, have been found to be highly disordered as compared with the ubiquitinating and DUB enzymes. These receptors contain several disorder-based protein binding regions, which are involved in protein–protein interaction/molecular recognition. Additionally, several PTMs are located within the disordered regions of these receptors. Furthermore, ubiquitinating enzymes, such as UBE2R2 (E2) and STUB1 (E3), considered important drug targets for cancer and neurogenerative diseases, were also found to be highly disordered, along with many MoRFs and PTMs in disordered regions. Therefore, this study identified highly disordered proteins or proteins with functional DPRs from UPS, which may have a crucial role in the progression of diseases.

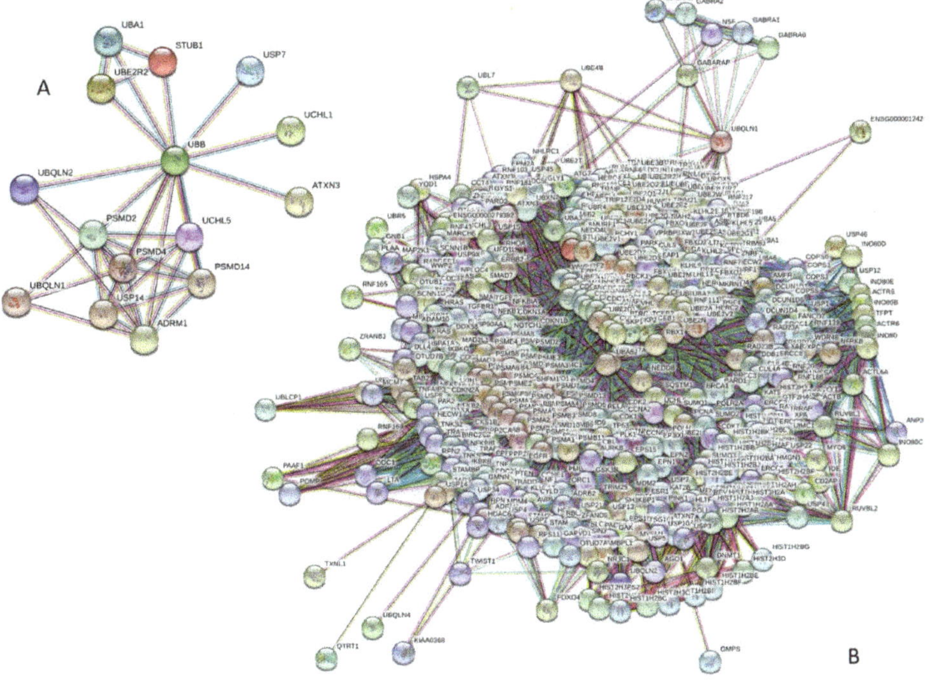

Figure 3. Evaluation of the intractability of 15 UPS proteins associated with human diseases by STRING platform. (**A**) Network of the inter-set PPI interactions (15 subunits, highest confidence level of 0.9). (**B**) PPI network centered at 15 UPS proteins associated with human diseases (515 proteins, highest confidence level of 0.9).

3.1.1. Intrinsic Disorder in Ubiquitin-Activating Enzyme (E1 Enzyme)

Two E1 enzymes have been identified in mammals, UBA1 and UBA6 [103], with UBA1 (UniProt ID: P22314) is a 1058-amino-acid-long protein most commonly participating in the Ub activation (first step of ubiquitination) in an ATP-dependent manner [64]. The reduced level of E1 enzyme has been reported in AD and HD [64,104]. Furthermore, another neurodegenerative disorder, SMAX2, was found to occur due to the mutation in UBA1 gene [21,105]. Here, according to our disorder analysis, the mean percent of predicted intrinsic disorder (PPID) calculated by averaging the outputs of five commonly used disorder predictors, such as PONDR® VLXT, PONDR® VSL2, PONDR® VL3, PONDR® FIT, and IUPred, was 9.74% (Table 2). Bhowmick et al. reported the presence of structural disorder in the human ubiquitinating enzyme, where they found 5.97% of average percentage of disordered residues in two E1 enzymes (UBA1 & UBA6) by IUPRED [58]. Our IUPRED analysis represents PPID of 8.13% for UBA1, which indicates UBA1 is more disordered than UBA6 enzyme. UBA1 is a multi-domain enzyme containing four domains important for interactions with its partners. The N-terminal half of UBA1 consists of an inactive adenylation domain (IAD) (residues 1–439) that surrounds the first catalytic cysteine half-domain (FCCD) (residues 204–309). We found one IDPR (residues 346–358) in IAD domain (Figure 4(a1)). The C-terminal half of UBA1 consists of an active adenylation domain (AAD) (residues 440–950) that surrounds the second catalytic cysteine half-domain (SCCD) (residues 626–891). One IDPR (residues 813–834) was found through our analysis in AAD and SCCD (Figure 4(a1)). The reactive cysteine residue (C632) is present in SCCD that binds ubiquitin, and our analysis shows that this residue (C632) is located within the ordered region of UBA1 protein. The ubiquitin fold domain (UFD) (residues 951–1058) is present at the C-terminal region of UBA1, which allows this protein to recognize and recruit E2 enzymes [64]. Only one short IDPR (residues 1018–1026) was found in this domain, as shown in Figure 4(a1). In addition, a MoRF region is predicted at the C-terminus of UBA1 by two predictors, MoRFCHiBi_Web (residues 1048–1057) and MoRFpred (residues1051–1058). These IDPRs (mean of five IDP predictors) and MoRFs (MoRFCHiBi_Web) are mapped in Figure 4(a1). Only two crystal structures for UBA1 are available in PDB (PDB ID: 6DC6 (residues 49–1058) [106] and 4P22 (residues 1–439) [107]). However, the N-terminal region containing residues 1–48 is missing in both the available crystal structures. Interestingly, according to our intrinsic disorder analysis, this region was found to be disordered. Furthermore, one disorder-based interaction site was also identified at the N-terminus by three predictors, MoRFCHiBi_Web (residues 1–13), MoRFpred (residues 5–12), and ANCHOR (residues 1–16). Some other IDPRs (residues 346–358, 618–623, 813–834) are represented in available crystal structures (Figure 5(a1)) and identified MoRF regions are listed in Table 3. According to evaluation of PTMs by D2P2, which is represented in Figure 4(a2), UBA1 has 41 phosphorylation sites, of which 25 are located in IDPRs, 10 acetylation sites, of which 5 lie in the IDPRs, and 29 ubiquitylation sites, of which 12 are located in the IDPRs. These results signify the significant role of intrinsic disorder in PTMs of E1 enzyme.

3.1.2. Intrinsic Disorder in Ubiquitin-Conjugating Enzyme (E2 Enzyme)

All E2 enzymes interact with one E1 and one or more E3 enzymes. Humans have around 40 E2 enzymes, which are involved in transfer of Ub or ubiquitin-like proteins such as SUMO and NEDD8 [108]. E2 enzyme has two main functions: transfer of Ub from thioester to a thiol group (trans thiolation) and transfer of Ub from thioester to amino group (aminolysis) [108]. It has one catalytic domain that is made up of ~150 amino acids. Here, the thioester bond is formed between E2 cysteine and the C-terminus of Ub. This domain has four α-helices and four-stranded β-sheets (see Figure 4(b1)). There is also an E3 binding domain that is made up of several loops. Commonly, E2 transfers its thioester-linked Ub to cysteine in the active site of HECT-type E3 ligase in thiolation reactions. Human E2 enzymes seem to show a propensity to transfer Ub to free lysine of RING-type E3 ligases in aminolysis reactions. However, there are a few exceptions [109,110], and Stewart et al., in 2016, reviewed some fundamentally different intrinsic activities of several E2 enzymes. Maximal E2 activity occurs only in the presence of E3 [111]. Most E2 enzymes have short N- and C-terminal extensions that often contain intrinsically disordered regions [112]. Our intrinsic disorder analysis of the UBE2R2 E2 enzyme (UniProt ID: Q712K3) has also predicted the residues at N-terminal (1–23) and C-terminal (198–238) to be disordered (see Figure 4b). Along with these, two short stretches of residues 102–113 and 153–162 also fall in the disordered regions. In total, the disordered regions account for 37.39%, as calculated by mean PPID. Previous studies reported the presence of structural disorder, with 17.74% of the average percentage of disordered residues in 29 E2 enzymes by IUPred [58]. Our IUPRED analysis of UBE2R2 represents PPID of 24.79%. Furthermore, the E2 enzyme also showed disorder binding residues in the C-terminal region, which are predicted by all MoRF predictors (Table 3). MoRFCHiBi_Web server predicted three short MoRF regions (residues 166–170, 203–217, 219–226) at the C-terminal region. DISOPRED3 and MoRFpred also predicted few MoRF residues (1–6 and 11–19, respectively) at the N-terminal region. The recently crystallized structure of E2 enzyme of 1.5 Å resolution has four long and four short helices, forming 33% helical and 6 β-strands constituting a 12% β-sheet structure [113,114]. The unstructured C-terminal also suggested to be disordered, which is in correlation with our disorder analysis. For example, Cdc34 is an E2 protein whose catalytic domain is similar to other E2 proteins, but the acidic C-terminal region (66 amino acids) is disordered, which interacts with Ub in complex. Removal of this interaction leaves Ub free for transfer [113,114]. According to the prediction from D2P2 (Figure 4(b2)), E2 enzyme has five phosphorylations, with all of them in IDPRs, and eight ubiquitylation sites, of which seven are located in the IDPR region. This

disordered-region-based analysis of E2 enzyme would be helpful to deeply understand the functioning
of this enzyme in the cell.

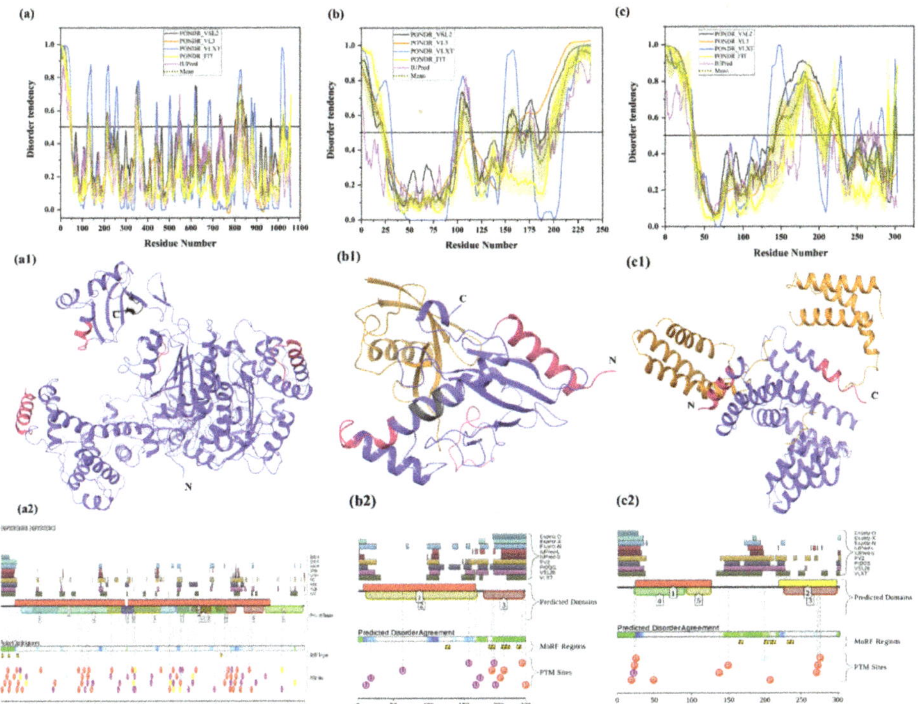

Figure 4. Intrinsic disorder in ubiquitinating enzymes E1 (UBA1), E2 (UBE2R2) and E3 (STUB1).
(a) ubiquitin-activating enzyme E1 (UniProt ID: P22314), (a1) crystal structure of the E1 enzyme
(PDB ID: 6DC6). (b) ubiquitin-conjugating enzyme E2 (UniProt ID: Q712K3), (b1) crystal structure
of E2 enzyme having residues 1–202 (PDB ID: 6NYO). (c) E3 ubiquitin ligase (UniProt ID: Q9UNE7),
(c1) crystal structure of E3 ubiquitin ligase (PDB ID: 4KBQ). In Plots (a–c), the outputs of PONDR®
VSL2, PONDR® VL3, PONDR® VLXT, PONDR® FIT, and IUPred are represented by black, orange,
blue, yellow, and purple lines, respectively. Mean disorder profile, calculated by averaging the outputs
of five predictor-specific per-residue disorder profiles, is depicted by olive color. Light-olive shadow
around the mean curve represents the error distribution for the mean. The light-yellow shadow around
the PONDR® FIT curve shows error distribution for PONDR® FIT. In (a1), crystal structure of the E1
enzyme (PDB ID: 6DC6) is represented in faded blue color, disordered residues are shown in salmon
pink color, and MoRF residues identified by MoRFCHiBi_Web server (1048–1057) are shown in grey
color. In (b1), E2 enzyme (1–202 length with missing residues 1–5 and 193–202 at N- and C-terminal,
respectively) is shown with Ubiquitin-60S ribosomal protein L40 (RPL40A; orange color). In (c1), Hsc70
Lid-Tail domains (orange color) in complex with E3 ubiquitin ligase (which is represented in faded
blue color); disordered residues in E3 are shown in salmon pink color. In (a2,b2,c2), functional disorder
profiles, MoRFs, and PTMs in E1, E2, and E3 enzymes using D2P2 server have been shown.

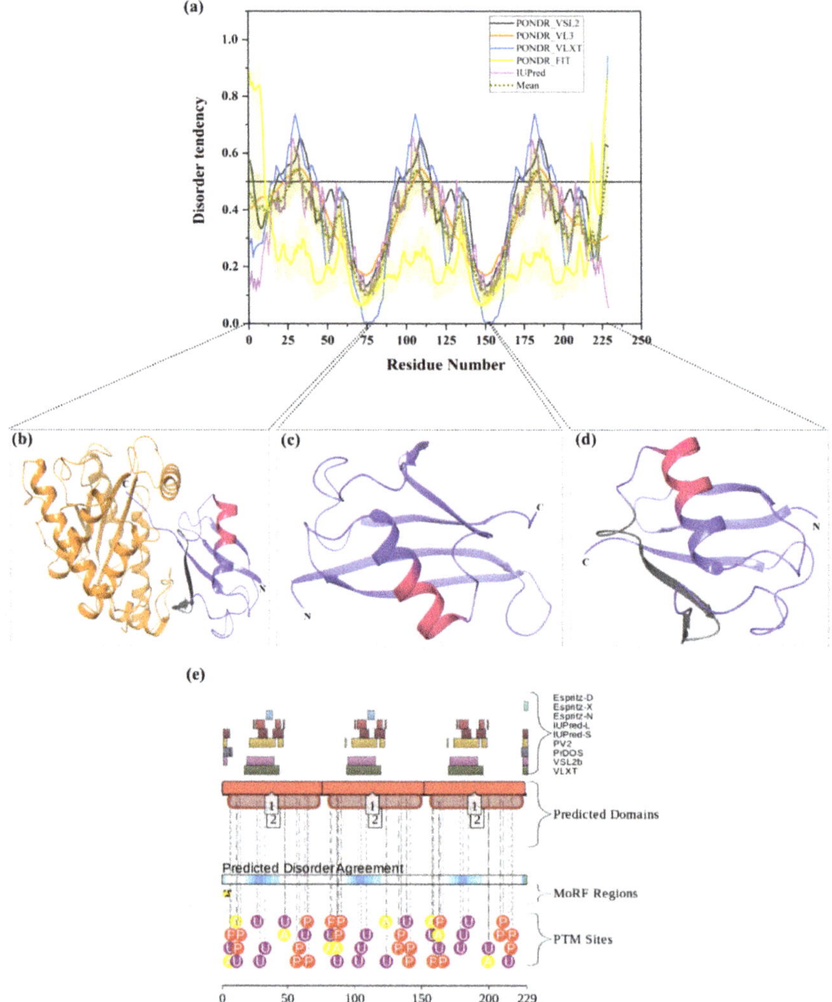

Figure 5. Analysis of intrinsic disorder predisposition in polyubiquitin-B (UBB) and structural characterization of mature ubiquitins. (**a**) Disorder analysis of human polyubiquitin-B (UniProt ID: P0CG47). (**b**) Crystal structure of chain B of Polyubiquitin-B (PDB ID: 6FDK). (**c**) Crystal structure of Chain D of Polyubiquitin-B (PDB ID: 6BYH). (**d**) Crystal structure of Chain A of Polyubiquitin-B (PDB ID: 4XOF). In Plot (**a**), the disorder profile obtained forms a set of disorder predictors such as PONDR® VSL2, PONDR® VL3, PONDR® VLXT, PONDR® FIT and IUPred, represented by black, orange, blue, yellow, and purple curves respectively. Mean disorder profile, which was calculated from average of five predictor-specific per-residue disorder profiles, is shown in olive color. Predicted disorder scores above 0.5 are considered as disordered residues/regions. The light-olive shadow around the mean curve represents the error distribution for the mean. The light-yellow shadow around the PONDR® FIT curve shows the error distribution for PONDR® FIT. In Plot (**b**), a structure of Ub protein (faded blue) in complex with *Chlamydia trachomatis* effector protein Cdu1 (orange color) is represented. In Plots (**b**), (**c**), and (**d**), Ub is shown in faded blue color, and disordered residues are shown in salmon pink color. The position of MoRFCHiBi_Web-server-identified MoRFs (residues 40–50, shown in PDB ID:6FDK, and residues 192–202, shown in PDB ID: 4XOF) are represented by grey color. (**e**) Functional disorder profile of the UBB protein, using the D2P2 server, is shown.

3.1.3. Intrinsic Disorder in Ubiquitin Ligase (E3 Enzyme)

E3 ligases have been grouped into the RING (really interesting new gene), the HECT (homologous to the E6AP carboxyl terminus), and the RBR (RING between RING) types [115]. RING E3s have a RING or U-box catalytic domain that directly transfers Ub to the target protein. This class of E3s has around six hundred predicted members. They possess a catalytic domain and a substrate-recruiting module, which can be present in a single polypeptide or in different subunits of the multicomponent E3 complex. To understand how E2 stimulates transfer of Ub from E2, crystal structures of two E3 ligases (RNF4 and BIRC7), when they were bound to E2–Ub complex, were studied [116,117]. RING E3 transfers Ub through aminolysis. It alters thioester linkage between the cysteine in the C-terminus of Ub and the cysteine in the active site of E2, thus hydrolyzing the bond. The RING domain is responsible for both binding E2 and stimulating transfer of Ub. It can exist either in the form of a monomer or a dimer. It adopts a cross-brace structure with two Zn^{2+} ions [118]. A subset of E3 ligases called the cullin RING ligases have three subunits: the cullin protein, which acts as the scaffold protein, the RING box protein, which contains the RING domain, and the F-box protein, which binds the substrate. The RING domain helps in positioning Ub on E2 in the correct orientation. A conserved asparagine residue (Asn77) has been proposed to stabilize the oxyanion intermediate [119]. Some RING E3s have additional structural domains, which have been reviewed by Berndson and Wolberger in 2014 [120]. HECT E3 ligases participate in two different reactions: transfer of Ub from Cys in the active site of E2 to Cys in the HECT domain, and HECT-Ub thioester is attacked by substrate lysine. The HECT domain has an N-lobe to bind E2 and a C-lobe that contains active site cysteine. Multiple configurations are made possible via a flexible tether between these two lobes. A conformational change, which is required to bring both the lobes together for successful transfer of Ub to substrate, was also revealed [121]. Similar to E2 enzyme, STUB1 E3 enzyme (Uniprot ID: Q9UNE7) is characterized by a mean PPID of 37.95%. Bhowmick et al. reported the presence of structural disorder in different human E3 ubiquitin ligase families, where they found 20.03% of the average percentage of disordered residues by IUPRED [58]. Furthermore, four disorder-based binding regions (residues 163–169, 198–204, 206–214, 230–239) were detected in this protein as per ANCHOR server (see Table 3). However, not a single MoRF was detected by MoRFCHiBi_Web, MoRFpred, or DISOPRED3. The MoRFs identified by ANCHOR do not contains any region at the N- or C-terminal and are mainly located at the middle region, signifying the role of middle-disordered regions of the E3 enzyme in molecular recognition. A long stretch of 35 residues (1–35) at N-terminal and residues 143–203 are predicted to be disordered, and another long IDPR is predicted by several algorithms in the middle of the protein (residues 147–227) (see Figure 4c). The D2P2 analysis (Figure 4(c2)) recognized several PTMs in disordered regions of E3 enzyme, such as 10 phosphorylation sites, out of which eight lie in IDPRs, one ubiquitylation site lies in IDPRs, and one nitrosylation site lies in IDPRs. For the E3 enzyme, a truncated structure for residues 21–154 is available in PDB (PDB ID: 4KBQ), where initial three residues are missing and other residues lying in disorder regions are mapped (see Figure 4(c1)).

3.1.4. Intrinsic Disorder in Polyubiquitin-B (UBB)

Ubiquitin (Ub) is a 76-amino-acid-long, highly conserved protein expressed in all eukaryotic cells. Ub is encoded by four different genes, where *UBA52* and *RPS27A* code for a single copy of Ub fused to the ribosomal proteins L40 and S27a, respectively, and where a polyubiquitin precursor with exact head-to-tail Ub repeats is encoded by the *UBB* and *UBC* genes, with the corresponding products, polyubiquitin-B and polyubiquitin-C, containing three and six Ub chains, respectively. Mature Ub exists either as a free protein or as a conjugated form that is covalently bound to various intracellular proteins, typically for their degradation through 26S proteasome [15,71]. However, the Ub conjugation to other cellular proteins controls numerous eukaryotic cell functions [122]. Importantly, alterations in UPS have been observed in various types of human cancers and neurodegenerative diseases. Ubiquitin must be supplied in an adequate amount and in a timely manner for conjugation to a variety of proteins. It has been reported that the growth of cancer cells requires ubiquitin, and its downregulation inhibits the ubiquitination of multiple proteins associated with oncogenic pathways [123]. As a result, uncontrolled UPS plays a crucial role in several cellular processes related to tumorigenesis [123]. Increased levels of ubiquitin with enhanced cellular proliferation and stress have been observed in many types of cancer cells [123,124]. Interestingly, protein aggregates associated with familial and sporadic AD often contain proteins other than those, which are generally linked with diseases. One of these proteins is a frameshift form of ubiquitin, UBB+1. UBB+1 is produced by molecular misreading of a wild-type ubiquitin gene, and the presence of this Ub form has been allied with several disorders. UBB+1 accumulation leads to aberrant UPS system activity, which increases the aggregation of toxic proteins leading to cell death [125]. Accumulation of UBB+1 has been reported in several neuronal diseases, such as AD, Pick's disease, and PolyQ diseases (including HD), as well as non-neuronal tissue diseases [125]. More specifically, UBB+1 accumulation has been reported in the neuritic plaques and neurofibrillary tangles of AD patients [126]. In this study, based on its overall disorder content, UBB (UniProt ID: P0CG47) is classified as a least-disordered or highly ordered protein. The mean PPID for UBB was found to be 10.04% (Table 2). Total 21 residues (28–34, 104–110, and 180–186) in the full-length UBB (229 residues) are predicted to be disordered (see Figure 5a). Furthermore, two predictors, MoRFCHiBi_Web (residues 40–50, 116–122, 192–202) and MoRFpred (residues 221–228), have predicted MoRF residues (Table 3). Similarly, D2P2 has also predicted least disorder and MoRF residues (Figure 5e). PTM analysis by D2P2 in Figure 5e displays 21 phosphorylation sites, of which 1 lies in IDPRs, 9 acetylation sites, of which 5 are in IDPRs, and 21 ubiquitylation sites, of which 14 lie in IDPRs. Multiple crystal structures are available for human Ub in PDB. Figure 5b–d represents three illustrative examples of these structures, where the position of predicted IDPR and MoRF regions are also shown.

3.1.5. Intrinsic Disorder in Ubiquilin 1 (UBQLN1)

Ubiquilin 1 (UBQLN1) (UniProt ID: Q9UMX0) is a 589-residue-long extraproteasomal ubiquitin receptor [127] that plays a crucial role in the regulation of the protein quality control system [6]. Its structure consists of an N-terminal UBL domain (residues 37–111) and a C-terminal Ub-associated (UBA) domain (residues 546–586). Importantly, from our analysis, both UBL and UBA domains were found to be highly disordered and contain many disorder-based binding sites (see Figure 6a,(a1–3)). The UBA domain is identified in ubiquitination-linked proteins, such as E2 and E3 enzymes. Interestingly, ubiquilin interacts, via its UBA domain, more efficiently with poly-Ub chain than with mono-Ub. Ko et al. have shown that ubiquilin interacts especially with the poly-Ub chains of ubiquitylated proteins via the UBA domain and with the subunit of 19S proteasome via the UBL domain [128]. It was also reported that the absence of UBQLN1 is associated with the destruction of protein synthesis and cell cycle arrest [129]. A study has shown that UBQLN1 is essential for the transport of mislocalized mitochondrial proteins to proteasome for their degradation [130]. For proteasomal degradation, the central portion of UBQLN1 is crucial to bind at the hydrophobic domains of mitochondrial proteins [129]. Additionally, in response to myocardial ischemia/reperfusion injury, UBQLN1 plays a significant role in cardiac ubiquitination-proteasome coupling [127]. In approximately 50% of human lung adenocarcinoma cases, UBQLN1 has been reported to be lost or under-expressed [131]. UBQLN1 is also known to regulate the activity and expression of Insulin-like growth factor-1 receptor (IGF1R), a receptor that regulates growth, proliferation, and survival [131]. As a result, UBQLN1 is associated with the pathophysiology of cancer and neurodegenerative diseases [6]. UBQLN1 polymorphism substantially increases the risk of AD, possibly due to its induction of alternative splicing in the brain. UBQLN1 induces Aβ production by affecting APP processing and trafficking. It also regulates the activity of γ-secretase complex by regulating presenilin 1 endoproteolysis within the γ-secretase complex, a protease that cleaves APP and generates Aβ peptide, which is responsible for AD pathogenesis [6]. Furthermore, UBQLN1 also controls the level of β-secretase, a rate-limiting enzyme in the production of Aβ peptides. Therefore, the reduced UBQLN1 level in AD brain may result in perturbed processing of APP and Aβ generation [6]. UBQLN1 acts as a molecular chaperone for APP by binding and preventing its aggregation, and a reduced level of UBQLN1 was found in the brains of AD patients [67]. Several studies have also reported that UBQLN-family proteins are linked to the pathogenesis of PolyQ diseases. Studies in cellular and animal HD models have reported that UBQLN1 suppresses PolyQ-induced protein aggregation and toxicity. Involvement of UBQLN1 in multiple diseases may be due to its highly disordered nature, since this protein is one of the most disordered proteins in a set of UPS-related proteins analyzed in this study. In fact, in our analysis, UBQLN1 was found to be highly disordered, with a mean PPID of 87.10% (Table 2). Figure 6a shows that all five predictors have predicted the presence of long IDPRs in this protein. There are also multiple short as well as long disorder-based binding regions in human UBQLN1 (Table 3). Despite high levels of predicted disorder, NMR solution structures were determined for the comparatively ordered N- and C-terminal regions of human UBQLN1 (Figure 6(a1,a2)). According to D2P2 (Figure 6(a3)), UBQLN1 has eight phosphorylation sites and four ubiquitylation sites, and all of them are located in the IDPR region of this protein. However, many PTMs sites are located at the UBL domain of the UBQLN1 protein. Since UBQLN1 has multiple interactions and functions, results from our analysis (presence of multiple IDPRs and disorder-based binding sites) signify the central role of intrinsic disorder in protein–protein interaction and protein degradation via UPS.

3.1.6. Intrinsic Disorder in Ubiquilin 2 (UBQLN2)

Ubiquilin-2 (UBQLN2) (UniProt ID: Q9UHD9) is a 624-residue-long protein present in cytosol, which is mostly expressed in the brain, liver, spleen, pancreas, heart, and other tissues [6,26]. Like UBQLN1, UBQLN2 is also actively involved in misfolded protein degradation via the ubiquitin-proteasome system [26]. In addition, it also plays an important role in the regulation of the progression of the cell cycle and cellular signaling [69]. Polyubiquitinated proteins that underwent a three-step enzymatic cascade are recognized by the UBL domain of UBQLN2 and transported for degradation to the S5a/PSMD4 receptor of the 26S proteasome [132]. UBQLN2 consists of an N-terminal ubiquitin-like domain (UBL) domain (residues 33–103), which interacts with the proteasome, and a C-terminal ubiquitin-associated (UBA) domain (residues 582–624), which is crucial for the UPS activity. Figure 6b,(b1,b2) shows the presence of significant intrinsic disorder and MoRFs in both domains (UBL and UBA) of UBQLN2. Additionally, UBQLN2 contains one proline-rich repeat domain containing 12 PXX repeats (490–535 AA) involved in protein–protein interactions and four stress-induced protein 1 (STI-1)-like motifs present in regions 178–247 and 379–462, which are responsible for the UBQLN2 interaction with autophagy mediators and HSPs [26]. Interestingly, our ANCHOR-based MoRF analysis found the presence of proline-rich repeat domain and STI-1 motifs in disorder-based binding regions (Table 3). *UBQLN2* gene mutations were reported in frontotemporal dementia (FTD) and amyotrophic lateral sclerosis (ALS). Abnormal UBQLN2 inclusions have been observed in the cytosol of degenerating motor neurons of ALS patients [26]. Furthermore, the disturbance of autophagic and proteasomal protein degradation was reported in ALS-linked mutations in UBQLN2 [133]. Furthermore, ALS-linked mutations in this protein are also linked to neuroinflammation, the formation of stress granules (SGs), and dysfunction of autophagy [26]. Hjerpe et al. reported that mutations in UBQLN2 are associated with defective chaperone binding, impaired aggregate clearance and cognitive deficits in mice and neurodegeneration in humans [134]. Interestingly, numerous mutations were found in the PXX domain of UBQLN2, and these mutations have been reported to provoke impairments in autophagy and 26S proteasome [26]. Studies in neuronal cells have reported that dysregulation of UBQLN2 in neurons may activate NF-κB and cytosolic TDP-43 aggregation [135]. Similar to UBQLN1, UBQLN2 is also predicted to be highly disordered (Figure 6b), with a mean PPID of 80.93%. D2P2 server-based analysis (Figure 6(b2)) provides a further illustration of the highly disordered nature of this protein and shows that UBQLN2 is heavily decorated by multiple PTMs and includes a very large number of MoRFs. According to D2P2 (Figure 6(b2)), UBQLN2 has nine phosphorylation sites located in the IDPRs, and six ubiquitylation sites out of which four lie in IDPR regions. Solution NMR structure was determined for the N-terminal region of human UBQLN2 (residues 1–103) containing UBL domain. Figure 6(b1) shows that this region is characterized by high structural dynamics and conformational flexibility.

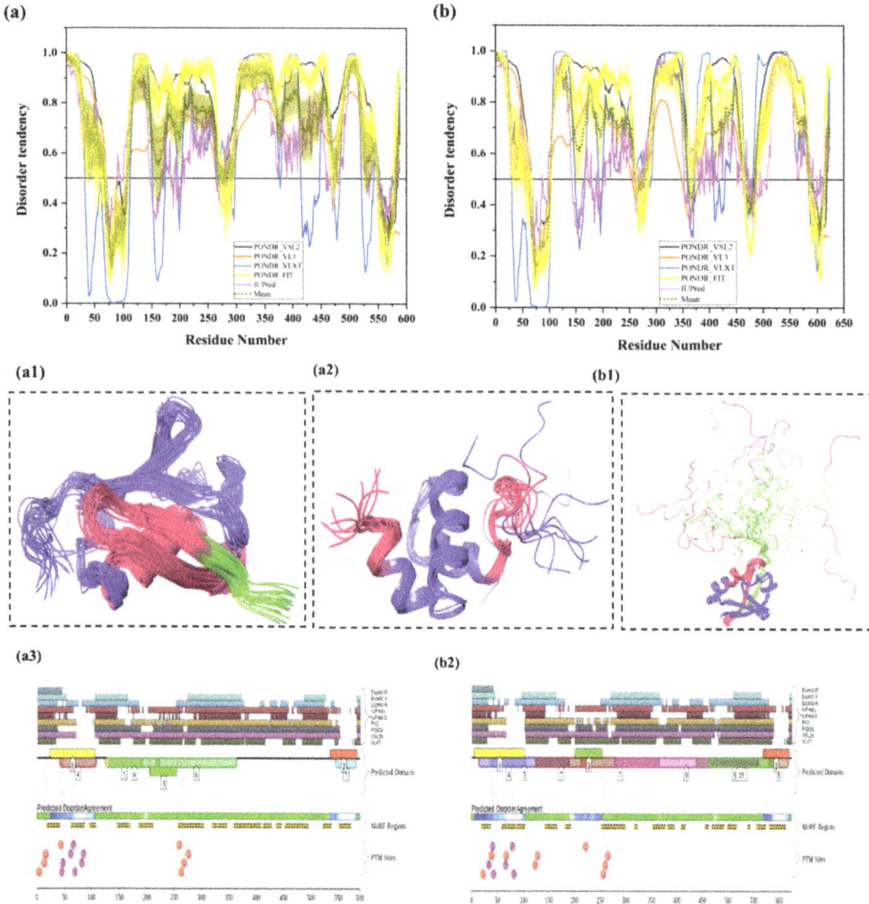

Figure 6. Intrinsic disorder predisposition and structural characterization of human UBQLN1 and UBQLN2. (**a**) Disorder profile of human UBQLN1 (UniProt ID: Q9UMX0). (**b**) Disorder profile of UBQLN2 (UniProt ID: Q9UHD9) (**a1**) NMR solution structure of the N-terminal UBL domain (residues 34–112) of UBQLN1 (PDB ID: 2KLC). (**a2**) NMR solution structure of the C-terminal UBA domain (residues 541–586) of UBQLN1 (PDB ID: 2JY5). (**b1**) NMR solution structure of the N-terminal UBL domain (residues 1–103) of human UBQLN2 (PDB ID: 1J8C). In Plots (**a**) and (**b**), intrinsic disorder profiles generated by disorder predictors, such as PONDR® VSL2, PONDR® VL3, PONDR® VLXT, PONDR® FIT and IUPred, are shown by black, orange, blue, yellow, and purple curves respectively. Mean disorder profile is calculated from the average of five predictor-specific per-residue disorder profiles, represented by the olive color curve. Predicted disorder scores above 0.5 are considered disordered residues/regions. The light-olive shadow around the mean curve represents error distribution for mean. The light-yellow shadow around PONDR® FIT curve shows error distribution for that predictor. In (**a1,a2,b1**), UBQLN 1 and 2 are represented by faded blue color, and disordered residues are shown in salmon pink color. The position of the MoRF region (residues 34–39) recognized by the MoRFCHiBi_web server in UBQLN1 is represented in (PDB ID: 2KLC), and MoRF region (residues 10–38) in UBQLN2 are represented in (PDB ID: 1J8C) as MoRFs lying in the IDP region by faded green color. (**a3,b2**) Functional disorder profile of UBQLN1 and UBQLN2 proteins, respectively, using the D2P2 server, is shown, depicting PTM sites.

3.1.7. Intrinsic Disorder in Ubiquitin C-Terminal Hydrolase Isozyme L1 (UCHL1)

UCHL1 (UniProt ID: P09936) is a small, 223-residue-long protein, which is highly abundant in the brain (it is estimated that UCHL1 accounts for 1%–5% of total neuronal protein) and is normally expressed exclusively in neurons and testis [31,70]. UCHL1 catalyzes hydrolysis of C-terminal ubiquityl esters and amides. It is a thiol protease that recognizes peptide bonds at the C-terminal glycine of Ub [136]. The overall structure of this protein (see Figure 7(a1)) resembles a structure typical for the papain family of cysteine proteases, with two lobes: one having five helices and the other having two helices. The identified disordered residues from our analysis are shown in salmon pink color. An active site consisting of a Cys, His, and Asp triad is situated between the two lobes. UCHL1 negatively regulates cytokines and induces NF-κB and STAT1 signaling [137]. Oxidation of a few residues in UCHL1 has been associated with AD. Mutations leading to increased activity of this enzyme have been associated with preserving cognitive functions in AD patients [138]. UCHL1 is essential for the maintenance of axonal integrity, and its dysfunction is associated with neurodegenerative disease [31]. Furthermore, UCHL1 downregulation has been observed in idiopathic AD, as well as in PD brains [139]. I93M mutation in UCHL1 occurs in four out of seven family members who developed PD. According to our analysis, this mutation is located in ordered regions of this protein (Figure 7a). Previous reports suggest that this mutation inhibits α-synuclein degradation via 26S proteasome [31,32]. In-vitro studies have described destabilization of the 3D structure of UCHL1 after deletion of a few amino acids from either the N- or C-terminus, which leads to the partial unfolding of this protein and toxic gain-of-function [31]. The regions consisting of residues 5–10 and 211–216 are involved in interaction with Ub [140]. Although UCHL1 is predicted to have rather low levels of intrinsic disorder, the aforementioned regions related to interaction of this protein with Ub are located within the disordered tails of UCHL1. Furthermore, region 215–220 was predicted to be a disorder-based binding region by the MoRFpred server (see Table 3). According to D2P2 (Figure 7(a2)), UCHL1 has 14 ubiquitylation sites, of which 13 are located in IDPRs. Moreover, it also comprises three phosphorylation sites, two acetylation sites, one nitrosylation site, and one prenylation site, and all of them are located in IDPRs of UCHL1.

3.1.8. Intrinsic Disorder in Ubiquitin C-terminal Hydrolase Isozyme L5 (UCHL5)

UCHL5 (UniProt ID: Q9Y5K5) ia also known as ubiquitin C-terminal hydrolase 37 (UCH37). It is a 329-residue-long DUB that binds to the 19S regulatory subunit of 26S proteasome and deubiquitinates polyubiquitinated proteins. It has a thiol-dependent Ub-specific cysteine protease activity. It is physically associated with a base component of 19S proteasome. It removes poly-Ub chain from the distal end. UCHL5 gets activated by binding of ADRM1, which interacts with the C-terminal tail of UCHL5, a region that is different from the UCH catalytic domain [33,141]. Ub-mediated degradation occurs in the absence of Hedgehog signaling. UCHL5 is a critical regulator of hedgehog signaling [141]. Ge et al. reported lower expression of this enzyme in the brain tissues of glioma patients [142]. In vitro analysis revealed that UCHL5 can inhibit migration and invasion of glioma cells mediated via a downregulation of SNRPF [142]. It was reported that UCHL5 deubiquitinates Tcf3, which helps it to fully activate the Wnt/β–catenin pathway [143]. Furthermore, the region of UCHL5 amino acid residues from 313–329 interacts with proteasomal receptor ADRM1 [144], and, according to our analysis, this ADRM1 binding region of UCHL5 is the part of disorder-based protein binding region (see Table 3). Human UCHL5 is predicted to have more intrinsic disorder than UCHL1 (PPID of 16.11% vs. 7.62%) (compare Figure 7a,b). D2P2 server (Figure 7(b2)) identified several PTMs in UCHL5. These include two phosphorylation sites, of which one is located in the IDPRs, two acetylation sites, of which one is located in the IDPRs, 14 ubiquitylation sites, of which 10 are located in IDPRs, one nitrosylation site in IDPR, and one mono-methylation site. Interestingly, some PTM sites are located in the ADRM1 binding region of this protein. These results signify the important role of intrinsic disorder in UCHL5 for interaction with proteasome receptor ADRM1 and further 26S proteasome-mediated degradation of target proteins.

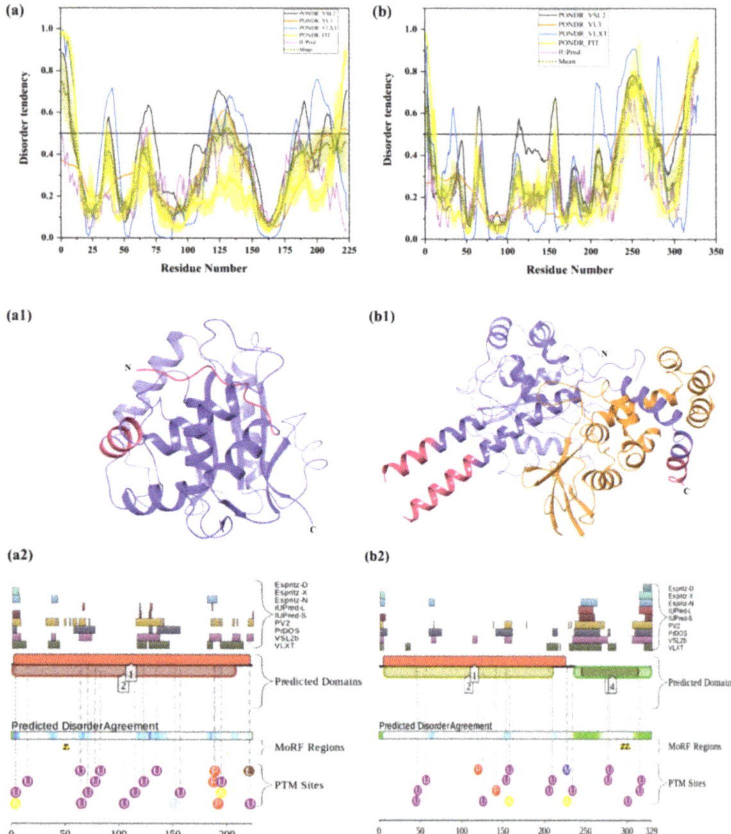

Figure 7. Intrinsic disorder predisposition and structural characterization of UCHL1 and UCHL5.
(**a**) Disorder profile of human UCHL1 (UniProt ID: P09936) (**b**) Disorder profile of human UCHL5
(UniProt ID: Q9Y5K5) (**a1**) Crystal structure of UCHL1 (PDB ID: 4JKJ). (**b1**) Crystal structure of UCHL5
(PDB ID: 4UEL). In (**a,b**), intrinsic disorder profiles obtained from disorder predictors, such as PONDR®
VSL2, PONDR® VL3, PONDR® VLXT, PONDR® FIT and IUPred, are depicted by black, orange,
blue, yellow, and purple lines respectively. Mean disorder profile, calculated from the average of five
predictor-specific per-residue disorder profiles, is represented by the olive color. Predicted disorder
scores above 0.5 are considered disordered residues/regions. The light-olive shadow around the mean
curve represents the error distribution for the mean. The light-yellow shadow around the PONDR® FIT
curve shows the error distribution for PONDR® FIT. In (**a1,b1**), UCHL1 and UCHL5 are represented
by faded blue color; in Plot (**b1**), the DEUBAD domain of the RPN13 protein and Ub in complex with
UCHL5 are shown in orange color. Disordered residues are shown in salmon pink color. In (**a2,b2**),
the functional disorder profile of UCHL1 and UCHL5 proteins using the D2P2 server have been shown.

3.1.9. Intrinsic Disorder in Ubiquitin-Specific-Processing Protease 7 (USP7)

USP7 (UniProt ID: Q93009) is also known as ubiquitin carboxyl-terminal hydrolase 7 or herpesvirus-
associated ubiquitin-specific protease (HAUSP), and is a member of the USP family of deubiquitylating
enzymes; it cleaves ubiquitin from polyubiquitin chains of substrate protein [10,145,146]. Figure 8(a1)
represents the crystal structure of USP7 (PDB ID: 4YOC), and IDPRs and MoRFs identified in this
study are denoted by salmon pink and grey color, respectively. This 135-kDa cellular protein is
associated with numerous cellular processes, such as oncogenesis and tumor suppression, immune
functions, DNA dynamics, epigenetic modulations, DNA damage and repair processes, regulation of

gene expression and protein function, and host–virus interactions. Dysfunctions in USP7 at different physiological conditions lead to the development of various pathological conditions, such as cancer, immune dysfunction, metabolic diseases, and neurological pathologies. USP7 is mainly recognized in cancers and virus-associated host–pathogen interactions [10]. The full-length USP7 consists of 1102 amino acids [147] and has several functional domains, including poly Q stretch (amino acid 4–10), tumor necrosis factor receptor-associated factor (TRAF)-like domain, C-terminal domain (CTD), and middle catalytic (CAT) domain [10,147]. The presence of poly-Q repeat in HSP7 specifies that it may have a link to the neurodegenerative disorders associated with the poly-Q expansion [10]. TRAF domain (amino acids 62–208) is required for protein–protein interactions and also plays an important role in nuclear localization of USP7 [147,148]. The TCAT domain (amino acids 208–560) is crucial for the catalytic activity. Some residues from this domain are located in the disorder-based binding regions (Table 3). This catalytic domain mediates ubiquitination and deubiquitination of substrate proteins. CTD (amino acids 560–1102) consists of five ubiquitin-like (UBL) folds that enable protein–protein interactions with other proteins, such as ICP0, ataxin-1, DNA (cytosine-5-)-methyltransferase 1 (DNMT1), and ubiquitin-like PHD and RING finger domain-containing protein 1 (UHRF1). MDM2 and p53 also interact with USP7 via CTD, which serves as a second site of interaction [10,147,149,150]. Interestingly, our analysis found MoRF regions (Table 3) at CTD of USP7 using three different servers: MoRFCHiBi_Web (residues 1077–1082 and 1090–1102), MoRFpred (residues 1094–1099), and DISOPRED3 (residues 1056–1061 and 1084–1093). In various pathological conditions, variation in the USP7 expression level has been observed in different organs. However, not a single mutation has been reported to date in the USP7 gene [10]. Being one of the longest proteins in the set analyzed in this study, USP7 is characterized by relatively low disorder content (see Figure 8a). In fact, the mean PPID of this protein is 11.62%. However, Table 3 and Figure 8(a2) show that human USP7 has several MoRFs and multiple PTM sites including 16 phosphorylation sites, of which nine are located in the IDPRs, six acetylation sites, of which four are located in the IDPRs, and 17 ubiquitylation sites, of which nine are located in the IDPRs. These results suggest the functional importance of intrinsic disorder in this protein.

3.1.10. Intrinsic Disorder in Ubiquitin Carboxyl-Terminal Hydrolase 14 (USP14)

USP14 (UniProt ID: P54578), is a proteasome-associated deubiquitinating enzyme, which is unique among known UBP enzymes [35,151]. Figure 8(b1) is the crystal structure of USP14 (PDB ID: 4GJQ), where IDPRs found in our analysis are represented in salmon pink color. It plays an important role in the development of synapses [152]. Furthermore, it is crucial for the degradation of ubiquitinated proteins, since proteasome activation occurs when the polyubiquitin chain binds to USP14, which further degrades substrate protein [153]. This reduces the degradation of toxic/pathogenic misfolded proteins by the proteasome, and studies show that USP14 inhibition enhances the degradation of toxic proteins associated with neurodegenerative diseases, such as AD, ALS, PD, HD, etc. [36]. Therefore, inhibitors of USP14 may provide good therapeutics for protein aggregation diseases. Full-length human USP14 consists of 494 amino acids, with ubiquitin-like (UBL) domain (9 kDa) (residues 1–90) at the N-terminus followed by the catalytic domain (45 kDa) (residues 91–494) [35], which is vital for the catalytic activity of USP14, whereas the C-terminal domain facilitates protein–protein interaction [35]. One disorder-based binding region (residues 66–75) was identified by DISOPRED3 in the UBL domain (Table 3). In addition, many PTM sites are also located in the C-terminal catalytic domain (Figure 8(b2)). The interaction of USP14 with the 19S regulatory particle of 26S proteasome through its UBL domain stimulates DUB's catalytic activity one-hundred-fold [154]. It directly interacts with the ATPase ring of the proteasome, which leads to the conformation changes in the 19S proteasome that allow proper substrate interaction. It negatively regulates proteasome activity. Furthermore, it negatively regulates autophagy since its genetic inhibition results in downregulation of autophagic flux. Most USP proteins consist of two ubiquitin binding domains; one for proximal Ub, which cleaves isopeptide linkage between two ubiquitin moieties, and the other for distal Ub. The catalytic core consists of Cys, His, and Asp/Asn residues. There are two Cys-X-X-Cys motifs in human USP14 that bind zinc, which may help in proper folding of the core [155]. Figure 8b shows that USP14 is predicted to have several IDPRs, possessing the relatively high levels of intrinsic disorder with a mean PPID of 23.48%, whereas Figure 8(b2) indicates that some of the IDPRs in this protein are used for protein–protein interactions (see also Table 3) and serve as a site of various PTMs, including 23 phosphorylation sites, of which 19 are located in the IDPRs, eight acetylation sites, of which 3 are located in the IDPRs, and 17 ubiquitylation sites, of which 10 are located in IDPRs.

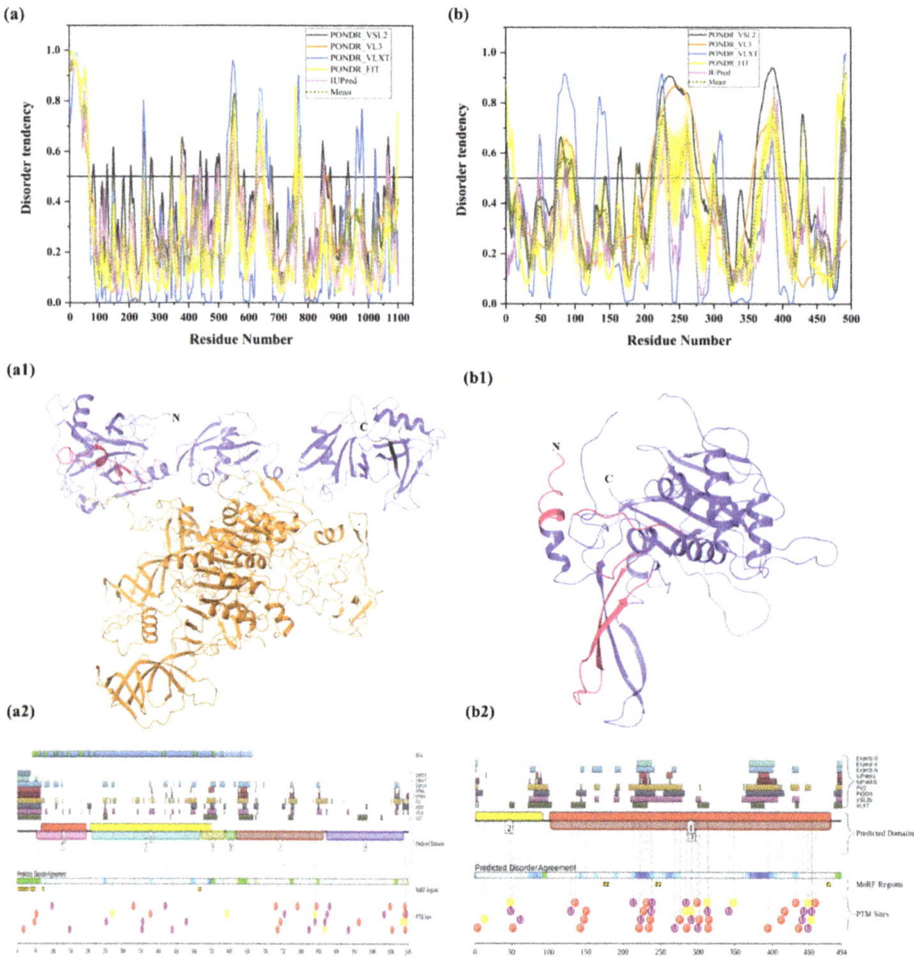

Figure 8. Intrinsic disorder predisposition and structural characterization of human USP7 and USP14. (a) Disorder profile of human USP7 (UniProt ID: Q93009). (b) Disorder profile of human USP14 (UniProt ID: P54578) (a1) crystal structure of USP7 (PDB ID: 4YOC). (b1) Crystal structure of USP14 (PDB ID: 4GJQ). In Plots (a) and (b), disorder profiles generated by sets of disorder predictors such as PONDR® VSL2, PONDR® VL3, PONDR® VLXT, PONDR® FIT, and IUPred are depicted by black, orange, blue, yellow, and purple curves respectively. The mean disorder profile calculated from the average of five predictor-specific per-residue disorder profiles is shown by the olive color curve. Predicted disorder scores above 0.5 are considered disordered residues/regions. The light-olive shadow around mean curve represents the error distribution for the mean. The light-yellow shadow around the PONDR® FIT curve shows the error distribution for PONDR® FIT. In Plots a1 and b1, USP7 and USP14 are represented by faded blue color; in Plot (a1), human DNA (cytosine-5)-methyltransferase 1 (DNMT1)

complexed with USP7 is shown in orange color. Disordered residues are shown by the salmon pink color. In Plot (**a1**), the positions of MoRFs (residues 1077–1082) predicted by the MoRFCHiBi_web server are shown by grey color in USP7 (PDB ID: 4YOC). (**a2,b2**) depict the PTM sites and MoRF regions obtained from the D2P2 server.

3.1.11. Intrinsic Disorder in Ataxin-3 (ATXN3)

ATXN3 (UniProt ID: P54252) is a 42-kDa ubiquitously expressed deubiquitinating enzyme involved in the degradation of misfolded chaperone substrates, transcription, cytoskeleton regulation, and maintenance of protein homeostasis [39,74,156]. Polyglutamine repeat expansion in the unstructured C-terminus of the human ataxin-3 protein leads to Spinocerebellar Ataxia Type 3 (SCA3), an age-related neurodegenerative disease [74]. CAG repeat expansion in the ATXN3 coding region is a crucial molecular defect reported in SCA3. Blount et al reported that ubiquitin-binding site 2 (UbS2) of ATXN3 interacts with Rad23 and prevents its degradation by proteasome [73]. Ataxin-3 is a ubiquitin-specific protease that mainly binds with the long polyubiquitin chains of unwanted proteins through its C-terminal UIMs and cleaves the ubiquitin from polyubiquitin-tagged proteins through its N-terminal ubiquitin protease domain just before they are degraded by 26 S proteasome so that the ubiquitin can be used again. However, reports have shown that ATXN3 shows weak or no activity for the chains with four or fewer ubiquitins [74,157]. Besides the important role of ATXN3 in the degradation of proteins, it has been associated with the regulation of transcriptional process. Interestingly, ATXN3 directly binds to DNA through a leucine zipper motif present in between 223 to 270 amino acids [39]. The structure of Ataxin-3 contains two ubiquitin-binding sites UbS1 (residues 77–78) and UbS2 (residue 87) on the catalytic domain. UbS2 mainly controls normal ATXN3 protein levels and turnover in cells. Next to UbS2, three ubiquitin-interacting motifs (UIM) are present, such as UIM1 (residues 224–243), UIM2 (residues 244–263) and UIM3 (residues 331–349), which bind to ubiquitin chains at least four moieties long [73]. Interestingly, from our analysis, all three UIM are present in disordered regions of ATXN3 (Figure 9a,c,d). Moreover, several disorder-based binding regions are present in three UIMs (Table 3). UIMs function in mediating high-affinity binding of ATXN3 to Ub chains. The IDPRs and MoRFs identified in our analysis are represented in Figure 9b,c on the crystal structure of the Josephin domain of ataxin-3 (PDB ID: 2AGA) and the NMR solution structure of the tandem UIM domain of ataxin-3 (PDB ID: 2KLZ). In addition, UIMs restrict the chain types that can be trimmed by the ATXN3 protein [158]. These UIMs are separated by the polyglutamine (polyQ) domain, which interacts with BECN1 via the deubiquitination of the Lys-402 of BECN1 that stabilizes BECN1 leads to starvation-induced autophagy [159]. The VCP binding site (residues 282–285) is present prior to the polyQ region, which was found to be located in disordered and MoRF regions of this protein [73]. Notably, the Arginine-rich region at the C-terminal of ATXN3 binds the AAA ATPase protein VCP [73]. A nuclear localization signal (NLS) is present in the region from 273 to 286 amino acids, which is also located in disordered and MoRF regions. Residues 1–27 and 237–286 are important for cellular localization, and both the regions are found to be MoRFs (see Table 3). Additionally, ATXN3 also contains six nuclear export signals (NES); among them, residues 77–99 and 141–158 are important NES since they show significant nuclear export activity. The large Josephin domain at the N-terminus of ATXN3 has low isopeptidase activity. UIMs along with the Josephin domain either rescue proteins from degradation or stimulate protein degradation by deubiquitinating protein and maintaining free reusable ubiquitin [39]. Finally, mutation in ATXN3 leads to consequences such as impaired autophagy, compromised axonal transport, mitochondrial dysfunction, transcriptional deregulation, and proteasomal dysfunction [39]. Figure 9a,d shows that the C-terminal half of human ATXN3 is predicted to be highly disordered, possessing several MoRF sites; however, no single PTM site was identified. Importantly, the vast majority of the aforementioned functional regions of this protein are concentrated within this IDPR, clearly indicating the importance of intrinsic disorder for the overall functionality of human ATXN3.

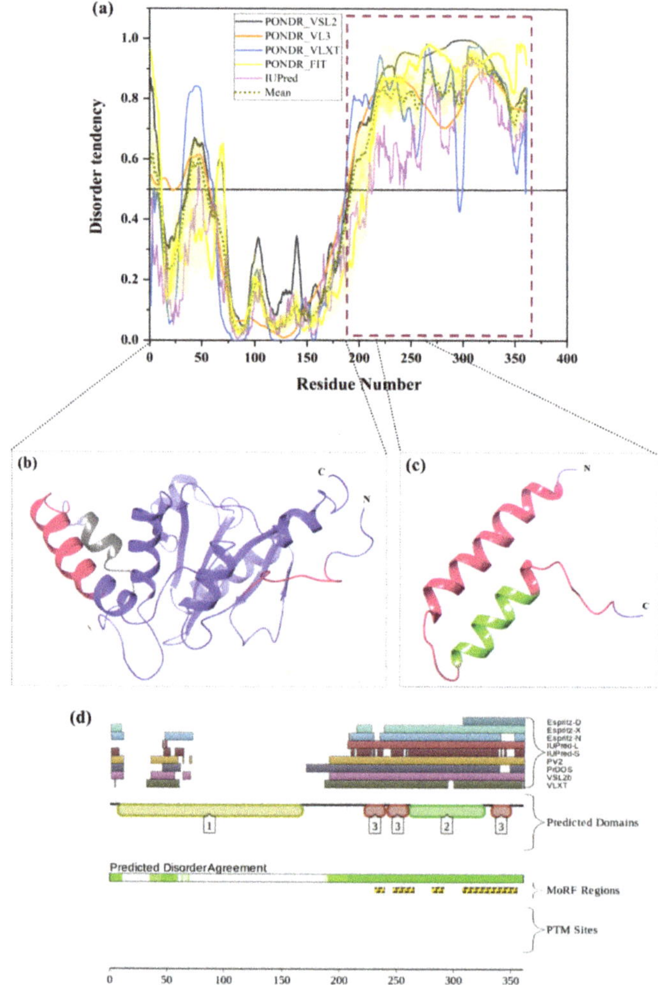

Figure 9. Intrinsic disorder predisposition of ataxin-3. (**a**) Intrinsic disorder profile generated for ataxin-3 (UniProt ID: P54252) by a set of per-residue disorder predictors, such as PONDR® VSL2, PONDR® VL3, PONDR® VLXT, PONDR® FIT, and IUPred. (**b**) Crystal structure of the Josephin domain of ataxin-3 (PDB ID: 2AGA). (**c**) NMR solution structure of the tandem UIM domain of ataxin-3 (PDB ID: 2KLZ). In Plot (**a**), disorder profiles generated by set of disorder predictors, such as PONDR® VSL2, PONDR® VL3, PONDR® VLXT, PONDR® FIT, and IUPred, are depicted by black, orange, blue, yellow, and purple curves respectively. A mean disorder profile calculated from average of five predictor-specific per-residue disorder profile is shown by the olive color curve. Predicted disorder scores above 0.5 are considered as disordered residues/regions. The light-olive shadow around the mean curve represents the error distribution for the mean. The light-yellow shadow around the PONDR® FIT curve shows the error distribution for PONDR® FIT. In Plots (**b**) and (**c**), ataxin-3 is represented by a faded blue color; disordered residues are shown by a salmon pink color. In (**b**), the position of MoRFs (residues 56–65) predicted by the MoRFCHiBi_web server is shown by a gray color (PDB ID: 2AGA). In (**c**), the position of MoRFs (residues 246–255) predicted by the MoRFCHiBi_Web server is represented as MoRFs lying in IDPRs by faded green color (PDB ID:2KLZ). (**d**) The D2P2 server-based functional disorder profile is shown.

3.1.12. Intrinsic Disorder in Adhesion-Regulating Molecule 1 (ADRM1)

ADRM1 (also known as proteasome regulatory particle non-ATPase 13, RPN13; UniProt ID: Q16186) is a 407-residue-long protein, which, in addition to being a component of the 26S proteasome, also has a role in cell adhesion. It is a proteasomal receptor that recognizes K-48-linked poly-Ub chains on the target proteins. It also acts as a receptor for a deubiquitinase, UCHL37. ADRM1 has a Pleckstrin-like a receptor for the ubiquitin domain located at its N-terminal region (residues 22–130), which binds to Ub. From our analysis, this domain is the part of the ordered region in ADRM1 (Figure 10a). Husnjak et al., in 2008, showed that blocking ADRM1 via SiRNA or RA190 triggers plasmacytoid dendritic cell, cytotoxic T lymphocyte, and natural killer cell-mediated lysis of multiple myeloma cells [160]. Jiang et al. demonstrated, in 2017, that levels of ADRM1 are elevated in high-grade ovarian serous carcinoma and serous tubal intraepithelial carcinoma [161]. Importantly, the inhibition of ADRM1 leads to the accumulation of poly-Ub-proteins, triggering apoptosis in cancer cells [161]. The NMR solution structure determined for full-length human ADRM1 (PDB ID: 2KR0) revealed that this protein contains high levels of intrinsic disorder, with the N- and C-terminally located ubiquitin- and UCH37-binding domains (residues 22–130 and 253–407, respectively) being packed against each other when ADRM1 is not incorporated into the proteasome [144]. Figure 10c represents this rather unusual structure, where a large central part of the protein is completely unstructured. In line with these observations, Figure 10a,b shows that the central 150-residue-long region is predicted to be highly disordered but contains multiple PTMs and MoRFs. According to D2P2 (Figure 10b), ADRM1 has 16 phosphorylation sites, of which 15 are located in the IDPRs, 2 acetylation sites in the IDPRs, 12 ubiquitylation sites, of which 9 are in IDPRs, and 2 mono-methylation sites, of which 1 is located in IDPRs. Fascinatingly, ADRM1 is the proteasomal ubiquitin receptor, which plays a vital role in recognition, recruitment, and eventually degradation of protein substrates through 26S proteasome in a well-controlled manner. These processes require interaction with biological partners and the identified dynamic and flexible regions may have crucial roles for protein degradation. Therefore, ADRM1 could be targeted for drug development against cancer as well as neurodegenerative diseases.

3.1.13. Intrinsic Disorder in 26S Proteasome Non-ATPase Regulatory Subunit 2 (PSMD2)

PSMD2 (also known as 26S proteasome regulatory subunit RPN1; UniProt ID: Q13200) is a 100-kDa protein composed of 909 amino acids. PSMD2 may have a role in presenting ubiquitinated substrates to the proteasome. Shi et al., demonstrated, in 2016, that PSMD2 has two binding sites, T1 and T2, in its toroid domain, which interact with Ub, UBL of shuttles, and the UBL domains of DUB and Ubp6, respectively [162]. It has a leucine-rich repeat like domain at its N-terminus, a horseshoe-shaped structure which has a β-sheet at its inner side that interacts with the UBL domains of Rad23 and Dsk2 shuttles and helps in unloading the substrate onto the proteasomal ATPases [163]. Furthermore, the knockdown of PSMD2-suppressed cell proliferation in breast cancer cell lines and also an upregulation of p21 and p27 was seen [164]. Figure 11a shows that, although human PSMD2 is mostly ordered, it contains several long IDPRs, including a 100-residue-long N-terminal region, which is heavily decorated by different PTMs (see Figure 11(a1)), including 17 phosphorylation sites of which 12 are located in the IDPRs, 27 ubiquitylation sites, of which 16 are located in IDPRs, and 1 acetylation site. In addition to PTM sites, the N-terminal region also comprises several disorder-based binding regions (Table 3) which were identified by three different predictors: MoRFCHiBi_Web (residues 1–13, 96–102), MoRFpred (residues 51–62), and ANCHOR (residues 1–30, 35–79). These regions may have a vital role in polyubiquitinated protein recognition on 19S RP of the 26S proteasome.

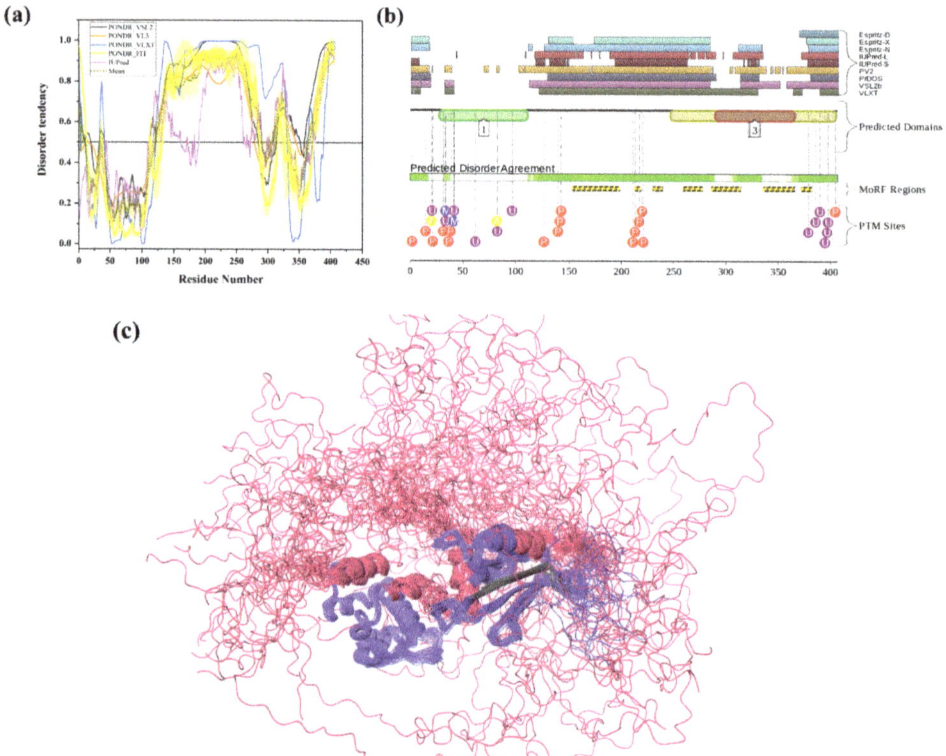

Figure 10. Intrinsic disorder predisposition of ADRM1. (**a**) Intrinsic disorder profile generated for ADRM1 (UniProt ID: Q16186) by a set of per-residue disorder predictors, such as PONDR® VSL2, PONDR® VL3, PONDR® VLXT, PONDR® FIT, and IUPred. (**b**) D2P2 server-based functional disorder profile for ADRM1. (**c**) Solution NMR structure of the ADRM1 with 20 different conformations (PDB ID: 2KR0). In Plot (**a**), disorder profiles generated by sets of disorder predictors, such as PONDR® VSL2, PONDR® VL3, PONDR® VLXT, PONDR® FIT, and IUPred, are depicted by black, orange, blue, yellow, and purple curves, respectively. The mean disorder profile, calculated from the average of five predictor-specific per-residue disorder profiles, is shown by the olive color curve. The predicted disorder score above 0.5 are considered as disordered residues/regions. The light-olive shadow around mean curve represents the error distribution for the mean. The light-yellow shadow around PONDR® FIT curve shows error distribution for PONDR® FIT. In (**c**), ADRM1 is represented by the faded blue color, and disordered residues are shown by the salmon pink color.

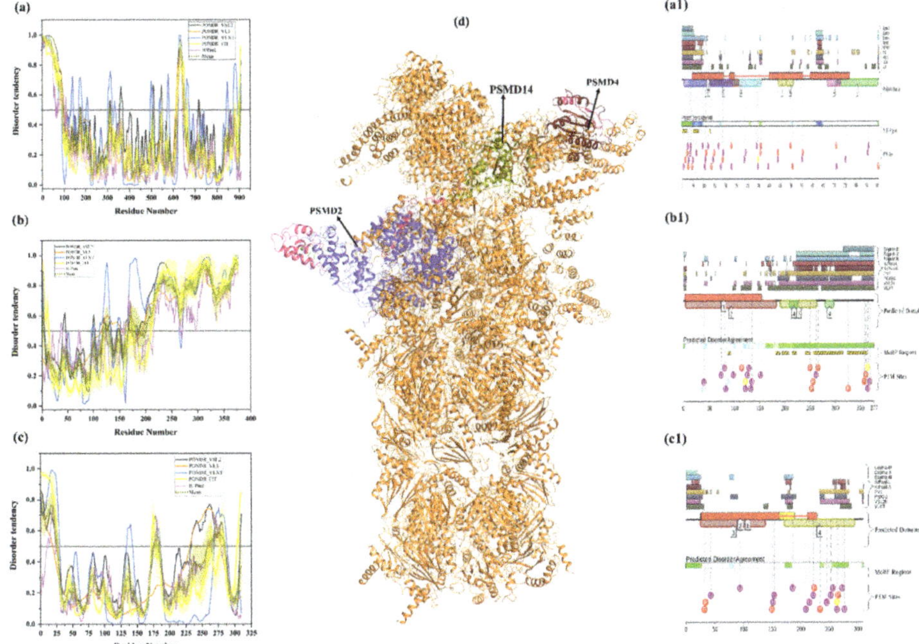

Figure 11. Intrinsic disorder predisposition of human PSMD2 (**a**), PSMD4 (**b**), and PSMD14 (**c**). Plots show the intrinsic disorder profile generated by a set of per-residue disorder predictors, such as PONDR® VSL2, PONDR® VL3, PONDR® VLXT, PONDR® FIT, and IUPred. The corresponding outputs are depicted by black, orange, blue, yellow, and purple curves respectively. Mean disorder profile is calculated from the average of five predictor-specific per-residue disorder profiles and shown by the olive color curve. Predicted disorder scores above 0.5 are considered disordered residues/regions. Light-olive shadow around the mean curve represents the error distribution for the mean. The light-yellow shadow around the PONDR® FIT curve shows the error distribution for that predictor. (**d**) portrays the cryo-EM structures of substrate-engaged human 26S proteasome in seven different conformational states (PDB ID: 6MSB); all three reported proteins PSMD2 (blue), PSMD4 (wine) and PSMD14 (olive) are shown in this complex. D2P2 server-based functional disorder profiles are shown in (**a1,b1,c1**).

3.1.14. Intrinsic Disorder in 26S Proteasome Non-ATPase Regulatory Subunit 4 (PSMD4)

PSMD4 is a non-ATPase 19S base component of the 26S proteasome complex (UniProt ID: P55036). Apart from this, it is also present in substantial amounts in free form. This is a 41-kDa protein composed of 377 amino acids. It is also called RPN10 and is responsible for recognizing Ub moieties on proteins. It has an N-terminal von Willebrand factor A domain (VWA) (residues 5–188) and two C-terminal helical UIM (residues 211–230 and 282–301). From our analysis, we observed that the N-terminal VWA domain is located in an ordered region, and the C-terminal UIM is located in disordered regions of PSMD4 (Figure 11b,(b1)). Moreover, several MoRF regions and PTMs are found in the UIM domain (Table 3 and Figure 11(b1)). Mice expressing PSMD4 lacking the UIM domains but having an intact VWA domain died in utero, whereas liver-specific deletion of UIMs resulted in the accumulation of ubiquitinated proteins; yet, these mice lived longer than PSMD4-null mice, suggesting that the VWA domain may act as a facilitator [165]. Jiang et al. demonstrated that knockdown of PSMD4 in hepatocellular carcinoma cell lines suppressed cell proliferation, which could be reversed by overexpressing AKT as PSMD4 promoted PTEN degradation [166]. Recently, Chen et al. found that UIM-2 of PSMD4 is capable of interacting with the UBL domain of UBQLN2 and prefers K11 and K48 Ub linkages in substrates [167]. They also resolved the structure of these interactions [167]. Figure 11b shows that, when not in a complex with the proteasome, a fragment of the PSMD4 protein (residues 196–306) containing two UIMs is characterized by high conformational flexibility (PDB ID: 1YX4) [168]. This is in line with the results of the prediction of intrinsic disorder predisposition of this protein (Figure 11b,(b1)), showing high levels of disorder in the C-terminal part of this protein. Similar to IDPRs in other proteins, this disordered C-terminal region of PSMD4 contains multiple PTMs and disordered binding regions. D2P2 analysis (Figure 11(b1)) predicted the presence of 8 phosphorylation sites in IDPRs, 14 ubiquitylation site, of which 11 are located in IDPRs, 2 acetylation sites in IDPRs, and 2 nitrosylation sites, of which 1 is located in a disordered region. Importantly, regions important for interaction with UBQLN1 (residues 197–262) or binding to Ub (residues 216–220 and 287–291) are all localized within this disordered half, as well as MoRF regions of this protein, indicating the crucial role of intrinsic disorder in PSMD4 function.

3.1.15. Intrinsic Disorder in 26S Proteasome Non-ATPase Regulatory Subunit 14 (PSMD14)

PSMD14 (also called Rpn11; UniProt ID: O00487) is made up of 310 amino acids present in the 19S cap and helps in substrate deubiquitination [169]. It is a zinc-dependent metallopeptidase that belongs to the JAMM family of proteases [170]. It is present in a heterodimeric complex with Rpn8. The cryo-EM structures of substrate-engaged human 26S proteasome in seven different conformational states (PDB ID: 6MSB) are shown in Figure 11d, where PSMD14 is represented in olive color. Downregulation of the PSMD14 gene has been associated with AD [171]. The knockdown of PSMD14 via RNAi has been reported to induce cell cycle arrest, ultimately leading to senescence in carcinoma cell lines [76]. So far, this enzyme has been implicated in cancer by many researchers. In cancer cells, RNAi of PSMD14 decreased proteasome activity and inhibited cell growth. Zhang et al. reported that the human ortholog of PSMD14, POH1, deubiquitinates pro-interleukin-1β, which helps in suppressing inflammasome activity in macrophages of mice [172]. Furthermore, the knockdown of POH1 inhibited tumor progression and induced apoptosis in mitochondria in vitro and RNAi of POH1 achieved similar results in vivo [173]. PSMD14 stabilizes the SNAIL protein by deubiquitination, which is associated with human esophageal squamous cell carcinoma [174]. Similarly, this enzyme has also been associated with breast cancer [38]. In an interesting study by Song et al., it was shown that PSMD14 is overexpressed in multiple myeloma cells, and its pharmacological inhibition helps these cells to overcome their resistance to the proteasome inhibitor bortezomib [175]. PSMD14 is involved in various biological processes, such as programmed cell death, DNA repair, and embryonic cell development and differentiation [38]. Mpr1p, Pad1p N-terminal (MPN) domain (residues 31–166) present at the N-terminal region of the protein is crucial for the functioning of PSMD14, which releases ubiquitin from ubiquitinated proteins [176], and our analysis identified that the MPN domain is located in the ordered region of PSMD14 protein (see Figure 11c,(c1)).

According to the multifactorial disorder analysis, human PSMD14 contains several IDPRs (see Figure 11c). Some of these IDPRs serve as disorder-based binding sites (see Table 3), whereas others are PTM sites (see Figure 11(c1)), which include six phosphorylation sites, of which four are located in IDPRs, 14 ubiquitylation sites, of which 8 are located in IDPRs, and one acetylation site in a disordered region. At the N-terminal region of PSMD14, a short MoRF region identified by three different servers—MoRFCHiBi_Web (residues 1–7), MoRFpred (residues 1–9), and DISOPRED3 (residues 1–12)—indicates participation of its N-terminal region in 26S proteasome-mediated degradation of polyubiquitinated target protein.

4. Conclusions

The ubiquitin proteasome system plays a key role in the pathogenesis of various types of cancers and neurodegenerative diseases. The components of UPS actively participate in the process of protein degradation, and any disturbance in this system leads to the occurrence of aforementioned diseases. Our intrinsic disorder and MoRF analysis of disease-associated UPS proteins recognized numerous functional IDPRs and disorder-based binding regions. Using five different IDP prediction tools, we found that UBA1 and UCHL1 are highly ordered; USP14, PSMD14, UCHL5, PSMD2, USP7, and UBB are moderately disordered; and UBQLN1, UBQLN2, ADRM1, PSMD4, ATXN3, STUB1, and UBE2R2 are highly disordered proteins. We also documented multiple post-translational modifications (PTMs) sites in all the proteins except ataxin-3, and most of the identified PTMs are located in the disordered regions of proteins. Since these proteins have to interact with their biological partners for the normal functioning of protein homeostasis, disorder-based binding regions may have an important role due to their flexibility and dynamic nature. For example, the 19S regulatory particle (RP) of the 26S proteasome contains receptors (such as ADRM1 and PSMD4) for interaction with the polyubiquitin chain of the target protein. Interestingly, in this study, these receptors are found to be highly disordered. Moreover, the domains of extraproteasomal polyubiquitin receptors (UBQLN1 and UBQLN2), which are responsible for interaction with proteasome as well as polyubiquitinated proteins for their degradation, are found to be highly disordered. These IDPRs in the components of UPS may have functional roles in maintaining cellular protein homeostasis, normal functioning of UPS, and aberrant UPS functionality in the pathogenesis of the diseases mentioned above. Further studies are required for complete elucidation of the roles of these identified IDPRs and disorder-based binding regions in the pathogenesis of diseases.

Author Contributions: R.G.: conception and design, interpretation of data, editing of the manuscript, and study supervision. V.N.U.: acquisition of data, interpretation of data, and editing of the manuscript. K.G., P.K., and S.K.K.: acquisition of data, interpretation of data, and writing of the manuscript. All authors have read and agreed to the published version of the manuscript.

Funding: This research was funded by the Department of Biotechnology (DBT), Government of India (grant number: BT/PR16871/NER/95/329/2015).

Acknowledgments: Authors would like to thank IIT Mandi for all the facilities. KG is grateful to DBT, India (BT/PR16871/NER/95/329/2015). PK was supported by an IIT Mandi-IIT Ropar-PGI Chandigarh, BioX consortium grant (IITM/INT/RG/18). SKK is thankful to DBT, India (BT/PR15453/BRB/10/1460/2015).

Conflicts of Interest: All the authors declare that there is no financial competing interest.

References

1. Etlinger, J.D.; Goldberg, A.L. A soluble ATP dependent proteolytic system responsible for the degradation of abnormal proteins in reticulocytes. *Proc. Natl. Acad. Sci. USA* **1977**, *74*, 54–58. [CrossRef] [PubMed]
2. Wilkinson, K.D. The discovery of ubiquitin-dependent proteolysis. *Proc. Natl. Acad. Sci. USA* **2005**, *102*, 15280–15282. [CrossRef] [PubMed]
3. Ciechanover, A.; Heller, H.; Elias, S.; Haas, A.L.; Hershko, A. U2 R ATP dependence1980 Ciechanover. *Proc. Natl. Acad. Sci. USA* **1980**, *77*, 1365–1368. [CrossRef] [PubMed]
4. Hershko, A.; Ciechanover, A.; Rose, I.A. Resolution of the ATP dependent proteolytic system from reticulocytes: A component that interacts with ATP. *Proc. Natl. Acad. Sci. USA* **1979**, *76*, 3107–3110. [CrossRef] [PubMed]
5. Goldberg, A.L. Protein degradation and protection against misfolded or damaged proteins. *Nature* **2003**, *426*, 895–899. [CrossRef] [PubMed]
6. Zhang, C.; Saunders, A.J. An emerging role for Ubiquilin 1 in regulating protein quality control system and in disease pathogenesis. *Discov. Med.* **2009**, *8*, 18–22.
7. He, M.; Zhou, Z.; Shah, A.A.; Zou, H.; Tao, J.; Chen, Q.; Wan, Y. The emerging role of deubiquitinating enzymes in genomic integrity, diseases, and therapeutics. *Cell Biosci.* **2016**, *6*, 1–15. [CrossRef]
8. Tu, Y.; Chen, C.; Pan, J.; Xu, J.; Zhou, Z.G.; Wang, C.Y. The ubiquitin proteasome pathway (UPP) in the regulation of cell cycle control and DNA damage repair and its implication in tumorigenesis. *Int. J. Clin. Exp. Pathol.* **2012**, *5*, 726–738.
9. Dwane, L.; Gallagher, W.M.; Chonghaile, T.N.; O'Connor, D.P. The emerging role of nontraditional ubiquitination in oncogenic pathways. *J. Biol. Chem.* **2017**, *292*, 3543–3551. [CrossRef]
10. Bhattacharya, S.; Chakraborty, D.; Basu, M.; Ghosh, M.K. Emerging insights into HAUSP (USP7) in physiology, cancer and other diseases. *Signal Transduct. Target. Ther.* **2018**, *3*, 1–12. [CrossRef]

11. Gadhave, K.; Bolshette, N.; Ahire, A.; Pardeshi, R.; Thakur, K.; Trandafir, C.; Istrate, A.; Ahmed, S.; Lahkar, M.; Muresanu, D.F.; et al. The ubiquitin proteasomal system: A potential target for the management of Alzheimer's disease. *J. Cell. Mol. Med.* **2016**, *20*, 1392–1407. [CrossRef] [PubMed]

12. Bedford, L.; Paine, S.; Sheppard, P.W.; Mayer, R.J.; Roelofs, J. Assembly, structure, and function of the 26S proteasome. *Trends Cell Biol.* **2010**, *20*, 391–401. [CrossRef] [PubMed]

13. Kulathu, Y.; Komander, D. Atypical ubiquitylation-the unexplored world of polyubiquitin beyond Lys48 and Lys63 linkages. *Nat. Rev. Mol. Cell Biol.* **2012**, *13*, 508–523. [CrossRef]

14. Li, W.; Ye, Y. Polyubiquitin chains: Functions, structures, and mechanisms. *Cell. Mol. Life Sci.* **2008**, *65*, 2397–2406. [CrossRef] [PubMed]

15. Tian, Y.; Ding, W.; Wang, Y.; Ji, T.; Sun, S.; Mo, Q.; Chen, P.; Fang, Y.; Liu, J.; Wang, B.; et al. Ubiquitin B in cervical cancer: Critical for the maintenance of cancer stem-like cell characters. *PLoS ONE* **2013**, *8*, e84457. [CrossRef] [PubMed]

16. Dennissen, F.J.A.; Kholod, N.; Hermes, D.J.H.P.; Kemmerling, N.; Steinbusch, H.W.M.; Dantuma, N.P.; van Leeuwen, F.W. Mutant ubiquitin (UBB+1) associated with neurodegenerative disorders is hydrolyzed by ubiquitin C-terminal hydrolase L3 (UCH-L3). *FEBS Lett.* **2011**, *585*, 2568–2574. [CrossRef]

17. Dennissen, F.J.A.; Kholod, N.; Steinbusch, H.W.M.; Van Leeuwen, F.W. Misframed proteins and neurodegeneration: A novel view on Alzheimer's and Parkinson's diseases. *Neurodegener. Dis.* **2010**, *7*, 76–79. [CrossRef]

18. Van Leeuwen, F.W.; De Kleijn, D.P.V.; Van Den Hurk, H.H.; Neubauer, A.; Sonnemans, M.A.F.; Sluijs, J.A.; Köycü, S.; Ramdjielal, R.D.J.; Salehi, A.; Martens, G.J.M.; et al. Frameshift mutants of β amyloid precursor protein and ubiquitin-B in Alzheimer's and Down patients. *Science* **1998**, *279*, 242–247. [CrossRef]

19. Fischer, D.F.; De Vos, R.A.I.; Van Dijk, R.; De Vrij, F.M.S.; Proper, E.A.; Sonnemans, M.A.F.; Verhage, M.C.; Sluijs, J.A.; Hobo, B.; Zouambia, M.; et al. Disease-specific accumulation of mutant ubiquitin as a marker for proteasomal dysfunction in the brain. *FASEB J.* **2003**, *17*, 2014–2024. [CrossRef]

20. Cook, J.C.; Chock, P.B. Isoforms of mammalian ubiquitin-activating enzyme. *J. Biol. Chem.* **1992**, *267*, 24315–24321.

21. Dlamini, N.; Josifova, D.J.; Paine, S.M.L.; Wraige, E.; Pitt, M.; Murphy, A.J.; King, A.; Buk, S.; Smith, F.; Abbs, S.; et al. Clinical and neuropathological features of X-linked spinal muscular atrophy (SMAX2) associated with a novel mutation in the UBA1 gene. *Neuromuscul. Disord.* **2013**, *23*, 391–398. [CrossRef]

22. David, Y.; Ziv, T.; Admon, A.; Navon, A. The E2 ubiquitin-conjugating enzymes direct polyubiquitination to preferred lysines. *J. Biol. Chem.* **2010**, *285*, 8595–8604. [CrossRef]

23. Alpi, A.F.; Chaugule, V.; Walden, H. Mechanism and disease association of E2-conjugating enzymes: Lessons from UBE2T and UBE2L3. *Biochem. J.* **2016**, *473*, 3401–3419. [CrossRef]

24. Miller, V.M.; Nelson, R.F.; Gouvion, C.M.; Williams, A.; Rodriguez-Lebron, E.; Harper, S.Q.; Davidson, B.L.; Rebagliati, M.R.; Paulson, H.L. CHIP suppresses polyglutamine aggregation and toxicity in vitro and in vivo. *J. Neurosci.* **2005**, *25*, 9152–9161. [CrossRef]

25. Marín, I. The ubiquilin gene family: Evolutionary patterns and functional insights. *BMC Evol. Biol.* **2014**, *14*, 63. [CrossRef]

26. Renaud, L.; Picher-Martel, V.; Codron, P.; Julien, J.-P. Key role of UBQLN2 in pathogenesis of amyotrophic lateral sclerosis and frontotemporal dementia. *Acta Neuropathol. Commun.* **2019**, *7*, 103. [CrossRef]

27. Ravid, T.; Hochstrasser, M. Diversity of degradation signals in the ubiquitin-proteasome system. *Nat. Rev. Mol. Cell Biol.* **2008**, *9*, 679–689. [CrossRef]

28. Jang, H.H. Regulation of Protein Degradation by Proteasomes in Cancer. *J. Cancer Prev.* **2018**, *23*, 153–161. [CrossRef]

29. Reyes-Turcu, F.E.; Ventii, K.H.; Wilkinson, K.D. Regulation and cellular roles of ubiquitin-specific deubiquitinating enzymes. *Annu. Rev. Biochem.* **2009**, *78*, 363–397. [CrossRef]

30. Amerik, A.Y.; Hochstrasser, M. Mechanism and function of deubiquitinating enzymes. *Biochim. Biophys. Acta* **2004**, *1695*, 189–207. [CrossRef]

31. Bishop, P.; Rocca, D.; Henley, J.M. Ubiquitin C-terminal hydrolase L1 (UCH-L1): Structure, distribution and roles in brain function and dysfunction. *Biochem. J.* **2016**, *473*, 2453–2462. [CrossRef]

32. Setsuie, R.; Wang, Y.L.; Mochizuki, H.; Osaka, H.; Hayakawa, H.; Ichihara, N.; Li, H.; Furuta, A.; Sano, Y.; Sun, Y.J.; et al. Dopaminergic neuronal loss in transgenic mice expressing the Parkinson's disease-associated UCH-L1 I93M mutant. *Neurochem. Int.* **2007**, *50*, 119–129. [CrossRef]

33. Yao, T.; Song, L.; Xu, W.; DeMartino, G.N.; Florens, L.; Swanson, S.K.; Washburn, M.P.; Conaway, R.C.; Conaway, J.W.; Cohen, R.E. Proteasome recruitment and activation of the Uch37 deubiquitinating enzyme by Adrm1. *Nat. Cell Biol.* **2006**, *8*, 994–1002. [CrossRef]

34. Kim, R.Q.; Sixma, T.K.; van Dijk, W.J.; Sixma, T.K.; Faesen, A.C.; Luna-Vargas, M.P.A.; Geurink, P.P.; Clerici, M.; Merkx, R.; van Dijk, W.J.; et al. Mechanism of USP7/HAUSP activation by its C-Terminal ubiquitin-like domain and allosteric regulation by GMP-synthetase. *Mol. Cell* **2016**, *281*, 147–159.

35. Hu, M.; Li, P.; Song, L.; Jeffrey, P.D.; Chenova, T.A.; Wilkinson, K.D.; Cohen, R.E.; Shi, Y. Structure and mechanisms of the proteasome-associated deubiquitinating enzyme USP14. *EMBO J.* **2005**, *24*, 3747–3756. [CrossRef]

36. Miller, S.; Ortuno, D.; Carlisle, H.J. Does inactivation of USP14 enhance degradation of proteasomal substrates that are associated with neurodegenerative diseases? *F1000Research* **2016**, *5*, 137.

37. Gomes, A.V. Genetics of proteasome diseases. *Scientifica* **2013**, *2013*, 637629. [CrossRef]

38. Luo, G.; Hu, N.; Xia, X.; Zhou, J.; Ye, C. RPN11 deubiquitinase promotes proliferation and migration of breast cancer cells. *Mol. Med. Rep.* **2017**, *16*, 331–338. [CrossRef]

39. Evers, M.M.; Toonen, L.J.A.; van Roon-Mom, W.M.C. Ataxin-3 protein and RNA toxicity in spinocerebellar ataxia type 3: Current insights and emerging therapeutic strategies. *Mol. Neurobiol.* **2014**, *49*, 1513–1531. [CrossRef]

40. Tanaka, K. The proteasome: Overview of structure and functions. *Proc. Japan Acad. Ser. B Phys. Biol. Sci.* **2009**, *85*, 12–36. [CrossRef]

41. Lu, X.; Nowicka, U.; Sridharan, V.; Liu, F.; Randles, L.; Hymel, D.; Dyba, M.; Tarasov, S.G.; Tarasova, N.I.; Zhao, X.Z.; et al. Structure of the Rpn13-Rpn2 complex provides insights for Rpn13 and Uch37 as anticancer targets. *Nat. Commun.* **2017**, *8*, 15540. [CrossRef]

42. Collins, G.A.; Goldberg, A.L. The Logic of the 26S Proteasome. *Cell* **2017**, *169*, 792–806. [CrossRef] [PubMed]

43. Ciechanover, A.; Schwartz, A.L. The ubiquitin system: Pathogenesis of human diseases and drug targeting. *Biochim. Biophys. Acta* **2004**, *1695*, 3–17. [CrossRef] [PubMed]

44. Kulkarni, P.; Uversky, V.N. Intrinsically Disordered Proteins in Chronic Diseases. *Biomolecules* **2019**, *9*, 147. [CrossRef] [PubMed]

45. Giri, R.; Morrone, A.; Toto, A.; Brunori, M.; Gianni, S. Structure of the transition state for the binding of c-Myb and KIX highlights an unexpected order for a disordered system. *Proc. Natl. Acad. Sci. USA* **2013**, *110*, 14942–14947. [CrossRef]

46. Gianni, S.; Morrone, A.; Giri, R.; Brunori, M. A folding-after-binding mechanism describes the recognition between the transactivation domain of c-Myb and the KIX domain of the CREB-binding protein. *Biochem. Biophys. Res. Commun.* **2012**, *428*, 205–209. [CrossRef]

47. Toto, A.; Camilloni, C.; Giri, R.; Brunori, M.; Vendruscolo, M.; Gianni, S. Molecular Recognition by Templated Folding of an Intrinsically Disordered Protein. *Sci. Rep.* **2016**, *6*, 21994. [CrossRef]

48. Sharma, N.; Fonin, A.V.; Shpironok, O.G.; Silonov, S.A.; Turoverov, K.K.; Uversky, V.N.; Kuznetsova, I.M.; Giri, R. Folding perspectives of an intrinsically disordered transactivation domain and its single mutation breaking the folding propensity. *Int. J. Biol. Macromol.* **2019**. [CrossRef]

49. Gadhave, K.; Giri, R. Amyloid formation by intrinsically disordered trans-activation domain of cMyb. *Biochem. Biophys. Res. Commun.* **2020**, *524*, 446–452. [CrossRef]

50. Habchi, J.; Tompa, P.; Longhi, S.; Uversky, V.N. Introducing protein intrinsic disorder. *Chem. Rev.* **2014**, *114*, 6561–6588. [CrossRef]

51. Perdigão, N.; Heinrich, J.; Stolte, C.; Sabir, K.S.; Buckley, M.J.; Tabor, B.; Signal, B.; Gloss, B.S.; Hammang, C.J.; Rost, B.; et al. Unexpected features of the dark proteome. *Proc. Natl. Acad. Sci. USA* **2015**, *112*, 15898–15903. [CrossRef] [PubMed]

52. Basile, W.; Salvatore, M.; Bassot, C.; Elofsson, A. Why do eukaryotic proteins contain more intrinsically disordered regions? *PLoS Comput. Biol.* **2019**, *15*, e1007186. [CrossRef] [PubMed]

53. Kumar, D.; Sharma, N.; Giri, R. Therapeutic interventions of cancers using intrinsically disordered proteins as drug targets: C-myc as model system. *Cancer Inform.* **2017**, *16*, 1176935117699408. [CrossRef] [PubMed]

54. Aarthy, M.; Kumar, D.; Giri, R.; Singh, S.K. E7 oncoprotein of human papillomavirus: Structural dynamics and inhibitor screening study. *Gene* **2018**, *658*, 159–177. [CrossRef] [PubMed]

55. Kumar, A.; Kumar, P.; Kumari, S.; Uversky, V.N.; Giri, R. Folding and structural polymorphism of p53 C-terminal domain: One peptide with many conformations. *Arch. Biochem. Biophys.* **2020**, *684*, 108342. [CrossRef] [PubMed]

56. Gadhave, K.; Gehi, B.R.; Kumar, P.; Xue, B.; Uversky, V.N.; Giri, R. The dark side of Alzheimer's disease: Unstructured biology of proteins from the amyloid cascade signaling pathway. *Cell. Mol. Life Sci.* **2020**, 1–46. [CrossRef]

57. Boomsma, W.; Nielsen, S.V.; Lindorff-Larsen, K.; Hartmann-Petersen, R.; Ellgaard, L. Bioinformatics analysis identifies several intrinsically disordered human E3 ubiquitin-protein ligases. *PeerJ* **2016**, *2016*, e1725. [CrossRef]

58. Bhowmick, P.; Pancsa, R.; Guharoy, M.; Tompa, P. Functional Diversity and Structural Disorder in the Human Ubiquitination Pathway. *PLoS ONE* **2013**, *8*, e65443. [CrossRef]

59. Guharoy, M.; Bhowmick, P.; Tompa, P. Design principles involving protein disorder facilitate specific substrate selection and degradation by the ubiquitin-proteasome system. *J. Biol. Chem.* **2016**, *291*, 6723–6731. [CrossRef]

60. Manasanch, E.E.; Orlowski, R.Z. Proteasome inhibitors in cancer therapy. *Nat. Rev. Clin. Oncol.* **2017**, *14*, 417–433. [CrossRef]

61. Lim, K.H. Diverse Misfolded Conformational Strains and Cross-seeding of Misfolded Proteins Implicated in Neurodegenerative Diseases. *Front. Mol. Neurosci.* **2019**, *12*, 158. [CrossRef] [PubMed]

62. Bateman, A. UniProt: A worldwide hub of protein knowledge. *Nucleic Acids Res.* **2019**, *47*, D506–D515.

63. Lee, I.; Schindelin, H. Structural insights into E1-catalyzed ubiquitin activation and transfer to conjugating enzymes. *Cell* **2008**, *134*, 268–278. [CrossRef]

64. Groen, E.J.N.; Gillingwater, T.H. UBA1: At the Crossroads of Ubiquitin Homeostasis and Neurodegeneration. *Trends Mol. Med.* **2015**, *21*, 622–632. [CrossRef]

65. Valimberti, I.; Tiberti, M.; Lambrughi, M.; Sarcevic, B.; Papaleo, E. E2 superfamily of ubiquitin-conjugating enzymes: Constitutively active or activated through phosphorylation in the catalytic cleft. *Sci. Rep.* **2015**, *5*, 14849. [CrossRef]

66. Kumar, P.; Ambasta, R.K.; Veereshwarayya, V.; Rosen, K.M.; Kosik, K.S.; Band, H.; Mestril, R.; Patterson, C.; Querfurth, H.W. CHIP and HSPs interact with beta-APP in a proteasome-dependent manner and influence Abeta metabolism. *Hum. Mol. Genet.* **2007**, *16*, 848–864. [CrossRef]

67. Stieren, E.S.; El Ayadi, A.; Xiao, Y.; Siller, E.; Landsverk, M.L.; Oberhauser, A.F.; Barral, J.M.; Boehning, D. Ubiquilin-1 is a molecular chaperone for the amyloid precursor protein. *J. Biol. Chem.* **2011**, *286*, 35689–35698. [CrossRef]

68. Williams, K.L.; Warraich, S.T.; Yang, S.; Solski, J.A.; Fernando, R.; Rouleau, G.A.; Nicholson, G.A.; Blair, I.P. UBQLN2/ubiquilin 2 mutation and pathology in familial amyotrophic lateral sclerosis. *Neurobiol. Aging* **2012**, *33*, 2527.e3–2527.e10. [CrossRef]

69. Zhang, K.Y.; Yang, S.; Warraich, S.T.; Blair, I.P. Ubiquilin 2: A component of the ubiquitin-proteasome system with an emerging role in neurodegeneration. *Int. J. Biochem. Cell Biol.* **2014**, *50*, 123–126. [CrossRef]

70. Das, C.; Hoang, Q.Q.; Kreinbring, C.A.; Luchansky, S.J.; Meray, R.K.; Ray, S.S.; Lansbury, P.T.; Ringe, D.; Petsko, G.A. Structural basis for conformational plasticity of the Parkinson's disease-associated ubiquitin hydrolase UCH-L1. *Proc. Natl. Acad. Sci. USA* **2006**, *103*, 4675–4680. [CrossRef]

71. Fang, Y.; Fu, D.; Tang, W.; Cai, Y.; Ma, D.; Wang, H.; Xue, R.; Liu, T.; Huang, X.; Dong, L.; et al. Ubiquitin C-terminal Hydrolase 37, a novel predictor for hepatocellular carcinoma recurrence, promotes cell migration and invasion via interacting and deubiquitinating PRP19. *Biochim. Biophys. Acta* **2013**, *1833*, 559–572. [CrossRef]

72. Kim, E.; Park, S.; Lee, J.H.; Mun, J.Y.; Choi, W.H.; Yun, Y.; Lee, J.; Kim, J.H.; Kang, M.-J.; Lee, M.J. Dual Function of USP14 Deubiquitinase in Cellular Proteasomal Activity and Autophagic Flux. *Cell Rep.* **2018**, *24*, 732–743. [CrossRef]

73. Blount, J.R.; Tsou, W.-L.; Ristic, G.; Burr, A.A.; Ouyang, M.; Galante, H.; Scaglione, K.M.; Todi, S. V Ubiquitin-binding site 2 of ataxin-3 prevents its proteasomal degradation by interacting with Rad23. *Nat. Commun.* **2014**, *5*, 4638. [CrossRef]

74. Tzvetkov, N.; Breuer, P. Josephin domain-containing proteins from a variety of species are active de-ubiquitination enzymes. *Biol. Chem.* **2007**, *388*, 973–978. [CrossRef]

75. Qiu, X.-B.; Ouyang, S.-Y.; Li, C.-J.; Miao, S.; Wang, L.; Goldberg, A.L. hRpn13/ADRM1/GP110 is a novel proteasome subunit that binds the deubiquitinating enzyme, UCH37. *EMBO J.* **2006**, *25*, 5742–5753. [CrossRef]

76. Byrne, A.; McLaren, R.P.; Mason, P.; Chai, L.; Dufault, M.R.; Huang, Y.; Liang, B.; Gans, J.D.; Zhang, M.; Carter, K.; et al. Knockdown of human deubiquitinase PSMD14 induces cell cycle arrest and senescence. *Exp. Cell Res.* **2010**, *316*, 258–271. [CrossRef]

77. Peng, K.; Radivojac, P.; Vucetic, S.; Dunker, A.K.; Obradovic, Z. Length-dependent prediction of protein intrinsic disorder. *BMC Bioinformatics* **2006**, *7*, 208. [CrossRef]

78. Peng, K.; Vucetic, S.; Radivojac, P.; Brown, C.J.; Dunker, A.K.; Obradovic, Z. Optimizing long intrinsic disorder predictors with protein evolutionary information. *J. Bioinform. Comput. Biol.* **2005**, *3*, 35–60. [CrossRef]

79. Romero, P.; Obradovic, Z.; Li, X.; Garner, E.C.; Brown, C.J.; Dunker, A.K. Sequence complexity of disordered protein. *Proteins Struct. Funct. Genet.* **2001**, *42*, 38–48. [CrossRef]

80. Xue, B.; Dunbrack, R.L.; Williams, R.W.; Dunker, A.K.; Uversky, V.N. PONDR-FIT: A meta-predictor of intrinsically disordered amino acids. *Biochim. Biophys. Acta* **2010**, *1804*, 996–1010. [CrossRef]

81. Dosztányi, Z.; Csizmok, V.; Tompa, P.; Simon, I. IUPred: Web server for the prediction of intrinsically unstructured regions of proteins based on estimated energy content. *Bioinformatics* **2005**, *21*, 3433–3434. [CrossRef]

82. Garg, N.; Kumar, P.; Gadhave, K.; Giri, R. The dark proteome of cancer: Intrinsic disorderedness and functionality of HIF-1α along with its interacting proteins. *Prog. Mol. Biol. Transl. Sci.* **2019**, *166*, 371–403.

83. Kumar, D.; Singh, A.; Kumar, P.; Uversky, V.N.; Rao, C.D.; Giri, R. Understanding the penetrance of intrinsic protein disorder in rotavirus proteome. *Int. J. Biol. Macromol.* **2019**, *144*, 892–908. [CrossRef]

84. Mishra, P.M.; Uversky, V.N.; Giri, R. Molecular Recognition Features in Zika Virus Proteome. *J. Mol. Biol.* **2018**, *430*, 2372–2388. [CrossRef]

85. Giri, R.; Kumar, D.; Sharma, N.; Uversky, V.N. Intrinsically Disordered Side of the Zika Virus Proteome. *Front. Cell. Infect. Microbiol.* **2016**, *6*, 144. [CrossRef]

86. Singh, A.; Kumar, A.; Yadav, R.; Uversky, V.N.; Giri, R. Deciphering the dark proteome of Chikungunya virus. *Sci. Rep.* **2018**, *8*, 5822. [CrossRef]

87. Malhis, N.; Jacobson, M.; Gsponer, J. MoRFchibi SYSTEM: Software tools for the identification of MoRFs in protein sequences. *Nucleic Acids Res.* **2016**, *44*, W488–W493. [CrossRef]

88. Dosztányi, Z.; Mészáros, B.; Simon, I. ANCHOR: Web server for predicting protein binding regions in disordered proteins. *Bioinformatics* **2009**, *25*, 2745–2746. [CrossRef]

89. Disfani, F.M.; Hsu, W.-L.; Mizianty, M.J.; Oldfield, C.J.; Xue, B.; Dunker, A.K.; Uversky, V.N.; Kurgan, L. MoRFpred, a computational tool for sequence-based prediction and characterization of short disorder-to-order transitioning binding regions in proteins. *Bioinformatics* **2012**, *28*, i75–i83. [CrossRef]

90. Jones, D.T.; Cozzetto, D. DISOPRED3: Precise disordered region predictions with annotated protein-binding activity. *Bioinformatics* **2015**, *31*, 857–863. [CrossRef]

91. Giri, R.; Bhardwaj, T.; Shegane, M.; Gehi, B.R.; Kumar, P.; Gadhave, K.; Oldfield, C.J.; Uversky, V.N. When Darkness Becomes a Ray of Light in the Dark Times: Understanding the COVID-19 via the Comparative Analysis of the Dark Proteomes of SARS-CoV-2, Human SARS and Bat SARS-Like Coronaviruses. *bioRxiv* **2020**. [CrossRef]

92. Singh, A.; Kumar, A.; Uversky, V.N.; Giri, R. Understanding the interactability of chikungunya virus proteins: Via molecular recognition feature analysis. *RSC Adv.* **2018**, *8*, 27293–27303. [CrossRef]

93. Oates, M.E.; Romero, P.; Ishida, T.; Ghalwash, M.; Mizianty, M.J.; Xue, B.; Dosztányi, Z.; Uversky, V.N.; Obradovic, Z.; Kurgan, L.; et al. D^{2}P^2: Database of disordered protein predictions. *Nucleic Acids Res.* **2013**, *41*, 508–516. [CrossRef]

94. Szklarczyk, D.; Gable, A.L.; Lyon, D.; Junge, A.; Wyder, S.; Huerta-Cepas, J.; Simonovic, M.; Doncheva, N.T.; Morris, J.H.; Bork, P.; et al. STRING v11: Protein-protein association networks with increased coverage, supporting functional discovery in genome-wide experimental datasets. *Nucleic Acids Res.* **2019**, *47*, 607–613. [CrossRef]

95. Ciechanover, A.; Schwartz, A.L. The ubiquitin-proteasome pathway: The complexity and myriad functions of proteins death. *Proc. Natl. Acad. Sci. USA* **1998**, *95*, 2727–2730. [CrossRef]

96. Wehmer, M.; Sakata, E. Recent advances in the structural biology of the 26S proteasome. *Int. J. Biochem. Cell Biol.* **2016**, *79*, 437–442. [CrossRef]

97. Toh-e, A. Structure and function of the yeast 26S proteasome. *Seikagaku* **1999**, *71*, 173–181.

98. Livneh, I.; Cohen-Kaplan, V.; Cohen-Rosenzweig, C.; Avni, N.; Ciechanover, A. The life cycle of the 26S proteasome: From birth, through regulation and function, and onto its death. *Cell Res.* **2016**, *26*, 869–885. [CrossRef]

99. Marshall, R.S.; Vierstra, R.D. Dynamic Regulation of the 26S Proteasome: From Synthesis to Degradation. *Front. Mol. Biosci.* **2019**, *6*, 40. [CrossRef]

100. Rajagopalan, K.; Mooney, S.M.; Parekh, N.; Getzenberg, R.H.; Kulkarni, P. A majority of the cancer/testis antigens are intrinsically disordered proteins. *J. Cell. Biochem.* **2011**, *112*, 3256–3267. [CrossRef]

101. Ishida, T.; Kinoshita, K. PrDOS: Prediction of disordered protein regions from amino acid sequence. *Nucleic Acids Res.* **2007**, *35*, 460–464. [CrossRef]

102. Walsh, I.; Martin, A.J.M.; Di Domenico, T.; Tosatto, S.C.E. ESpritz: Accurate and fast prediction of protein disorder. *Bioinformatics* **2012**, *28*, 503–509. [CrossRef]

103. Liu, X.; Sun, L.; Gursel, D.B.; Cheng, C.; Huang, S.; Rademaker, A.W.; Khan, S.A.; Yin, J.; Kiyokawa, H. The non-canonical ubiquitin activating enzyme UBA6 suppresses epithelial-mesenchymal transition of mammary epithelial cells. *Oncotarget* **2017**, *8*, 87480–87493. [CrossRef]

104. Liu, H.-Y.; Pfleger, C.M. Mutation in E1, the ubiquitin activating enzyme, reduces Drosophila lifespan and results in motor impairment. *PLoS ONE* **2013**, *8*, e32835. [CrossRef]

105. Ramser, J.; Ahearn, M.E.; Lenski, C.; Yariz, K.O.; Hellebrand, H.; von Rhein, M.; Clark, R.D.; Schmutzler, R.K.; Lichtner, P.; Hoffman, E.P.; et al. Rare missense and synonymous variants in UBE1 are associated with X-linked infantile spinal muscular atrophy. *Am. J. Hum. Genet.* **2008**, *82*, 188–193. [CrossRef]

106. Lv, Z.; Williams, K.M.; Yuan, L.; Atkison, J.H.; Olsen, S.K. Crystal structure of a human ubiquitin E1-ubiquitin complex reveals conserved functional elements essential for activity. *J. Biol. Chem.* **2018**, *293*, 18337–18352. [CrossRef]

107. Xie, S.T. Expression, purification, and crystal structure of N-terminal domains of human ubiquitin-activating enzyme (E1). *Biosci. Biotechnol. Biochem.* **2014**, *78*, 1542–1549. [CrossRef]

108. Stewart, M.D.; Ritterhoff, T.; Klevit, R.E.; Brzovic, P.S. E2 enzymes: More than just middle men. *Cell Res.* **2016**, *26*, 423–440. [CrossRef]

109. Pruneda, J.N.; Littlefield, P.J.; Soss, S.E.; Nordquist, K.A.; Chazin, W.J.; Brzovic, P.S.; Klevit, R.E. Structure of an E3:E2~Ub complex reveals an allosteric mechanism shared among RING/U-box ligases. *Mol. Cell* **2012**, *47*, 933–942. [CrossRef]

110. Wenzel, D.M.; Lissounov, A.; Brzovic, P.S.; Klevit, R.E. UBCH7 reactivity profile reveals parkin and HHARI to be RING/HECT hybrids. *Nature* **2011**, *474*, 105–108. [CrossRef]

111. Cohen, I.; Wiener, R.; Reiss, Y.; Ravid, T. Distinct activation of an E2 ubiquitin-conjugating enzyme by its cognate E3 ligases. *Proc. Natl. Acad. Sci. USA* **2015**, *112*, 625–632. [CrossRef]

112. Schelpe, J.; Monté, D.; Dewitte, F.; Sixma, T.K.; Rucktooa, P. Structure of UBE2Z Enzyme Provides Functional Insight into Specificity in the FAT10 Protein Conjugation Machinery. *J. Biol. Chem.* **2016**, *291*, 630–639. [CrossRef]

113. Williams, K.M.; Qie, S.; Atkison, J.H.; Salazar-Arango, S.; Alan Diehl, J.; Olsen, S.K. Structural insights into E1 recognition and the ubiquitin-conjugating activity of the E2 enzyme Cdc34. *Nat. Commun.* **2019**, *10*, 3296. [CrossRef]

114. Spratt, D.E.; Shaw, G.S. Association of the disordered C-terminus of CDC34 with a catalytically bound ubiquitin. *J. Mol. Biol.* **2011**, *407*, 425–438. [CrossRef]

115. Uchida, C.; Kitagawa, M. RING-, HECT-, and RBR-type E3 Ubiquitin Ligases: Involvement in Human Cancer. *Curr. Cancer Drug Targets* **2016**, *16*, 157–174. [CrossRef]

116. Plechanovová, A.; Jaffray, E.G.; Tatham, M.H.; Naismith, J.H.; Hay, R.T. Structure of a RING E3 ligase and ubiquitin-loaded E2 primed for catalysis. *Nature* **2012**, *489*, 115–120. [CrossRef]

117. Dou, H.; Buetow, L.; Sibbet, G.J.; Cameron, K.; Huang, D.T. BIRC7-E2 ubiquitin conjugate structure reveals the mechanism of ubiquitin transfer by a RING dimer. *Nat. Struct. Mol. Biol.* **2012**, *19*, 876–883. [CrossRef]

118. Lorick, K.L.; Jensen, J.P.; Fang, S.; Ong, A.M.; Hatakeyama, S.; Weissman, A.M. RING fingers mediate ubiquitin-conjugating enzyme (E2)-dependent ubiquitination. *Proc. Natl. Acad. Sci. USA* **1999**, *96*, 11364–11369. [CrossRef]

119. Berndsen, C.E.; Wiener, R.; Yu, I.W.; Ringel, A.E.; Wolberger, C. A conserved asparagine has a structural role in ubiquitin-conjugating enzymes. *Nat. Chem. Biol.* **2013**, *9*, 154–156. [CrossRef]

120. Berndsen, C.E.; Wolberger, C. New insights into ubiquitin E3 ligase mechanism. *Nat. Struct. Mol. Biol.* **2014**, *21*, 301–307. [CrossRef]

121. Kamadurai, H.B.; Qiu, Y.; Deng, A.; Harrison, J.S.; MacDonald, C.; Actis, M.; Rodrigues, P.; Miller, D.J.; Souphron, J.; Lewis, S.M.; et al. Mechanism of ubiquitin ligation and lysine prioritization by a HECT E3. *Elife* **2013**, *2*, e00828. [CrossRef]

122. Pickart, C.M. Echanisms ndPickart, C.M. Echanisms nderlying ubiquitination. *Annu. Rev. Biochem.* **2001**, *70*, 503–533. [CrossRef]

123. Oh, C.; Park, S.; Lee, E.K.; Yoo, Y.J. Downregulation of ubiquitin level via knockdown of polyubiquitin gene Ubb as potential cancer therapeutic intervention. *Sci. Rep.* **2013**, *3*, 2623. [CrossRef]

124. Finch, J.S.; St John, T.; Krieg, P.; Bonham, K.; Smith, H.T.; Fried, V.A.; Bowden, G.T. Overexpression of three ubiquitin genes in mouse epidermal tumors is associated with enhanced cellular proliferation and stress. *Cell Growth Differ.* **1992**, *3*, 269–278.

125. Tank, E.M.H.; True, H.L. Disease-associated mutant ubiquitin causes proteasomal impairment and enhances the toxicity of protein aggregates. *PLoS Genet.* **2009**, *5*, e1000382. [CrossRef]

126. Chen, X.; Petranovic, D. Role of frameshift ubiquitin B protein in Alzheimer's disease. *Wiley Interdiscip. Rev. Syst. Biol. Med.* **2016**, *8*, 300–313. [CrossRef]

127. Fang, X.; Trexler, C.; Chen, J. Ushering in the cardiac role of Ubiquilin1. *J. Clin. Invest.* **2018**, *128*, 5195–5197. [CrossRef]

128. Ko, H.S.; Uehara, T.; Tsuruma, K.; Nomura, Y. Ubiquilin interacts with ubiquitylated proteins and proteasome through its ubiquitin-associated and ubiquitin-like domains. *FEBS Lett.* **2004**, *566*, 110–114.

129. Whiteley, A.M.; Prado, M.A.; Peng, I.; Abbas, A.R.; Haley, B.; Paulo, J.A.; Reichelt, M.; Katakam, A.; Sagolla, M.; Modrusan, Z.; et al. Ubiquilin1 promotes antigen-receptor mediated proliferation by eliminating mislocalized mitochondrial proteins. *Elife* **2017**, *6*, 26435. [CrossRef]

130. Itakura, E.; Zavodszky, E.; Shao, S.; Wohlever, M.L.; Keenan, R.J.; Hegde, R.S. Ubiquilins Chaperone and Triage Mitochondrial Membrane Proteins for Degradation. *Mol. Cell* **2016**, *63*, 21–33. [CrossRef]

131. Kurlawala, Z.; Dunaway, R.; Shah, P.P.; Gosney, J.A.; Siskind, L.J.; Ceresa, B.P.; Beverly, L.J. Regulation of insulin-like growth factor receptors by Ubiquilin1. *Biochem. J.* **2017**, *474*, 4105–4118. [CrossRef] [PubMed]

132. McKinnon, C.; Tabrizi, S.J. The ubiquitin-proteasome system in neurodegeneration. *Antioxid. Redox Signal.* **2014**, *21*, 2302–2321. [CrossRef]

133. Osaka, M.; Ito, D.; Suzuki, N. Disturbance of proteasomal and autophagic protein degradation pathways by amyotrophic lateral sclerosis-linked mutations in ubiquilin 2. *Biochem. Biophys. Res. Commun.* **2016**, *472*, 324–331. [CrossRef] [PubMed]

134. Hjerpe, R.; Bett, J.S.; Keuss, M.J.; Solovyova, A.; McWilliams, T.G.; Johnson, C.; Sahu, I.; Varghese, J.; Wood, N.; Wightman, M.; et al. UBQLN2 Mediates Autophagy-Independent Protein Aggregate Clearance by the Proteasome. *Cell* **2016**, *166*, 935–949. [CrossRef] [PubMed]

135. Picher-Martel, V.; Dutta, K.; Phaneuf, D.; Sobue, G.; Julien, J.-P. Ubiquilin-2 drives NF-κB activity and cytosolic TDP-43 aggregation in neuronal cells. *Mol. Brain* **2015**, *8*, 71. [CrossRef] [PubMed]

136. Leroy, E.; Boyer, R.; Auburger, G.; Leube, B.; Ulm, G.; Mezey, E.; Harta, G.; Brownstein, M.J.; Jonnalagada, S.; Chernova, T.; et al. The ubiquitin pathway in Parkinson's disease. *Nature* **1998**, *395*, 451–452. [CrossRef]

137. Gu, Y.; Ding, X.; Huang, J.; Xue, M.; Zhang, J.; Wang, Q.; Yu, H.; Wang, Y.; Zhao, F.; Wang, H.; et al. The deubiquitinating enzyme UCHL1 negatively regulates the immunosuppressive capacity and survival of multipotent mesenchymal stromal cells. *Cell Death Dis.* **2018**, *9*, 459. [CrossRef]

138. Rydning, S.L.; Backe, P.H.; Sousa, M.M.L.; Iqbal, Z.; Øye, A.-M.; Sheng, Y.; Yang, M.; Lin, X.; Slupphaug, G.; Nordenmark, T.H.; et al. Novel UCHL1 mutations reveal new insights into ubiquitin processing. *Hum. Mol. Genet.* **2017**, *26*, 1031–1040. [CrossRef]

139. Choi, J.; Levey, A.I.; Weintraub, S.T.; Rees, H.D.; Gearing, M.; Chin, L.-S.; Li, L. Oxidative modifications and down-regulation of ubiquitin carboxyl-terminal hydrolase L1 associated with idiopathic Parkinson's and Alzheimer's diseases. *J. Biol. Chem.* **2004**, *279*, 13256–13264. [CrossRef]

140. Boudreaux, D.A.; Maiti, T.K.; Davies, C.W.; Das, C. Ubiquitin vinyl methyl ester binding orients the misaligned active site of the ubiquitin hydrolase UCHL1 into productive conformation. *Proc. Natl. Acad. Sci. USA* **2010**, *107*, 9117–9122. [CrossRef]

141. Zhou, Z.; Yao, X.; Pang, S.; Chen, P.; Jiang, W.; Shan, Z.; Zhang, Q. The deubiquitinase UCHL5/UCH37 positively regulates Hedgehog signaling by deubiquitinating Smoothened. *J. Mol. Cell Biol.* **2018**, *10*, 243–257. [CrossRef] [PubMed]

142. Ge, J.; Hu, W.; Zhou, H.; Yu, J.; Sun, C.; Chen, W. Ubiquitin carboxyl-terminal hydrolase isozyme L5 inhibits human glioma cell migration and invasion via downregulating SNRPF. *Oncotarget* **2017**, *8*, 113635–113649. [CrossRef] [PubMed]

143. Han, W.; Lee, H.; Han, J.-K. Ubiquitin C-terminal hydrolase37 regulates Tcf7 DNA binding for the activation of Wnt signalling. *Sci. Rep.* **2017**, *7*, 42590. [CrossRef]

144. Chen, X.; Lee, B.-H.; Finley, D.; Walters, K.J. Structure of proteasome ubiquitin receptor hRpn13 and its activation by the scaffolding protein hRpn2. *Mol. Cell* **2010**, *38*, 404–415. [CrossRef]

145. Everett, R.D. A novel ubiquitin-specific protease is dynamically associated with the PML nuclear domain and binds to a herpesvirus regulatory protein. *EMBO J.* **1997**, *16*, 1519–1530. [CrossRef]

146. Holowaty, M.N.; Sheng, Y.; Nguyen, T.; Arrowsmith, C.; Frappier, L. Protein Interaction Domains of the Ubiquitin-specific Protease, USP7/HAUSP. *J. Biol. Chem.* **2003**, *278*, 47753–47761. [CrossRef]

147. Wang, Z.; Kang, W.; You, Y.; Pang, J.; Ren, H.; Suo, Z.; Liu, H.; Zheng, Y. USP7: Novel Drug Target in Cancer Therapy. *Front. Pharmacol.* **2019**, *10*, 427. [CrossRef]

148. Zapata, J.M.; Pawlowski, K.; Haas, E.; Ware, C.F.; Godzik, A.; Reed, J.C. A diverse family of proteins containing tumor necrosis factor receptor-associated factor domains. *J. Biol. Chem.* **2001**, *276*, 24242–24252. [CrossRef]

149. Pfoh, R.; Lacdao, I.K.; Georges, A.A.; Capar, A.; Zheng, H.; Frappier, L.; Saridakis, V. Crystal Structure of USP7 Ubiquitin-like Domains with an ICP0 Peptide Reveals a Novel Mechanism Used by Viral and Cellular Proteins to Target USP7. *PLoS Pathog.* **2015**, *11*, e1004950. [CrossRef]

150. Kim, R.Q.; Sixma, T.K. Regulation of USP7: A High Incidence of E3 Complexes. *J. Mol. Biol.* **2017**, *429*, 3395–3408. [CrossRef]

151. Kim, H.T.; Goldberg, A.L. The deubiquitinating enzyme Usp14 allosterically inhibits multiple proteasomal activities and ubiquitin-independent proteolysis. *J. Biol. Chem.* **2017**, *292*, 9830–9839. [CrossRef] [PubMed]

152. Hanpude, P.; Bhattacharya, S.; Dey, A.K.; Maiti, T.K. Deubiquitinating enzymes in cellular signaling and disease regulation. *IUBMB Life* **2015**, *67*, 544–555. [CrossRef]

153. Kim, H.T.; Goldberg, A.L. UBL domain of Usp14 and other proteins stimulates proteasome activities and protein degradation in cells. *Proc. Natl. Acad. Sci. USA* **2018**, *115*, 11642–11650. [CrossRef] [PubMed]

154. Leggett, D.S.; Hanna, J.; Borodovsky, A.; Crosas, B.; Schmidt, M.; Baker, R.T.; Walz, T.; Ploegh, H.; Finley, D. Multiple associated proteins regulate proteasome structure and function. *Mol. Cell* **2002**, *10*, 495–507. [CrossRef]

155. Ye, Y.; Scheel, H.; Hofmann, K.; Komander, D. Dissection of USP catalytic domains reveals five common insertion points. *Mol. Biosyst.* **2009**, *5*, 1797–1808. [CrossRef] [PubMed]

156. Li, F.; Macfarlan, T.; Pittman, R.N.; Chakravarti, D. Ataxin-3 is a histone-binding protein with two independent transcriptional corepressor activities. *J. Biol. Chem.* **2002**, *277*, 45004–45012. [CrossRef]

157. Berke, S.J.S.; Chai, Y.; Marrs, G.L.; Wen, H.; Paulson, H.L. Defining the role of ubiquitin-interacting motifs in the polyglutamine disease protein, ataxin-3. *J. Biol. Chem.* **2005**, *280*, 32026–32034. [CrossRef]

158. Todi, S.V.; Winborn, B.J.; Scaglione, K.M.; Blount, J.R.; Travis, S.M.; Paulson, H.L. Ubiquitination directly enhances activity of the deubiquitinating enzyme ataxin-3. *EMBO J.* **2009**, *28*, 372–382. [CrossRef]

159. Ashkenazi, A.; Bento, C.F.; Ricketts, T.; Vicinanza, M.; Siddiqi, F.; Pavel, M.; Squitieri, F.; Hardenberg, M.C.; Imarisio, S.; Menzies, F.M.; et al. Polyglutamine tracts regulate beclin 1-dependent autophagy. *Nature* **2017**, *545*, 108–111. [CrossRef]

160. Husnjak, K.; Elsasser, S.; Zhang, N.; Chen, X.; Randles, L.; Shi, Y.; Hofmann, K.; Walters, K.J.; Finley, D.; Dikic, I. Proteasome subunit Rpn13 is a novel ubiquitin receptor. *Nature* **2008**, *453*, 481–488. [CrossRef] [PubMed]

161. Jiang, R.T.; Yemelyanova, A.; Xing, D.; Anchoori, R.K.; Hamazaki, J.; Murata, S.; Seidman, J.D.; Wang, T.-L.; Roden, R.B.S. Early and consistent overexpression of ADRM1 in ovarian high-grade serous carcinoma. *J. Ovarian Res.* **2017**, *10*, 53. [CrossRef] [PubMed]

162. Shi, Y.; Chen, X.; Elsasser, S.; Stocks, B.B.; Tian, G.; Lee, B.-H.; Shi, Y.; Zhang, N.; de Poot, S.A.H.; Tuebing, F.; et al. Rpn1 provides adjacent receptor sites for substrate binding and deubiquitination by the proteasome. *Science* **2016**, *351*, aad9421. [CrossRef] [PubMed]

163. Elsasser, S.; Gali, R.R.; Schwickart, M.; Larsen, C.N.; Leggett, D.S.; Müller, B.; Feng, M.T.; Tübing, F.; Dittmar, G.A.G.; Finley, D. Proteasome subunit Rpn1 binds ubiquitin-like protein domains. *Nat. Cell Biol.* **2002**, *4*, 725–730. [CrossRef] [PubMed]

164. Li, Y.; Huang, J.; Zeng, B.; Yang, D.; Sun, J.; Yin, X.; Lu, M.; Qiu, Z.; Peng, W.; Xiang, T.; et al. PSMD2 regulates breast cancer cell proliferation and cell cycle progression by modulating p21 and p27 proteasomal degradation. *Cancer Lett.* **2018**, *430*, 109–122. [CrossRef]

165. Hamazaki, J.; Sasaki, K.; Kawahara, H.; Hisanaga, S.-I.; Tanaka, K.; Murata, S. Rpn10-Mediated Degradation of Ubiquitinated Proteins Is Essential for Mouse Development. *Mol. Cell. Biol.* **2007**, *27*, 6629–6638. [CrossRef]

166. Jiang, Z.; Zhou, Q.; Ge, C.; Yang, J.; Li, H.; Chen, T.; Xie, H.; Cui, Y.; Shao, M.; Li, J.; et al. Rpn10 promotes tumor progression by regulating hypoxia-inducible factor 1 alpha through the PTEN/Akt signaling pathway in hepatocellular carcinoma. *Cancer Lett.* **2019**, *447*, 1–11. [CrossRef]

167. Chen, X.; Ebelle, D.L.; Wright, B.J.; Sridharan, V.; Hooper, E.; Walters, K.J. Structure of hRpn10 Bound to UBQLN2 UBL Illustrates Basis for Complementarity between Shuttle Factors and Substrates at the Proteasome. *J. Mol. Biol.* **2019**, *431*, 939–955. [CrossRef]

168. Wang, Q.; Young, P.; Walters, K.J. Structure of S5a bound to monoubiquitin provides a model for polyubiquitin recognition. *J. Mol. Biol.* **2005**, *348*, 727–739. [CrossRef]

169. Schweitzer, A.; Aufderheide, A.; Rudack, T.; Beck, F.; Pfeifer, G.; Plitzko, J.M.; Sakata, E.; Schulten, K.; Förster, F.; Baumeister, W. Structure of the human 26S proteasome at a resolution of 3.9 Å. *Proc. Natl. Acad. Sci. USA* **2016**, *113*, 7816–7821. [CrossRef]

170. Mansour, W.; Nakasone, M.A.; von Delbrück, M.; Yu, Z.; Krutauz, D.; Reis, N.; Kleifeld, O.; Sommer, T.; Fushman, D.; Glickman, M.H. Disassembly of Lys11 and mixed linkage polyubiquitin conjugates provides insights into function of proteasomal deubiquitinases Rpn11 and Ubp6. *J. Biol. Chem.* **2015**, *290*, 4688–4704. [CrossRef]

171. Puthiyedth, N.; Riveros, C.; Berretta, R.; Moscato, P. Identification of Differentially Expressed Genes through Integrated Study of Alzheimer's Disease Affected Brain Regions. *PLoS ONE* **2016**, *11*, e0152342. [CrossRef] [PubMed]

172. Zhang, L.; Liu, Y.; Wang, B.; Xu, G.; Yang, Z.; Tang, M.; Ma, A.; Jing, T.; Xu, X.; Zhang, X.; et al. POH1 deubiquitinates pro-interleukin-1β and restricts inflammasome activity. *Nat. Commun.* **2018**, *9*, 4225. [CrossRef] [PubMed]

173. Wang, C.-H.; Lu, S.-X.; Liu, L.-L.; Li, Y.; Yang, X.; He, Y.-F.; Chen, S.-L.; Cai, S.-H.; Wang, H.; Yun, J.-P. POH1 Knockdown Induces Cancer Cell Apoptosis via p53 and Bim. *Neoplasia* **2018**, *20*, 411–424. [CrossRef]

174. Zhu, R.; Liu, Y.; Zhou, H.; Li, L.; Li, Y.; Ding, F.; Cao, X.; Liu, Z. Deubiquitinating enzyme PSMD14 promotes tumor metastasis through stabilizing SNAIL in human esophageal squamous cell carcinoma. *Cancer Lett.* **2018**, *418*, 125–134. [CrossRef]

175. Song, Y.; Li, S.; Ray, A.; Das, D.S.; Qi, J.; Samur, M.K.; Tai, Y.-T.; Munshi, N.; Carrasco, R.D.; Chauhan, D.; et al. Blockade of deubiquitylating enzyme Rpn11 triggers apoptosis in multiple myeloma cells and overcomes bortezomib resistance. *Oncogene* **2017**, *36*, 5631–5638. [CrossRef]

176. Maytal-Kivity, V.; Reis, N.; Hofmann, K.; Glickman, M.H. MPN+, a putative catalytic motif found in a subset of MPN domain proteins from eukaryotes and prokaryotes, is critical for Rpn11 function. *BMC Biochem.* **2002**, *3*, 28. [CrossRef]

Article

Nickel and GTP Modulate *Helicobacter pylori* UreG Structural Flexibility

Annalisa Pierro [1], Emilien Etienne [1], Guillaume Gerbaud [1], Bruno Guigliarelli [1], Stefano Ciurli [2], Valérie Belle [1], Barbara Zambelli [2,*] and Elisabetta Mileo [1,*]

[1] Aix Marseille Univ, CNRS, BIP, Bioénergétique et Ingénierie des Protéines, IMM, Marseille, France; apierro@imm.cnrs.fr (A.P.); eetienne@imm.cnrs.fr (E.E.); ggerbaud@imm.cnrs.fr (G.G.); guigliar@imm.cnrs.fr (B.G.); belle@imm.cnrs.fr (V.B.)

[2] Laboratory of Bioinorganic Chemistry, Department of Pharmacy and Biotechnology, University of Bologna, 40127 Bologna, Italy; stefano.ciurli@unibo.it

* Correspondence: barbara.zambelli@unibo.it (B.Z.); emileo@imm.cnrs.fr (E.M.)

Received: 2 June 2020; Accepted: 10 July 2020; Published: 16 July 2020

Abstract: UreG is a P-loop GTP hydrolase involved in the maturation of nickel-containing urease, an essential enzyme found in plants, fungi, bacteria, and archaea. This protein couples the hydrolysis of GTP to the delivery of Ni(II) into the active site of apo-urease, interacting with other urease chaperones in a multi-protein complex necessary for enzyme activation. Whereas the conformation of *Helicobacter pylori* (*Hp*) UreG was solved by crystallography when it is in complex with two other chaperones, in solution the protein was found in a disordered and flexible form, defining it as an intrinsically disordered enzyme and indicating that the well-folded structure found in the crystal state does not fully reflect the behavior of the protein in solution. Here, isothermal titration calorimetry and site-directed spin labeling coupled to electron paramagnetic spectroscopy were successfully combined to investigate *Hp*UreG structural dynamics in solution and the effect of Ni(II) and GTP on protein mobility. The results demonstrate that, although the protein maintains a flexible behavior in the metal and nucleotide bound forms, concomitant addition of Ni(II) and GTP exerts a structural change through the crosstalk of different protein regions.

Keywords: intrinsically disordered proteins; EPR spectroscopy; isothermal titration calorimetry; protein-ligand interaction; site-directed spin labeling; protein structural dynamics

---

1. Introduction

The discovery of antimicrobials to defeat bacterial pathogens is among the most important medical advances of the last century. However, antimicrobial resistance (AMR) has impaired the efficacy of antibiotics against infections in the last decades, and is considered by the World Health Organization (WHO) as one of the most important threats to public health for the next future [1,2]. In 2017, WHO listed the twelve most important resistant bacteria at a global level for which there is urgent need for new therapies [3]. Ten of them produce the virulence factor urease, a nickel-enzyme that hydrolyzes urea to produce ammonia and carbamate, thus leading to pH increase [4]. This event provides a suitable environment for host colonization, both by producing a micro-environment compatible with bacterial growth and by supplying nitrogen sources. For example, *Staphylococcus aureus* urease activity determines biofilm formation and is required for bacterial persistence [5,6], while for *Proteus mirabilis* [7], *Staphylococcus saprophyticus* [8] and *Ureaplasma ureolyticum* [9] urease activity plays a central role for infection and urea stones formation in the urinary tract. Several of these pathogens are involved in bacterial infections of the respiratory apparatus. It is remarkable that half of patients who died of the recent CoViD19 epidemics in Wuhan (China) became co-infected with bacteria in the lungs and also required antibiotics [10]. Therefore, urease is an attractive target for the development of innovative antibacterial molecules, acting both as antibiotics, as well as preventive anti-virulence drugs or adjuvants for bacterial eradication.

One of the best-known pathogens that exploits the enzymatic activity of urease is the Gram-negative bacterium *Helicobacter pylori*, a widespread microbe, infecting the stomach of up to 50% and 80% of adults in industrialized and developing countries, respectively [11]. The infection causes chronic inflammation of the gastric mucosa, which can slowly progress to gastric ulcer and, through the premalignant stages of atrophic gastritis, to gastric adenocarcinoma or gastric mucosa-associated lymphoid tissue (MALT) lymphoma. In 1994, the WHO classified *H. pylori* as a class I carcinogen. The neutralization of pH driven by urease is required by *H. pylori* for the colonization of the gastric niche [12], while the generated ammonia and induced platelet activation also plays a critical role in the inflammatory response of the host and in the progress of the disease [13].

Bacterial ureases are generally heteropolymeric proteins with a quaternary structure $(\alpha\beta\gamma)_3$ [4,14,15]. In the genus *Helicobacter*, the trimer is of the type $(\alpha\beta)_3$, with the β subunits corresponding to the fused β and γ subunits normally found in other bacteria. The protein also presents a higher level of oligomerization with a $[(\alpha\beta)_3]_4$ quaternary structure [16]. Despite the different oligomeric organization, the structure of the known urease enzymes is fully conserved, and they present a substantially identical active site found in the α subunit. This site contains two Ni(II) ions bridged by the carboxylate group of a carbamylated lysine, essential to maintain the ions at the correct distance for catalysis, and by a hydroxide ion, the nucleophile in the hydrolysis reaction [4,14,15].

Although previous studies identified several molecules that bind urease and inhibit it competitively or uncompetitively, none of them is generally used in therapy, due to their severe side effects or limited ability to pass the bacterial membrane [14,15]. Recently, an alternative strategy to design urease inhibitors has been proposed by targeting, instead of the enzyme, the process that delivers nickel ions into the enzyme active site, precluding enzyme maturation to the active Ni(II)-loaded urease [17]. This activation process is governed by the interplay of at least four accessory proteins, named UreD, UreE, UreF, and UreG, coded by genes belonging to a single operon together with the structural genes [4]. UreE acts as the metallo-chaperone of the system that delivers Ni(II) into urease [18], through tunnels that pass across a complex formed by UreD, UreF, and UreG, the last acting as a molecular chaperone that prepares urease to incorporate the metal ion [19]. Precluding urease maturation by blocking delivery of Ni(II) into its active site could thus represent a novel approach to enzyme inhibition.

The central player of the urease chaperone activation network is UreG, a GTPase that couples the energy obtained from GTP hydrolysis to urease maturation [20]. *Hp*UreG interacts either with *Hp*UreE, forming a heterodimeric $Hp\text{UreG}_2\text{E}_2$ complex [18], or with *Hp*UreF and *Hp*UreD, forming a ternary $Hp\text{UreG}_2\text{F}_2\text{D}_2$ complex [21]. The multiplicity of partners of UreG is reflected in its folding flexibility: while the structure of the protein has been reported for the GDP-bound *Hp*UreG in the $Hp\text{UreG}_2\text{F}_2\text{D}_2$ complex [21], and for the GMPPNP (guanylyl imidodiphosphate)-bound *Klebsiella pneumonia* (Kp) UreG [22], in solution, both *Hp*UreG and *Kp*UreG feature high flexibility in solution, as shown by NMR spectroscopy, suggesting that the single conformation determined by X-ray crystallography does not reflect the flexible behavior of the protein in solution. This behavior is more generally observed in $^1\text{H},^{15}\text{N}$-HSQC NMR spectra of a plethora of UreG homologues from bacteria, archaea, and plants, which show broad signals with limited spread in the ^1H dimension, indicating a backbone mobility in the intermediate exchange regime [23–26]. Native mass spectrometry and site-directed spin labeling coupled to electron paramagnetic resonance (SDSL-EPR) confirmed the presence of a heterogeneous conformational landscape for *Sporosarcina pasteurii* (Sp) UreG in the gas phase and in solution respectively, with at least two conformers with different degree of folding that coexist in equilibrium [27,28].

The selection of the binding partner is defined by the nucleotide bound state of the protein: GTP binding facilitates the formation of the $Hp\text{UreG}_2\text{E}_2$ [29], while the GDP-bound form preferentially interacts in the $Hp\text{UreG}_2\text{F}_2\text{D}_2$ complex [21]. In addition, the concomitant presence of Ni(II) and GTP drives UreG dimerization in solution [21]. Ni(II) binds to a conserved Cys-Pro-His (CPH) motif, located on the protein interaction surface [30], while GTP binds on the opposite side of the protein [29].

A comparison of the crystal structure of the GMPPNP-bound *Kp*UreG and the GDP-bound UreG in the *Hp*UreG$_2$F$_2$D$_2$ complex suggested that the presence of GTP drives an allosteric modulation of the Ni(II) binding site, which assumes a square planar geometry able to accommodate Ni(II) [22], suggesting that the two protein regions that bind Ni(II) and GTP/GDP communicate by allostery to drive the necessary conformational changes for UreG to function. However, such allosteric effect has not been proven in solution. The present study addresses this point, combining multiple biophysical approaches: SDSL-EPR, isothermal titration calorimetry (ITC), and static and dynamic light scattering (MALS-QELS).

In particular, we targeted three different regions of the protein with nitroxide-based spin labels and we performed both continuous wave and pulsed EPR spectroscopy in the presence of Ni(II) and GTP in order to determine their effect on the structural dynamics of *Hp*UreG in solution, as well as to investigate the structural crosstalk of different protein regions occurring by flexibility modulation. The results obtained were complemented by ITC and MALS-QELS. Altogether, the results show that the concomitant addition of both Ni(II) and GTP induces a modification of the structure and mobility in two regions of the protein.

2. Materials and Methods

2.1. Protein Expression and Purification

The purification of *Hp*UreG and its mutants was performed using a protocol previously reported [20]. We improved the yield of the protein expression growing the cells into auto-induction medium containing glycerol (5 g/L), glucose (25 g/L), and lactose (100 g/L), instead of LB combined with IPTG induction used in the previous work. The cells were grown 3 h at 37 °C and 18 h at 28 °C. At the last step of purification, the proteins were in 20 mM TrisHCl pH 8 buffer, containing NaCl 150 mM and TCEP 1 mM. Protein concentration was estimated using absorbance at 280 nm and an extinction coefficient ε_{280} = 10,032 M^{-1}cm^{-1}.

2.2. Isothermal Titration Calorimetry

Ni(II) binding titrations of wild-type and C66A mutant HpUreG were performed at 25 °C using a high-sensitivity VP-ITC microcalorimeter (MicroCal, Norcross, GA, USA). The protein and the metal ion salt (NiSO$_4$) were diluted to 40–80 μM and 1.0 mM respectively into a solution of 20 mM TrisHCl pH 8, containing 150 mM NaCl and 1 mM TCEP, in the absence or in the presence of 150 μM of the non-hydrolyzable GTP analogue, GTPγS. A reference cell was filled with deionized water. Before each experiment, the baseline stability was verified. An interval of 5 min was applied between the injections to allow the system to reach thermal equilibrium. Control experiments were conducted by titration of the metal ion solution into the buffer alone under identical conditions, and the heat of dilution was negligible. The solution containing the protein was loaded into a sample cell (1.4093 mL) and was titrated with 55 × 5 μL injections with the Ni(II) solution. The raw data were processed and fitted using Affinimeter software, with a nonlinear least-squares minimization algorithm to theoretical titration curves with stoichiometric binding schemes. For Ni(II) titration over apo-HpUreG and C66A mutant, restriction of the binding parameters had to be made by fixing the stoichiometry of 2 and 1 respectively, as the low affinity binding did not provide an optimal sigmoidicity of the curve with a clear inflection point. Attempts to fit with stoichiometry of 1, 2, 3, and 4 were made, and the chosen stoichiometry was the one that provided the best fit of the experimental data.

2.3. HpUreG Mutants Design

The cysteine mutations were introduced into *Hp*UreG gene from Hp26695 strain urease operon (NCBI code NC000915) cloned into the pET15b expression vector (Novagen, Madison, WI, USA) in a previous work [20].

In order to relate and compare this work with previous studies, Cys66, wen mutated, was replaced by Alanine [18,21,29]. We decided to perform a Cysteine–Serine mutation for position 48 to preserve the surface charge of the protein, and a Cysteine–Alanine mutation in position 7 which is more buried in the crystal structure [22].

Mutants containing a single cysteine available for labeling were obtained by double mutation ("Site-directed mutagenesis" in Supplementary Materials Section S3). They were named as following: *Hp*UreG-C7A-C48S: C66^{Proxyl}; *Hp*UreG-C7A-C66A: C48^{Proxyl}; *Hp*UreG-C66A-C48S: C7^{Proxyl}.

Variants containing two labeling sites and thus needing only one cysteine mutated were designed to perform distances measurements by DEER-EPR ("Site-directed mutagenesis" in Supplementary Materials Section S3) were named as following: *Hp*UreG-C7A: C48^{Proxyl}/C66^{Proxyl}; *Hp*UreG-C66A: C7^{Proxyl}/C48^{Proxyl}; *Hp*UreG-C48S: C7^{Proxyl}/C66^{Proxyl}.

2.4. GTP Hydrolase Activity Assays

*Hp*UreG GTP hydrolyzing activity was measured by the SensoLyte® MG Phosphate Assay Kit (AnaSpec, Fremont, CA, USA), based on the colorimetric reaction involving malachite green reagent, molybdate, and orthophosphate.

Each sample was prepared mixing the reagents in order to obtain 20 µM of protein, 400 µM of GTP, and 2 mM of $MgSO_4$ in a final volume of 250 µL of buffer. The reaction mixture (RM) was incubated for 2 h at 37 °C. Every 30 min, 40 µL from the RM were incubated with 40 µL of Malachite Green Mix for 10 min in a final volume of 300 µL of buffer. After incubation, the absorbance at 600 nm was recorded. All the experiments were reproduced two times before estimating the Kcat values.

2.5. Protein Labeling with Nitroxide Spin Label

As *Hp*UreG variants are purified in presence of TCEP 1 mM, before labeling reaction, in order to avoid the reduction of the nitroxide spin label, the reductant removal is necessary. In general, a gel filtration using a PD-10 desalting column (GE Healthcare, Chicago, IL, USA) is sufficient. The labeling procedure is normally performed on 100 nmol of protein in a reductant-free buffer (Tris 20 mM, pH = 8, NaCl 150 mM) in the presence of a 10-fold excess of nitroxide spin label, the maleimido-Proxyl (Sigma-Aldrich, St. Luis, MO, USA). A 20-fold excess was used for double Cys variants. The mixture is then incubated at 4 °C, in the dark for 4 h under gentle stirring and continuous flow of argon. The excess of unbound label is removed by a second gel filtration with a PD-10 desalting column. The labeled protein is concentrated by using ultrafiltration (Vivaspin 5 kDa, Sartorius, Göttingen, Germany). The concentration of the labeled protein is evaluated by measuring the OD at 280 nm. The labeling yield of mono-labeled variants analyzed was between 80% and 100%, 150–170% for double-labeled ones.

2.6. EPR Spectroscopy

X-band room temperature (298 K) continuous wave EPR measurements were recorded on an Elexsys500 Bruker spectrometer equipped with a Super High Q sensitivity resonator operating at X band (9.9 GHz). The microwaves power was 10 mW, the magnetic field modulation amplitude was 0.1 mT, the field sweep was 15 mT, the receiver gain was 60 dB. All the samples were analyzed in quartz capillaries whose sensible volume was 40 µL.

The spin concentration was obtained by double integration of the EPR signal obtained under non-saturating conditions and the labeling yield was evaluated comparing the spin concentration with that one of a standard solution. For all variants, high labeling yields were obtained ranging from 80% to 100% for mono labeled samples and 150–170% for double-labeled samples.

X-band cw EPR spectra at room temperature were recorded at 50 µM of protein concentration in Tris 20 mM, pH = 8, NaCl = 150 mM. When present, Ni(II) was 2.5 mM ($NiSO_4$), GTP/GDP 3 mM (Sigma-Aldrich).

The EPR spectra were simulated using SimLabel program [31], a Matlab graphical user interface using the Easyspin toolbox [32].

2.7. DEER Measurements

Inter-label distance distributions were obtained using the four-pulse DEER sequence [33]. Experiments were performed on a Bruker ELEXSYS E580 spectrometer at Q-band using the standard EN 5107D2 resonator. The system was equipped with an Oxford helium temperature regulation unit and the data were acquired at 60 K. This temperature was optimized according to the relaxation times measured at variable temperatures in the range of 20–100 K with 10 K steps. All the measurements were performed on 20 μL of sample loaded into quartz capillaries. DEER samples were flash frozen in liquid nitrogen. Distance distribution were extracted from DEER data through a Tikhonov regularization after baseline correction, using DeerAnalysis2019 software (http://www.epr.ethz.ch/software/index Jeschke G. 2011. DeerAnalysis. ETH, Zürich, Switzerland). Distance distributions measured were compared with the distance distributions predicted, analyzing the crystal of HpUreG (PDB 4HI0) [22] using the MMM software [34].

3. Results and Discussion

3.1. Cys Variants Were Generated to Selectively Label Distinct Regions of HpUreG

SDSL-EPR spectroscopy is a non-destructive technique that provides details on protein structure and flexibility over a wide-range of temperatures and timescales [35–38]. Proteins can be studied in their native environment, that is in membranes, in cellular extract and also inside cells [39]. SDSL-EPR involves the grafting of a paramagnetic label, generally a thiol-specific nitroxide, on the protein of interest and the determination of the dynamic properties of the attached nitroxide by continuous wave (CW)-EPR spectroscopy [40,41]. Changes in the nitroxide spectrum are thoroughly related to the mobility of the nitroxide side-chain and to the local backbone motion, which can thus be used to follow protein structural changes and to reveal interaction sites in complexes in solution and at room temperature [40,42–45]. Distance distributions between two spin labels can be measured by pulsed double electron–electron resonance (DEER) techniques relying on their dipole–dipole coupling [46,47]. Inter-label distance distributions can be investigated between 15 and 80 Å, but in specific experimental conditions, 160 Å can be reached [48]. DEER experiments are usually carried out at cryogenic temperature (60 K), the low temperature being required to slow down the otherwise fast relaxation of nitroxide labels at higher temperature. As for all the other techniques requiring a freezing step, it is assumed that the conformational ensemble of the sample is captured.

Since most of the available nitroxide-based spin labels can specifically react with the thiol group of Cysteine, site-directed mutagenesis is often used to introduce Cys residues at specific locations in the primary structure of the protein of interest.

*Hp*UreG has three naturally occurring Cys, located in different regions of the protein (Figure 1A): (i) the conserved P-loop-motif, involved in GTP binding [20], accommodates Cys7; (ii) Helix 2, involved in GTP-dependent conformational changes, contains Cys48 [21]; (iii) the fully conserved CPH motif, involved in Ni(II) binding, includes Cys66 [20]. These positions allow, in principle, to monitor the mobility of three functionally important regions of the protein in solution. Consequently, six mutants containing one or two Cys residues were designed and labeled with the MA-Proxyl nitroxide to dissect the protein conformational landscape (Figure 1B): three double variants feature a unique position available for labeling (C7^{prox}, corresponding to the Cys48Ser/Cys66Ala labeled mutant; C48^{prox} corresponding to the Cys7Ala/Cys66Ala labeled mutant; C66^{prox}, corresponding to the Cys7Ala/Cys48Ser labeled mutant), while three single variants possess two positions available for labeling for distance measurements (C7^{prox}/C48^{prox}, corresponding to the Cys66Ala labeled mutant; C7^{prox}/C66^{prox}, corresponding to the Cys48Ser labeled mutant; C48^{prox}/C66^{prox}, corresponding to the Cys7Ala labeled mutant). Note that "WT^{prox}" corresponds to the wild-type protein (WT) labeled in the three naturally occurring Cys residues. The labeling reactions were checked by mass spectroscopy (see Supplementary Materials Figure S1). Any possible perturbation of the global structure and of the folding of *Hp*UreG mutations was excluded by controlling the global folding by circular dichroism (CD, see Figure S2). Similarly, the catalytic activity of *Hp*UreG was monitored for the wild-type protein labeled in the three Cys positions (WT^{prox}) (see Figure S1A), indicating no significant differences between the unlabeled and labeled proteins (k_{cat} = 0.027 min^{-1} and k_{cat} = 0.023 min^{-1}, respectively).

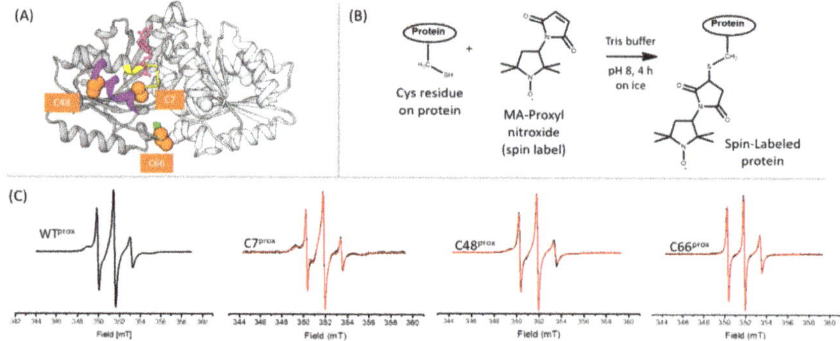

Figure 1. (**A**) *Hp*UreG structure in the presence of GDP (pink) from the *Hp*UreG$_2$F$_2$D$_2$ crystal (PDB 4HI0). The position of the three natural cysteine residues is highlighted in orange: Cys7 in the P-loop (yellow), Cys48 in the Helix 2 (purple), and Cys66 in the CPH Nickel binding site (green). (**B**) Labeling reaction scheme with MA-PROXYL nitroxide label. (**C**) Room temperature X-band EPR spectra of 50 µM of wild-type *Hp*UreG and its variants labeled with MA-PROXYL nitroxide in Tris buffer (black trace) and superimposed simulated spectra (red traces).

3.2. The Thermodynamics of Ni(II) and GTP-Driven Dimerization of HpUreG Was Characterized

Previous studies of Ni(II) binding to *Hp*UreG entailed the use of ITC, which showed that the isolated protein interacts with two Ni(II) ions per monomer with an exothermic reaction and a dissociation constant K_d = 10 µM [20]. Ni(II) binding was also monitored using the gradual increase of absorption peak at 337 nm, assigned to ligand-to-metal-charge transfer [29]. Differently from the ITC experiments, the latter approach did not detect any Ni(II) binding activity for the isolated protein, whereas metal binding occurred when GTP was added to the protein solution, under which condition *Hp*UreG was found to bind 0.5 equivalents of Ni(II) per protein monomer, with K_d = 0.33 µM, and to undergo dimerization upon metal binding [29].

Calorimetric titration of Ni(II) over a freshly purified *Hp*UreG sample, performed here for comparison with all other ITC data on *Hp*UreG mutants and labeled forms described in the present study, confirmed the results previously obtained by ITC, showing negative peaks following each metal additions, indicative of an exothermic binding event (Figure 2A, left panel). The integrated heat data generated a binding isotherm with a single inflection point, and a mild slope (Figure 2A, right panel and Table 1). The fit of the obtained data, performed using the AFFINImeter software [49] and a model involving a single set of sites, showed that two Ni(II) ions bind per *Hp*UreG monomer with similar affinity (K_d = 72 µM), a favorable enthalpic contribution and a minor entropic impact (Table 1). Previously reported studies on *Hp*UreG mutants indicated that at least one Ni(II) binding site is located on the CPH motif [20,29]. These observations outline two possible scenarios: (i) both Ni(II) ions bind to identical sites located in the region of CPH, or (ii) one Ni(II) ion binds to the CPH motif while the second binds to the different site; in the latter case, a possibility is represented by the Mg(II) binding sites close to the GTP binding pocket, as previously suggested [29]. The difference in Ni(II) affinity thus measured for *Hp*UreG (72 µM) with the previously reported value obtained by ITC (10 µM) [20] is likely due to the experimental conditions: for the ITC titrations, the relatively weak metal–protein affinity caused the value of the c-parameter, namely the product of the concentration of the protein in the cell by the binding constant, to be close to its lowest acceptable limit [20], rendering the calculated affinity constants less accurate.

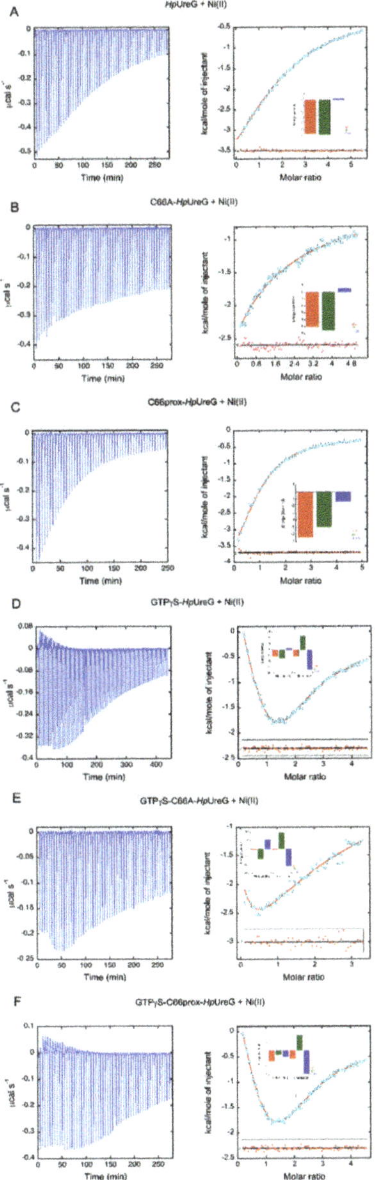

Figure 2. Isothermal Titration Calorimetry (ITC) of Ni(II) over *Hp*UreG and its variants in the presence and in the absence of GTPγS in the sample cell. The panels on the left report the heat flow response for the injections of the metal ion over the protein solutions. The panels on the right show the respective integrated heat data (blue circles) and their best fit obtained using a single set of binding sites (**A–C**) or a model involving protein dimerization upon metal binding (**D–F**); the insets show the binding signatures (ΔG, ΔH, −ΔT, ΔS) associated to each curve; the bottom plots show the residuals of the fitting procedure.

Table 1. Thermodynamic parameters of Ni(II) titrations over *Hp*UreG and its variants, in the absence and in the presence of GTPγS.

Sample in the ITC Cell	N	K_a (M^{-1})	K_d (μM)	ΔH (kcal mol^{-1})	ΔS (cal mol^{-1}K^{-1})
*Hp*UreG	2	$1.38 \pm 0.02 \times 10^4$	72 ± 1	-5.90 ± 0.06	-0.86
C66A-*Hp*UreG	1	$4.23 \pm 0.02 \times 10^3$	236 ± 1	-5.52 ± 0.01	-1.93
WT^{Prox}-*Hp*UreG	1	$4.33 \pm 0.04 \times 104$	23.0 ± 0.3	-4.99 ± 0.02	4.47
*Hp*UreG-GTPγS	HpUreG-Ni	$3.45 \pm 0.02 \times 10^4$	30.0 ± 0.2	-7.14 ± 0.06	-4.99
	HpUreG$_2$-Ni	$1.40 \pm 0.04 \times 10^4$	71 ± 2	$+11.2 \pm 0.3$	$+56.77$
C66A*Hp*UreG-GTPγS	C66AHpUreG-Ni	$6.1 \pm 0.8 \times 10^3$	160 ± 20	-61 ± 7	-187
	C66AHpUreG$_2$-Ni	$4.5 \pm 0.7 \times 10^4$	22 ± 3	$+99 \pm 9$	$+353$
WT^{Prox}-*Hp*UreG-GTPγS	WT^{Prox}-HpUreG-Ni	$4.20 \pm 0.04 \times 10^4$	23.8 ± 0.2	-2.63 ± 0.03	$+12.3$
	WT^{Prox}-HpUreG$_2$-Ni	$3.8 \pm 0.3 \times 10^3$	260 ± 20	$+9.1 \pm 0.5$	$+46.8$

Mutation of the Ni(II) binding residue Cys66 to Ala was previously reported to fully abolish Ni(II) binding capability of the protein [29]. Here, the ITC titration instead revealed that the Cys66Ala-*Hp*UreG mutant is still able to bind one Ni(II) ion per protein monomer, with an exothermic reaction (Figure 2B, left panel) and one order of magnitude lower affinity ($Kd = 236$ μM), with substantially invariant enthalpic and entropic contributions (Figure 2B, right panel and Table 1). This observation suggests that the two Ni(II) binding sites per monomer, observed in wild-type *Hp*UreG, are distinct, with one site, involving Cys66 in the CPH motif, that is abrogated by the Cys66-to-Ala mutation, while the second is maintained. The decreased affinity for the latter site ($K_d = 236$ μM vs. 72 μM for the mutated and WT protein, respectively) suggests the presence of cooperativity between the two metal binding sites. The triply labeled WT^{Prox}, while showing the same exothermic effect (Figure 2C, left panel) and the same stoichiometry as the Cys66Ala mutant, features an higher affinity for Ni(II) ($K_d = 23$ μM), a similar enthalpic value and positive entropic contribution (right panel of Figure 2C and Table 1), as compared to the Cys66Ala mutant ($K_d = 236$ μM), indicating that the Cys labeling with the nitroxide moiety still abolishes one of the two Ni(II) binding sites observed for the unlabeled WT protein, but, differently from the Cys66Ala mutation, maintains a similar affinity for the second Ni(II) binding event, supporting the idea of cooperativity between the two Ni-binding sites in the WT protein.

Ni(II)- and GTPγS -driven dimerization [29] was verified under the ITC experimental conditions using size-exclusion chromatography coupled to multi-angle light scattering (SEC-MALS, Figure S3). The obtained results confirmed that *Hp*UreG undergoes dimerization when both Ni(II) and GTPγS, a non-hydrolyzable GTP analogue, are added to the protein solution, while it remains monomeric in the presence of either Ni(II) or GTPγS alone.

The thermodynamics of Ni(II) and GTPγS-driven protein dimerization was therefore addressed using ITC. Ni(II) ions were titrated over *Hp*UreG in the presence of GTPγS in the sample cell. Negative peaks followed each injection of Ni(II) into the protein solution, indicating the occurrence of an exothermic reaction (Figure 2D, left panel). The observation of a slower endothermic effect following each injection, which terminates when one equivalent of metal is added to the protein solution, suggested the existence of another process, in addition to metal binding, similarly to what had been previously observed for the Ni(II)-sensor *Hp*NikR [50]. This type of ITC trace can be interpreted either as a conformational modification or as a change in the oligomerization state of the protein. As dimerization was demonstrated by light scattering experiments (Figure S3) [29], and the shape of the binding isotherm clearly indicated two inflection points suggestive of two reactions occurring upon metal titration, the data were analyzed using a model involving two successive equilibria, with protein dimerization following the binding of one Ni(II) per protein dimer (Scheme 1).

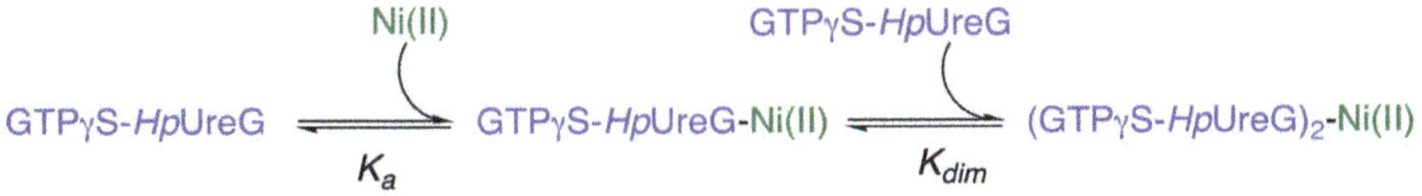

$$\text{GTP}\gamma\text{S-}\textit{Hp}\text{UreG} \;\underset{K_a}{\overset{\text{Ni(II)}}{\rightleftharpoons}}\; \text{GTP}\gamma\text{S-}\textit{Hp}\text{UreG-Ni(II)} \;\underset{K_{dim}}{\overset{\text{GTP}\gamma\text{S-}\textit{Hp}\text{UreG}}{\rightleftharpoons}}\; (\text{GTP}\gamma\text{S-}\textit{Hp}\text{UreG})_2\text{-Ni(II)}$$

Scheme 1. Equation defining the model used to fit the ITC data for Ni(II) titration over GTPγS-HpUreG, involving metal binding and protein dimerization.

Fitting of the binding isotherm (Figure 2D, right panel and Table 1) indicated that one Ni(II) ion binds per protein monomer in the presence of GTPγS, with a dissociation constant two times smaller than that reported for the apo-protein (K_d = 30 μM), while protein dimerization occurs with K_{dim} = 71 μM. In this case, Ni(II) binding occurs with thermodynamic parameters similar to the ones observed for the apo-protein, with favorable enthalpy and a negative entropic contribution (Table 1). On the other hand, dimerization is an entropy-driven endothermic process as expected (Table 1). The entire two-step process is characterized by a global dissociation constant of K_d = 2 nM. This value should be compared with that obtained by absorbance spectroscopy for Ni(II) binding to HpUreG in the presence of GTPγS (K_d = 0.33 μM) [29]. The difference between these two values could be attributed to the different pH at which the measurements were carried out in the present (8.0) and in the previous (7.2) work. The apparent decrease in affinity at lower pH is consistent with a proton dissociation event occurring upon metal binding, which possibly involves a cysteine residue (Cys66 in this case), as previously observed in the case of the nickel-dependent transcription factor HpNikR [50].

Ni(II) titration over the Cys66Ala mutant in the presence of GTPγS produced negative peaks indicative of an exothermic binding of Ni(II), but no endothermic effect was visible (Figure 2E, left panel), suggesting that a dimerization is either not occurring in this case or is occurring with much lower affinity, resulting in the absence of detectable endothermic heat. The binding isotherm (Figure 2E, right panel) showed two inflection points, and the same model reported in Scheme 1 was used to treat the data. According to the fit, the first event of Ni(II) binding occurs with one order of magnitude lower affinity (K_d = 160 μM) as observed for the Cys66Ala mutant in the absence of GTPγS (Figure 2A), while the second equilibrium shows a similar constant (K_d = 22 μM). In this case, the thermodynamic parameters associated to the metal binding step and to the second process (Table 1) are unusually high compared to all other similar data in this study, suggesting that additional phenomena other than metal binding and dimerization are occurring in the case of this mutant. It is worth noticing that, during sample manipulation, the Cys66Ala mutant was prone to precipitation, especially in the presence of Ni(II), suggesting that at least part of the protein sample undergoes aggregation upon Ni(II) titration, which might be the second process evidenced in the binding isotherm.

Ni(II) titration over the triply labeled WT[Prox] protein in the presence of GTPγS (Figure 2F, left panel) produced a bipartite binding isotherm (Figure 2F, right panel), whose analysis, performed according to Scheme 1, indicated that Ni(II) ion binding to the protein dimer occurs with an affinity similar to the wild type protein (K_d = 23.8 μM), and favorable enthalpic and entropic contributions (Table 1). On the other hand, the second process is less favorable for WT[Prox], occurring with a lower equilibrium constant (K_d = 260 μM) as compared to the dimerization of the WT protein. If the second process also involves dimerization for WT[Prox] (as suggested by the values of ΔH and ΔS, see Table 1), this decreased value could be due to a steric effect of the nitroxide label.

3.3. HpUreG Shows Distinct Flexibility in Different Protein Regions

The EPR spectrum of WT *Hp*UreG labeled with MA-Proxyl nitroxide (WT[Prox], Figure 1C) arises from the contribution of spin labels simultaneously grafted onto the three Cys residues. To separately dissect the conformational flexibility of different regions of HpUreG, the EPR spectrum of the nitroxide-labeled HpUreG variants that contain a single labeled cysteine (C7[Prox], C48[Prox], and C66[Prox]) were performed in solution and at room temperature (Figure 1C). Qualitatively, when the nitroxide mobility decreases, a broadening of the EPR spectral line shape is expected. For the single labelled *Hp*UreG variants, the spectra show different mobility: the line shape becomes sharper going from C7^{prox} to C48^{prox} and then to C66^{prox}, reflecting an increased mobility of the nitroxide moiety and, consequently, of the protein structural motif to which the label is attached (Figure 1C). A quantitative view of the nitroxide dynamics in terms of the rotational correlation time (τ_c) and of the magnetic parameters (*g*-factor and hyperfine A-tensors) was obtained by simulating the EPR spectra with SimLabel [31] (a MatLab graphical user interface using the Easyspin toolbox [32]) (see Table 2 and Supplementary Materials Section S2). As very often found in SDSL-EPR studies [27], the spectra of *Hp*UreG could be simulated by two components, which represent populations of spin labels characterized by different dynamics. These populations can be related either to rotameric states of the spin label or to structural sub-states of the protein in conformational equilibrium. In the case of *Hp*UreG, the large difference in the dynamics of the two components (see τ_c in Table 2), generally not observed for rotamers [51], suggests that they reflect distinct protein conformational states [51,52]. This conclusion is consistent with similar phenomena reported for *Sp*UreG [27], and is further supported by the observation that the *Hp*UreG EPR spectrum of C66^{prox} is modified by the addition of glycerol, a protective osmolyte: in this case, the spectrum can be simulated by increasing the contribution of the slower component, which changes from 44 to 70% (Table 2 and Figure S4). Protective osmolytes are indeed known to modify the conformational equilibria among different conformational states of the protein [53], stabilizing the protein structure toward a more folded conformation [52].

Table 2. Electron Paramagnetic Resonance (EPR) spectra simulation parameters: rotational correlation times (τ_c in ns) and proportion (%) of the simulated multi-components of the spectra of *Hp*UreG variants under various conditions.

Labeled Site	Component	Apo-Form		+Ni(II)		+Ni(II) and GTP	
		Weight %	τc (ns)	Weight %	τc (ns)	Weight %	τc (ns)
C66^{prox}	Fast	56	0.6	34	0.6	28	0.7
	Slow	44	2.4	66	2.4	72	3.4
C66^{prox} +*gly 30%*	Fast	30	0.6	22	0.6	*19*	0.8
	Slow/Rigid *	70	2.4	78	2.4	81 *	4.4 *
C48^{prox}	Fast	37	0.6	39	0.6	21	0.6
	Rigid	63	4.9	61	4.9	79	4.9
C7^{prox}	Fast	20	0.3	15	0.3	14	0.3
	Rigid	80	6.1	85	6.1	86	6.1

The symbol (*) indicates the "rigid" component.

Note that, in the following paragraphs, we named "fast" all spectral components characterized by τ_c values included between 0.3–0.7 ns, "slow" those characterized by τ_c values included between 0.7–4.0 ns, and "rigid" components characterized by τ_c values included between 4.0–6.2 ns. The sharpest EPR line shape is observed for the nitroxide grafted to Cys66. This spectrum is constituted by two components having almost the same proportion, one with τ_c = 0.6 ns ("fast") and the other with τ_c = 2.4 ns ("slow") (Table 2 and Figure 3). The fact that the "fast" component shows a mobility close to that normally observed for a spin label attached to loops or intrinsically disordered protein fragments [27,44,54] demonstrates that this region is highly flexible. This dynamic behaviour is similar to that observed for the *Sp*UreG orthologue containing the nitroxide label grafted onto the corresponding cysteine residue, which features two conformers with similar correlation times (τ_c "fast" = 0.3 ns and τ_c "slow" = 3.6 ns) (see Figure S5) [27].

Similarly, two components with different degrees of flexibility and comparable relative abundance are observed for C48^{prox}: one features a "fast" behavior (τ_c = 0.6 ns), while the other ("rigid") is consistent with a less flexible dynamic (τ_c = 4.9 ns) (Table 2 and Figure 3). On the other hand, a "rigid" component (τ_c = 6.1 ns) is dominant (80%) in the case of C7^{prox}, for which a less abundant component (20%) shows a "fast" behavior (τ_c = 0.3 ns) (Table 2 and Figure 3). The latter case can be explained by considering that the nitroxide moiety resides in a well-structured region or in a buried site. The high yield of labeling reached for this site (80–100%, Figure S1) suggests that Cys7 is accessible, so the observed rigid behavior is indicative of the presence of a highly rigid protein segment.

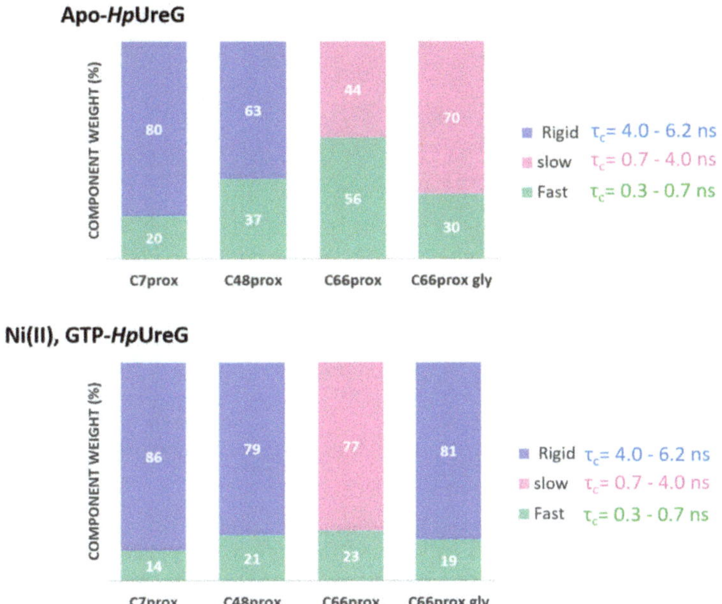

Figure 3. Conformational distribution of *Hp*UreG variants plotted in terms of the relative amount of the conformations obtained from simulation of EPR spectra. The different conformational ensembles are indicated using different colors and the relative correlation times are reported.

All these data experimentally confirm previously reported molecular dynamic simulations on *Hp*UreG, which suggested substantial rigidity in the protein regions involved in catalysis, justifying the residual catalytic activity of the isolated proteins, while evidencing the large dynamic flexibility for the protein portions involved in protein–protein interactions, which contain the residues in the conserved CPH motif [55].

To further investigate the structural dynamics of *Hp*UreG in the apo-state, double electron-electron resonance (DEER) experiments were applied. DEER experiments allow to measure the dipolar coupling between spin pairs, yielding distance distribution between the coupled spins. Three double-Cys variants were constructed and labeled (C7^{prox}/C48^{prox}, C7^{prox}/C66^{prox}, C48^{prox}/C66^{prox}) and their CW EPR spectra are reported in Supplementary Materials Figure S7. For all the DEER data shown in this section, the error on distance distribution results was calculated with the validation tool of DeerAnalysis (see Figure S8) [56]. DEER data of the C7^{prox}/C48^{prox} variant (Figure 4A) showed well-resolved distance distribution with 2 peaks centered at 2.4 and 3.5 nm, while the one obtained for C7^{proxyl}/C66^{proxyl} (Figure 4B) and for C48^{proxyl}/C66^{proxyl} (Figure 4C) displayed broad distance distribution. These results confirm the presence of considerable conformational heterogeneity in the protein sample, supporting the highly flexible behavior of *Hp*UreG, as already observed by previous NMR studies [20] and by the CW EPR data described above.

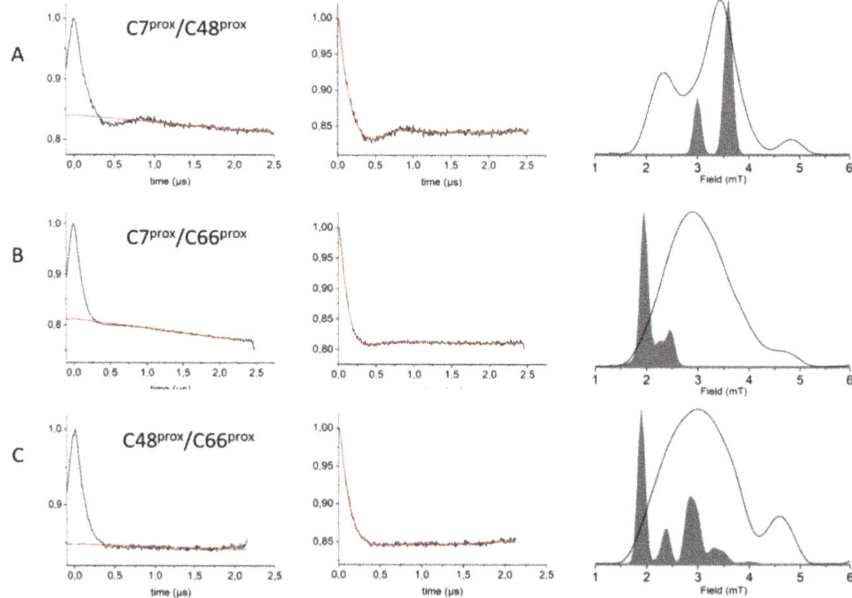

Figure 4. Inter-label distance distributions. (**Left panel**): experimental Q-band DEER traces recorded at 60 K for (**A**) C7^{prox}/C48^{prox}, (**B**) C7^{prox}/C66^{prox}, and (**C**) C48^{prox}/C66^{prox}. Red lines indicate the baseline used for background correction. (**Central panel**): corrected Double Electron-Electron Resonance (DEER) traces (black) with superimposed fits derived from Tikhonov regularization (red). (**Right panel**): Tikhonov derived distance distributions obtained using DeerAnalysis (black) [56] superimposed with distance distributions calculated by MMM software (gray curves) [34].

In all cases, the average distance distributions measured by DEER span a broader range than those predicted by an MMM [34] analysis based on the *Hp*UreG crystal structure (PDB: 4HI10) [21] and a library of rotamers for the MA-Proxyl spin label (Figure 4). This observation confirms that the experimental conditions used in the crystallization experiments (solutes and salts acting as precipitants) likely favored a more compact and rigid fold, as already shown for other biological systems [57], and that the protein structure observed in the crystal is different from the conformation that the protein assumes in solution.

3.4. Ni(II) Ions and GTP Binding Produce Changes in the Structural Dynamics of Different Protein Regions

Figure 5 shows the EPR spectra of the WT^{prox} in the presence of either GTP or Ni(II), and in the presence of both cofactors. No significant spectral changes were detected after the addition of either GTP or Ni(II), while addition of both cofactors resulted in a clear change of the EPR line shape, corresponding to an increase of the broader components of the spectrum. This indicates that both ligands are necessary to induce structural changes in *Hp*UreG.

To investigate the source of this spectral modifications, and in particular to sort out which of the three nitroxide labels bound to the protein contributes to the Ni(II) and GTP-driven line broadening, the experiment was repeated using the single-Cys variants of *Hp*UreG.

Addition of Ni(II) did not produce significant changes of the EPR line shapes for C7^{prox} (Figure 6A) and C48^{prox} (Figure 6B), while for the C66^{prox} spectrum induced a line broadening (Figure 6C), suggesting a reduction of the spin label mobility (Table 2 and Figure 3). Indeed, we observe a change in the weight of the two components (Table 2 and Figure 3), with the less flexible species becoming most abundant (from 44 to 66%).

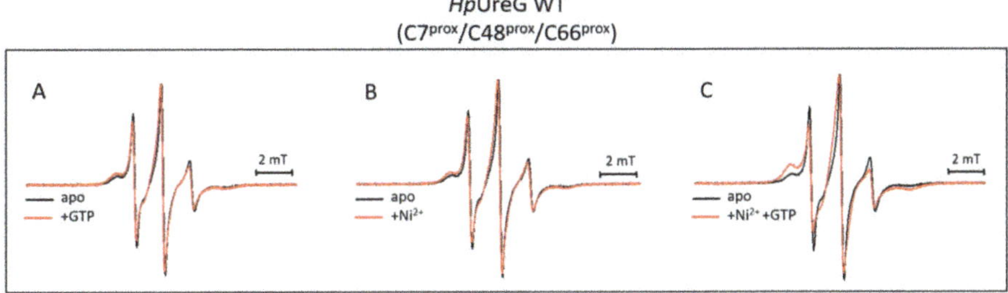

Figure 5. X-band CW EPR spectra of 50 μM *Hp*UreG labeled at the three naturally occurring Cys residues, recorded in solution and at room temperature. The apo-form spectra (black line) are compared to spectra recorded in presence of either 3 mM GTP and/or 2.5 mM Ni(II) (red line).

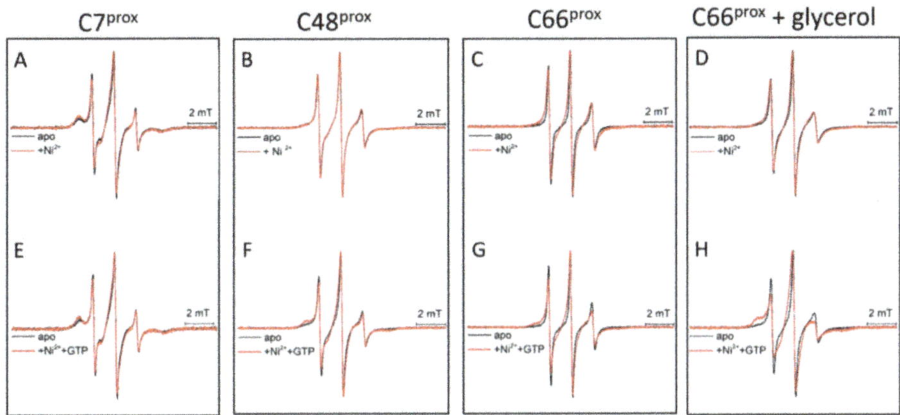

Figure 6. X-band CW EPR spectra of *Hp*UreG variants at room temperature, in the absence and in the presence of glycerol. In the upper panels (**A–D**), the apo-form spectra are in black, those obtained in presence of 2.5 mM of NiSO$_4$ are in red. In the lower panels (**E–H**), the apo-form spectra (**black line**) are compared to spectra recorded in presence of either 3 mM GTP and/or 2.5 mM Ni(II) (**red line**).

Differently, no significant spectral changes were detected on all protein variants after the addition of GTP or GDP alone (Figures S9 and S10). We also tested the effect of Mg(II) to protein samples containing GTP or GDP. The addition of Mg(II) in equimolar concentration with respect to GTP, did not affect the spectral shape of the protein (Supplementary Materials Figure S11).

Concomitant addition of both Ni(II) and GTP did not significantly affect the spectrum of C7^{prox} (Figure 6E), while it changed the line shape of C48^{prox} (Figure 6F), with a conformational shift toward the "rigid" component, which increased from 63% to 79% (Table 2 and Figure 3). In the case of C66^{prox}, additional spectral changes were observed when GTP and Ni(II) bind to the protein; the τ_c of both components increases, suggesting an induced structuration of this protein region (Table 2 and Supplementary Materials Section S2) or the dimerization of the protein upon both GTP and metal binding. These changes became more pronounced if the experiments were performed in the presence of glycerol, with a drastic change in the dynamics of the less flexible specie (Figure 6H, red trace), whose τ_c varies from 2.4 ns to 4.4 ns (Table 2). These results were confirmed for the *Hp*UreG C48^{prox}/C66^{prox} variant, which showed a similar behavior as found for C66^{prox} in glycerol (Figure S12).

The observed spectral changes indicate that Ni(II) and GTP induce structural and dynamics modifications in the Ni(II)-binding region of the protein where C66 is located, and in the region around Helix 2, containing C48. This suggests the occurrence of an allosteric communication between the protein regions proximal to the Ni(II) and the GTP binding sites, whereas the region around the Cys7 residue is not affected by the presence of either Ni or GTP, or both.

To further investigate the structural effect of ligand binding on *Hp*UreG fold, DEER experiments would have been of great interest. However, the decrease in the T_m (phase memory time) value (Figure S13A) associated with the field sweep (FS) intensity loss (Figure S13B) in the presence of Ni(II) prevented from obtaining properly exploitable DEER traces. Works are currently in progress to improve the quality of DEER experiments.

4. Conclusions

In this work, the structural dynamics of *Hp*UreG was explored in the absence and in the presence of its physiological cofactors, Ni(II) and GTP. In solution, *Hp*UreG fluctuates between different sub-states, this plasticity likely being a key factor to allow the protein to perform protein–protein and protein–metal ion interactions needed for Ni(II) ions delivery into the urease active site. ITC determined the conditions and the thermodynamic parameters of Ni(II) and GTP-driven protein dimerization, supported by light scattering data. SDSL-EPR demonstrated that the degree of structural flexibility changes along the protein backbone, with the region involved in GTP binding and the one involved in metal and protein interactions being more rigid and more flexible, respectively. EPR also revealed that concomitant addition of both Ni(II) and GTP is necessary for a structural transition in these two parts of the protein, located on opposite sides of the tertiary structure, with a shift of the conformational equilibrium towards a more structured state. Differently, addition of either the metal ions or the nucleotide produces only minor perturbations of the conformational equilibrium, indicating that both ligands are necessary to exert a significant conformational response. These observations suggest that binding of GTP in its pocket is propagated, along the protein backbone, to the metal binding site where Ni(II) is bound, and vice versa. The induced shift of the conformational ensemble of UreG likely regulates the protein function, possibly allowing the protein to shuttle Ni(II) ions from UreE to the UreD$_2$-UreF$_2$ complex and, eventually, to urease.

Overall, the present work represents an important contribution for the characterization of the dynamics of UreG and its role in the network of the urease chaperone proteins. In the perspective to extend this original approach, involving SDSL-EPR spectroscopy, to the study of urease network directly inside bacterial cells, the results presented here provide important insights useful in the research on molecules with anti-bacterial activities to overcome anti-microbial resistance (AMR).

Supplementary Materials: The following can be found at http://www.mdpi.com/2218-273X/10/7/1062/s1: Supporting figures and tables as well as detailed descriptions of all experimental procedures, comprising EPR spectra simulations, site-directed mutagenesis, measurement of UreG GTPase activity, Size exclusion chromatography and light scattering data, are provided with the manuscript. Section S1: Supplementary figures (S1–S13), Section S2: CW spectra simulation with SimLabel program, Section S3: Methods summary.

Author Contributions: Conceptualization, B.Z. and E.M.; Data curation, G.G. and B.Z.; Formal analysis, E.E., B.G., S.C., V.B. and B.Z.; Funding acquisition, S.C. and E.M.; Investigation, A.P. and G.G.; Methodology, A.P.; Project administration, B.Z. and E.M.; Software, E.E.; Validation, B.G. and S.C.; Writing—original draft, S.C., V.B., B.Z. and E.M. All authors have read and agreed to the published version of the manuscript.

Funding: We acknowledge financial support from the "Agence Nationale de la Recherche" (ANR-18-CE11-0007-01) and from the "Conseil Régional Région Sud" (A.P. PhD Fellowship EJD-2018-2021). We also thank the COST Action MOBIEU (CA15126) for supporting this work via a Short-Term Scientific Mission to B.Z. and E.M. The University of Bologna contributed to this study through funds for basic research and for travel fellowships awarded to A.P. The "Consorzio Interuniversitario di Risonanze Magnetiche di Metallo-Proteine" (CIRMMP) is acknowledged for its financial support to B.Z. and S.C.

Acknowledgments: We thank K.C. Tamburrini for the labeling of *Sp*UreG protein. We thank V. Fourmond for helpful discussion. We are grateful to the EPR facilities available at the national EPR network RENARD (IR 3443), and at the Aix-Marseille EPR center. We thank P. Mansuelle and R. Lebrun from Aix-Marseille Université and CNRS IMM (FR 3479), Plate-forme de Protéomique, 31 chemin Joseph Aiguier, 13402 Marseille, France, for mass spectrometry measurements and analyses.

Conflicts of Interest: The authors declare no conflict of interest.

References

1. Laxminarayan, R.; Duse, A.; Wattal, C.; Zaidi, A.K.; Wertheim, H.F.; Sumpradit, N.; Vlieghe, E.; Hara, G.L.; Gould, I.M.; Goossens, H.; et al. Antibiotic resistance-the need for global solutions. *Lancet Infect. Dis.* **2013**, *13*, 1057–1098. [CrossRef]

2. Tornimbene, B.; Eremin, S.; Escher, M.; Griskeviciene, J.; Manglani, S.; Pessoa-Silva, C.L. WHO Global Antimicrobial Resistance Surveillance System early implementation 2016–17. *Lancet Infect. Dis.* **2018**, *18*, 241–242. [CrossRef]

3. Global priority list of antibiotic-resistant bacteria to guide research, discovery, and development of new antibiotics. Available online: https://www.who.int/medicines/publications/global-priority-list-antibiotic-resistant-bacteria/en/ (accessed on 27 February 2017).

4. Zambelli, B.; Musiani, F.; Benini, S.; Ciurli, S. Chemistry of Ni2+ in urease: Sensing, trafficking, and catalysis. *Acc. Chem. Res.* **2011**, *44*, 520–530. [CrossRef] [PubMed]

5. Beenken, K.E.; Dunman, P.M.; McAleese, F.; Macapagal, D.; Murphy, E.; Projan, S.J.; Blevins, J.S.; Smeltzer, M.S. Global gene expression in Staphylococcus aureus biofilms. *J. Bacteriol.* **2004**, *186*, 4665–4684. [CrossRef]

6. Zhou, C.; Bhinderwala, F.; Lehman, M.K.; Thomas, V.C.; Chaudhari, S.S.; Yamada, K.J.; Foster, K.W.; Powers, R.; Kielian, T.; Fey, P.D. Urease is an essential component of the acid response network of Staphylococcus aureus and is required for a persistent murine kidney infection. *PLoS Pathog.* **2019**, *15*, e1007538. [CrossRef]

7. Jones, B.D.; Lockatell, C.V.; Johnson, D.E.; Warren, J.W.; Mobley, H.L. Construction of a urease-negative mutant of Proteus mirabilis: Analysis of virulence in a mouse model of ascending urinary tract infection. *Infect. Immun.* **1990**, *58*, 1120–1123. [CrossRef]

8. Gatermann, S.; Marre, R. Cloning and expression of Staphylococcus saprophyticus urease gene sequences in Staphylococcus carnosus and contribution of the enzyme to virulence. *Infect. Immun.* **1989**, *57*, 2998–3002. [CrossRef]

9. Ligon, J.V.; Kenny, G.E. Virulence of ureaplasmal urease for mice. *Infect. Immun.* **1991**, *59*, 1170–1171. [CrossRef]

10. Zhou, F.; Yu, T.; Du, R.; Fan, G.; Liu, Y.; Liu, Z.; Xiang, J.; Wang, Y.; Song, B.; Gu, X.; et al. Clinical course and risk factors for mortality of adult inpatients with COVID-19 in Wuhan, China: A retrospective cohort study. *Lancet* **2020**, *395*, 1054–1062. [CrossRef]

11. Testerman, T.L.; Morris, J. Beyond the stomach: An updated view of Helicobacter pylori pathogenesis, diagnosis, and treatment. *W. J. Gastroenterol.* **2014**, *20*, 12781–12808. [CrossRef]

12. Eaton, K.A.; Brooks, C.L.; Morgan, D.R.; Krakowka, S. Essential role of urease in pathogenesis of gastritis induced by Helicobacter pylori in gnotobiotic piglets. *Infect. Immun.* **1991**, *59*, 2470–2475. [CrossRef]

13. D'Elios, M.M.; Czinn, S.J. Immunity, inflammation, and vaccines for Helicobacter pylori. *Helicobacter* **2014**, *19*, 19–26. [CrossRef]

14. Maroney, M.J.; Ciurli, S. Nonredox nickel enzymes. *Chem. Rev.* **2014**, *114*, 4206–4228. [CrossRef] [PubMed]

15. Mazzei, L.; Musiani, F.; Ciurli, S. CHAPTER 5 Urease. In *The Biological Chemistry of Nickel*; The Royal Society of Chemistry: London, UK, 2017; pp. 60–97.

16. Ha, N.-C.; Oh, S.-T.; Sung, J.Y.; Cha, K.A.; Lee, M.H.; Oh, B.-H. Supramolecular assembly and acid resistance of Helicobacter pylori urease. *Nat. Struct. Biol.* **2001**, *8*, 505–509. [CrossRef]

17. Tarsia, C.; Danielli, A.; Florini, F.; Cinelli, P.; Ciurli, S.; Zambelli, B. Targeting Helicobacter pylori urease activity and maturation: In-cell high-throughput approach for drug discovery. *Biochim. Biophys. Acta Gen. Subj.* **2018**, *1862*, 2245–2253. [CrossRef]

18. Bellucci, M.; Zambelli, B.; Musiani, F.; Turano, P.; Ciurli, S. Helicobacter pylori UreE, a urease accessory protein: Specific Ni(2+)- and Zn(2+)-binding properties and interaction with its cognate UreG. *Biochem. J.* **2009**, *422*, 91–100. [CrossRef] [PubMed]

19. Musiani, F.; Gioia, D.; Masetti, M.; Falchi, F.; Cavalli, A.; Recanatini, M.; Ciurli, S. Protein Tunnels: The Case of Urease Accessory Proteins. *J. Chem. Theory Computat.* **2017**, *13*, 2322–2331. [CrossRef] [PubMed]

20. Zambelli, B.; Turano, P.; Musiani, F.; Neyroz, P.; Ciurli, S. Zn2+-linked dimerization of UreG from Helicobacter pylori, a chaperone involved in nickel trafficking and urease activation. *Proteins Struct. Funct. Bioinform.* **2009**, *74*, 222–239. [CrossRef]

21. Fong, Y.H.; Wong, H.C.; Yuen, M.H.; Lau, P.H.; Chen, Y.W.; Wong, K.-B. Structure of UreG/UreF/UreH Complex Reveals How Urease Accessory Proteins Facilitate Maturation of Helicobacter pylori Urease. *PLoS Biol.* **2013**, *11*, e1001678. [CrossRef] [PubMed]

22. Yuen, M.H.; Fong, Y.H.; Nim, Y.S.; Lau, P.H.; Wong, K.-B. Structural insights into how GTP-dependent conformational changes in a metallochaperone UreG facilitate urease maturation. *Proc. Natl. Acad. Sci. USA* **2017**, *114*, E10890–E10898. [CrossRef] [PubMed]

23. Miraula, M.; Ciurli, S.; Zambelli, B. Intrinsic disorder and metal binding in UreG proteins from Archae hyperthermophiles: GTPase enzymes involved in the activation of Ni(II) dependent urease. *J. Biol. Inorg. Chem.* **2015**, *20*, 739–755. [CrossRef] [PubMed]

24. Real-Guerra, R.; Staniscuaski, F.; Zambelli, B.; Musiani, F.; Ciurli, S.; Carlini, C.R. Biochemical and structural studies on native and recombinant Glycine max UreG: A detailed characterization of a plant urease accessory protein. *Plant Mol. Biol.* **2012**, *78*, 461–475. [CrossRef] [PubMed]

25. Zambelli, B.; Musiani, F.; Savini, M.; Tucker, P.; Ciurli, S. Biochemical studies on Mycobacterium tuberculosis UreG and comparative modeling reveal structural and functional conservation among the bacterial UreG family. *Biochemistry* **2007**, *46*, 3171–3182. [CrossRef]

26. Zambelli, B.; Stola, M.; Musiani, F.; De Vriendt, K.; Samyn, B.; Devreese, B.; Van Beeumen, J.; Turano, P.; Dikiy, A.; Bryant, D.A.; et al. UreG, a chaperone in the urease assembly process, is an intrinsically unstructured GTPase that specifically binds Zn2+. *J. Biol. Chem.* **2005**, *280*, 4684–4695. [CrossRef] [PubMed]

27. Palombo, M.; Bonucci, A.; Etienne, E.; Ciurli, S.; Uversky, V.N.; Guigliarelli, B.; Belle, V.; Mileo, E.; Zambelli, B. The relationship between folding and activity in UreG, an intrinsically disordered enzyme. *Sci. Rep.* **2017**, *7*, 5977. [CrossRef]

28. D'Urzo, A.; Santambrogio, C.; Grandori, R.; Ciurli, S.; Zambelli, B. The conformational response to Zn(II) and Ni(II) binding of Sporosarcina pasteurii UreG, an intrinsically disordered GTPase. *J. Biol. Inorg. Chem.* **2014**, *19*, 1341–1354. [CrossRef]

29. Yang, X.; Li, H.; Lai, T.-P.; Sun, H. UreE-UreG Complex Facilitates Nickel Transfer and Preactivates GTPase of UreG in Helicobacter pylori. *J. Biol. Chem.* **2015**, *290*, 12474–12485. [CrossRef] [PubMed]

30. Merloni, A.; Dobrovolska, O.; Zambelli, B.; Agostini, F.; Bazzani, M.; Musiani, F.; Ciurli, S. Molecular landscape of the interaction between the urease accessory proteins UreE and UreG. *Biochim. Biophys. Acta* **2014**, *1844*, 1662–1674. [CrossRef]

31. Etienne, E.; Le Breton, N.; Martinho, M.; Mileo, E.; Belle, V. SimLabel: A graphical user interface to simulate continuous wave EPR spectra from site-directed spin labeling experiments. *Magn. Reson. Chem. MRC* **2017**, *55*, 714–719. [CrossRef]

32. Stoll, S.; Schweiger, A. EasySpin, a comprehensive software package for spectral simulation and analysis in EPR. *J. Magn. Reason.* **2006**, *178*, 42–55. [CrossRef]

33. Pannier, M.; Veit, S.; Godt, A.; Jeschke, G.; Spiess, H.W. Dead-Time Free Measurement of Dipole–Dipole Interactions between Electron Spins. *J. Magn. Reason.* **2000**, *142*, 331–340. [CrossRef] [PubMed]

34. Jeschke, G. MMM: A toolbox for integrative structure modeling. *Protein Sci. Publ. Protein Soc.* **2018**, *27*, 76–85. [CrossRef]

35. Jeschke, G. The contribution of modern EPR to structural biology. *Emerg. Topics Life Sci.* **2018**. [CrossRef]

36. Bordignon, E. EPR Spectroscopy of Nitroxide Spin Probes. In *eMagRes*; Harris, R.K., Wasylishen, R.L., Eds.; John Wiley & Sons, Ltd.: Chichester, UK, 2017.

37. Roser, P.; Schmidt, M.J.; Drescher, M.; Summerer, D. Site-directed spin labeling of proteins for distance measurements in vitro and in cells. *Organ. Biomol. Chem.* **2016**, *14*, 5468–5476. [CrossRef] [PubMed]

38. Potapov, A.; Yagi, H.; Huber, T.; Jergic, S.; Dixon, N.E.; Otting, G.; Goldfarb, D. Nanometer-Scale Distance Measurements in Proteins Using Gd3+ Spin Labeling. *J. Am. Chem. Soc.* **2010**, *132*, 9040–9048. [CrossRef]

39. Bonucci, A.; Ouari, O.; Guigliarelli, B.; Belle, V.; Mileo, E. In-Cell EPR: Progress towards Structural Studies Inside Cells. *Chem. Bio. Chem.* **2020**, *21*, 451–460. [CrossRef]

40. Hubbell, W.L.; McHaourab, H.S.; Altenbach, C.; Lietzow, M.A. Watching proteins move using site-directed spin labeling. *Structure* **1996**, *4*, 779–783. [CrossRef]

41. Hubbell, W.L.; Cafiso, D.S.; Altenbach, C. Identifying conformational changes with site-directed spin labeling. *Nat. Struct. Biol.* **2000**, *7*, 735–739. [CrossRef] [PubMed]

42. Hubbell, W.L.; Lopez, C.J.; Altenbach, C.; Yang, Z. Technological advances in site-directed spin labeling of proteins. *Curr. Opin. Struct. Biol.* **2013**, *23*, 725–733. [CrossRef]

43. Le Breton, N.; Martinho, M.; Mileo, E.; Etienne, E.; Gerbaud, G.; Guigliarelli, B.; Belle, V. Exploring intrinsically disordered proteins using site-directed spin labeling electron paramagnetic resonance spectroscopy. *Front. Mol. Biosci.* **2015**, *2*, 21. [CrossRef]

44. Belle, V.; Rouger, S.; Costanzo, S.; Liquiere, E.; Strancar, J.; Guigliarelli, B.; Fournel, A.; Longhi, S. Mapping alpha-helical induced folding within the intrinsically disordered C-terminal domain of the measles virus nucleoprotein by site-directed spin-labeling EPR spectroscopy. *Proteins* **2008**, *73*, 973–988. [CrossRef] [PubMed]

45. Martinho, M.; Fournier, E.; Le Breton, N.; Mileo, E.; Belle, V. Nitroxide spin labels: Fabulous spy spins for biostructural EPR applications. In *Electron Paramagnetic Resonance: Volume 26*; The Royal Society of Chemistry: London, UK, 2019; Volume 26, pp. 66–88.

46. Jeschke, G. DEER distance measurements on proteins. *Ann. Rev. Phys. Chem.* **2012**, *63*, 419–446. [CrossRef] [PubMed]

47. Schiemann, O.; Prisner, T.F. Long-range distance determinations in biomacromolecules by EPR spectroscopy. *Quart. Rev. Biophys.* **2007**, *40*, 1–53. [CrossRef]

48. Schmidt, T.; Walti, M.A.; Baber, J.L.; Hustedt, E.J.; Clore, G.M. Long Distance Measurements up to 160 A in the GroEL Tetradecamer Using Q-Band DEER EPR Spectroscopy. *Angew. Chem. Int. Ed. Engl.* **2016**, *55*, 15905–15909. [CrossRef]

49. Pineiro, A.; Munoz, E.; Sabin, J.; Costas, M.; Bastos, M.; Velazquez-Campoy, A.; Garrido, P.F.; Dumas, P.; Ennifar, E.; Garcia-Rio, L.; et al. AFFINImeter: A software to analyze molecular recognition processes from experimental data. *Anal. Biochem.* **2019**, *577*, 117–134. [CrossRef] [PubMed]

50. Zambelli, B.; Bellucci, M.; Danielli, A.; Scarlato, V.; Ciurli, S. The Ni2+ binding properties of Helicobacter pylori NikR. *Chem. Commun.* **2007**, *21*, 3649–3651. [CrossRef]

51. López, C.J.; Fleissner, M.R.; Guo, Z.; Kusnetzow, A.K.; Hubbell, W.L. Osmolyte perturbation reveals conformational equilibria in spin-labeled proteins. *Protein Sci. Publ. Protein Soc.* **2009**, *18*, 1637–1652. [CrossRef]

52. Flores Jiménez, R.H.; Do Cao, M.A.; Kim, M.; Cafiso, D.S. Osmolytes modulate conformational exchange in solvent-exposed regions of membrane proteins. *Protein Sci. Publ. Protein Soc.* **2010**, *19*, 269–278. [CrossRef]

53. Bolen, D.W. Effects of naturally occurring osmolytes on protein stability and solubility: Issues important in protein crystallization. *Methods* **2004**, *34*, 312–322. [CrossRef]

54. Mileo, E.; Lorenzi, M.; Erales, J.; Lignon, S.; Puppo, C.; Le Breton, N.; Etienne, E.; Marque, S.R.; Guigliarelli, B.; Gontero, B.; et al. Dynamics of the intrinsically disordered protein CP12 in its association with GAPDH in the green alga Chlamydomonas reinhardtii: A fuzzy complex. *Mol. Biosyst.* **2013**, *9*, 2869–2876. [CrossRef]

55. Musiani, F.; Ippoliti, E.; Micheletti, C.; Carloni, P.; Ciurli, S. Conformational fluctuations of UreG, an intrinsically disordered enzyme. *Biochemistry* **2013**, *52*, 2949–2954. [CrossRef] [PubMed]

56. Jeschke, G.; Chechik, V.; Ionita, P.; Godt, A.; Zimmermann, H.; Banham, J.E.; Timmel, C.R.; Hilger, D.; Jung, H. DeerAnalysis2006—A Comprehensive Software Package for Analyzing Pulsed ELDOR Data. *Appl. Magn. Reson.* **2006**, *30*, 473–498. [CrossRef]

57. Sarver, J.L.; Zhang, M.; Liu, L.; Nyenhuis, D.; Cafiso, D.S. A Dynamic Protein-Protein Coupling between the TonB-Dependent Transporter FhuA and TonB. *Biochemistry* **2018**, *57*, 1045–1053. [CrossRef] [PubMed]

Article

Protein–Protein Interactions Mediated by Intrinsically Disordered Protein Regions Are Enriched in Missense Mutations

Eric T. C. Wong, Victor So, Mike Guron, Erich R. Kuechler, Nawar Malhis, Jennifer M. Bui and Jörg Gsponer *

Michael Smith Laboratories, University of British Columbia, Vancouver, BC V6T 1Z4, Canada;
eric_wong@live.ca (E.T.C.W.); victor.kaiming.so@gmail.com (V.S.); mike_guron@alumni.ubc.ca (M.G.);
erich.kuechler@gmail.com (E.R.K.); nmalhis@msl.ubc.ca (N.M.); jennbui@gmail.com (J.M.B.)
* Correspondence: gsponer@msl.ubc.ca; Tel.: +1-604-827-4731

Received: 13 June 2020; Accepted: 20 July 2020; Published: 24 July 2020

Abstract: Because proteins are fundamental to most biological processes, many genetic diseases can be traced back to single nucleotide variants (SNVs) that cause changes in protein sequences. However, not all SNVs that result in amino acid substitutions cause disease as each residue is under different structural and functional constraints. Influential studies have shown that protein–protein interaction interfaces are enriched in disease-associated SNVs and depleted in SNVs that are common in the general population. These studies focus primarily on folded (globular) protein domains and overlook the prevalent class of protein interactions mediated by intrinsically disordered regions (IDRs). Therefore, we investigated the enrichment patterns of missense mutation-causing SNVs that are associated with disease and cancer, as well as those present in the healthy population, in structures of IDR-mediated interactions with comparisons to classical globular interactions. When comparing the different categories of interaction interfaces, division of the interface regions into solvent-exposed rim residues and buried core residues reveal distinctive enrichment patterns for the various types of missense mutations. Most notably, we demonstrate a strong enrichment at the interface core of interacting IDRs in disease mutations and its depletion in neutral ones, which supports the view that the disruption of IDR interactions is a mechanism underlying many diseases. Intriguingly, we also found an asymmetry across the IDR interaction interface in the enrichment of certain missense mutation types, which may hint at an increased variant tolerance and urges further investigations of IDR interactions.

Keywords: intrinsically disordered proteins; single nucleotide variants; protein–protein interactions; interface core and rim; human disease

---

1. Introduction

Driven by the goal of understanding genetic diversity in the human population and how this diversity affects disease likelihood, efforts in sequencing human genomes have provided vast amounts of sequence variants, also known as single nucleotide variants (SNVs). If SNVs are located in protein-coding regions of the genome and are non-synonymous, they can result in premature stop codons (i.e., nonsense mutations) or substitutions of amino acids (i.e., missense mutations), both of which could impact the biological function of the encoded proteins. As a result, SNVs can be categorized as deleterious, benign, or even beneficial for human health. Studies have shown correlations between the phenotypic effects of SNVs and their localization to different functional regions of proteins [1–4].

Protein structural data is critical for mapping SNVs to functional regions and for understanding the molecular mechanism through which they lead to functional alterations. Each residue contributes differently to protein folding and function, and the different constraints on the residues are mirrored in the localization patterns of the SNVs observed in the protein structure. This relationship first became evident when disease-associated SNVs, specifically those that cause missense mutations, were mapped onto protein structures [5,6]. Mutations associated with diseases were shown to be enriched at active sites and buried regions that provide structural stability [2,7]. Disease-associated missense mutations were also found enriched at protein–protein interaction (PPI) interface regions, which does not come as a surprise given that the majority of proteins require interactions with other proteins to perform their functions properly. Importantly, the enrichment of disease-associated mutations at the protein interface, relative to the non-interface surface, shows much greater contrast when focusing on the residues at the core of the interface [2]. PPI interfaces can be divided into core and rim regions using the protein complex structures, where the core residues become buried upon binding while the rim residues remain relatively solvent-exposed and typically form the perimeter of the interface [8,9]. Disease-associated missense mutations are more common in the interface core, especially at the hotspot residues that contribute most to protein interaction affinity [2].

The enrichment of disease-associated missense mutations at interface regions clearly suggests that the disruption of PPIs is likely a common mechanism for altered biological function and disease [10]. An extensive mutagenesis study provides strong support for this hypothesis [1]. This study revealed that the majority of the tested disease-associated mutations disrupted PPIs, and the interaction-disrupting mutations can be divided into quasi-null and edgetic mutations. A quasi-null mutation abolishes all of a protein's interactions, likely through destabilization of the protein, while an edgetic mutation removes a specific subset of interactions. These edgetic mutations are more often found in interface regions, altering their binding properties. Furthermore, disease-associated mutations in different interface regions of the same protein can lead to different diseases, providing an explanation for the pleiotropic effects of disease-associated genes [10].

Structural mapping has also been exploited to investigate the functional impact of SNVs associated with specific classes of diseases, most prominently cancer. Somatic SNVs associated with cancer are of particular interest because some of them drive the propagation of the tumor cells, which contrasts the broader disruptiveness of germline disease-associated SNVs [6,11]. Somatic cancer-associated SNVs from tumor tissues were also found enriched at functional regions, but studies have reported higher enrichment at the protein surface compared to the buried regions [6,11]. Importantly, the properties of the SNVs depend on the native and cancer-associated functions of the proteins. Cancer-associated SNVs in oncoproteins cause gain-of-function and are more commonly found on the surfaces of proteins as well as clustered and recurrent in specific sequence positions [4,12,13]. While SNVs that can activate oncoproteins are limited to a select few residues, these oncoprotein SNVs appear to be under stronger positive selection in tumor cells, highlighting their key roles in driving oncogenesis [14]. Cancer-associated SNVs in tumor suppressor proteins are found enriched at the buried regions and are more often scattered across the sequence, resulting in destabilizing effects similar to the typical germline disease-associated SNV [4].

Recent sequencing efforts have also enabled the identification of SNVs present in the healthy human population. These SNVs and their observed frequency, as annotated in databases such as gnomAD [15], provide a first glimpse of the natural sequence variations in human populations. They are not directly associated with diseases and are depleted from functionally critical protein regions such as protein interfaces while their enrichment is inversely correlated with evolutionary sequence conservation [2,16], sharply contrasting the trends seen in disease-associated SNVs. This finding is unsurprising as the tolerance for sequence variation is likely to be higher at non-functional, less-conserved protein regions.

Studies of SNVs from the healthy population often make a distinction between common and rare variants. Common SNVs, which are often defined as those with greater than 1% frequency in the population, are typically functionally neutral [17,18]. Some common SNVs may even provide a selective advantage and may be beneficial for the adaptation of the population to environmental changes or stressors [19,20]. The localization patterns of these common SNVs most strongly contrast with disease-associated SNVs. On the other hand, rare SNVs account for the majority of variants in the population [21]. Rare SNVs consist of mutations under negative selection as well as novel mutations and thus are enriched in deleterious mutations [22–24]. These mutations also tend to have greater effects on function, which was demonstrated by a study on human height distribution that found an inverse correlation between frequency and phenotypic effect [25]. Therefore, common and rare SNV datasets contain SNVs with low and medium levels of deleteriousness on average, which provides contrast with the highly deleterious disease-associated SNVs.

Structural analyses of SNVs, most frequently on the SNVs causing missense mutations, have previously focused on independently-folding (i.e., globular) domains. However, many PPIs are mediated by protein regions that are not confined in a single folded conformation prior to binding, namely intrinsically disordered regions (IDRs) that participate in PPIs (interacting IDRs) [26–28]. IDRs are underrepresented in interaction and structural datasets [28–30], but IDRs are increasingly recognized for their prevalence and their critical roles in regulatory intermolecular interactions [31]. It has been hypothesized that some traits make IDRs particularly suitable for interactions involved in signaling and regulation [31], complementing globular domains that more often perform catalytic functions. IDRs contribute large interaction surfaces in the form of compact interaction modules such as shorter peptide motifs and longer molecular recognition features (MoRFs) [26,28,32]. It has been estimated that IDRs in the human proteome contain ~132,000 binding motifs [28]. Peptide motifs and MoRFs can be used in combinatorial ways due to alternative splicing and the modulation of their interaction propensities via post-translational modifications [33,34]. Moreover, the flexibility of IDRs allows multivalent and fuzzy, often promiscuous interactions with multiple partners as well as fast binding kinetics and low-affinity high-specificity partnerships [35–37]. Given these traits, it is not surprising that IDRs are a common feature of hubs in PPIs, which are proteins that make the largest number of interactions and thus greatly influence the connectivity of PPI networks [38–40].

Given the significance of IDR-mediated protein interactions, it is pertinent to know whether disease, common, and rare SNVs are similarly enriched/depleted at the interfaces of IDR-mediated interactions (IDR interactions) as at interfaces between folded domains (globular interactions). It has been established that IDRs are generally less enriched in disease SNVs compared to other protein regions and that they exhibit higher evolutionary rates that could be attributed to weaker structural constraints [41,42]. Nonetheless, with an estimated 22% of disease SNVs located in IDRs and a higher concentration of these mutations in IDRs that are involved in PPIs, the importance of understanding the mutations in IDRs should not be understated [43]. A study of sequence motifs revealed enrichment of disease-associated SNVs compared to benign SNVs [44], suggesting that the function-disrupting substitutions are concentrated in the interaction-mediating elements residing in IDRs. In the light of this finding and the fact that IDRs interact mostly with folded domains, although IDR–IDR interactions have been reported, one could expect the interfaces of the globular partners of IDRs (IDR-partners) to exhibit the familiar trends in SNV enrichment/depletion that have been observed in globular interaction interfaces. However, IDR interactions and globular interactions exhibit differences in both structure and function [27,45–48], and we have previously found IDR-partner interfaces to have distinctive physicochemical and geometric properties [49]. Moreover, IDR-partner interfaces bind to inherently dynamic IDRs that are potentially more accommodating to changes in the interface. Therefore, the IDR-partner interfaces may exhibit distinctive mutation enrichment patterns, which demand a closer inspection.

In this work, we built on previous studies analyzing the localization of several categories of missense mutations in protein complex structures, but we focused on IDR interactions. Specifically, we analyzed the bias of missense mutations among interface residues as well as buried and surface non-interface residues (Figure 1). Importantly, we separated protein interfaces into core and rim regions because we and others have demonstrated characteristic differences in the two interface regions, including residue composition and SNV enrichment [2,9,46,49]. To calculate mutation enrichments, we mapped disease-associated SNVs from SwissVar, somatic cancer-associated SNVs from COSMIC, and SNVs from gnomAD that cause missense mutations onto available structures of IDR interactions, as well as globular interactions which serve as a control [15,50,51]. Our analyses reveal that interface regions of interacting IDRs are at least as enriched in disease-associated SNVs as globular interactions and exhibit depletion of gnomAD SNVs, especially at the interface core regions. Notably, IDR-partner interfaces exhibit a strong presence of disease-associated SNVs. However, our analyses may also provide preliminary evidence of a greater tolerance for common gnomAD SNVs at IDR-partner interfaces, which deserves further investigation. Overall, our findings are concordant with studies that have associated IDRs with numerous diseases, especially cancer [52–55].

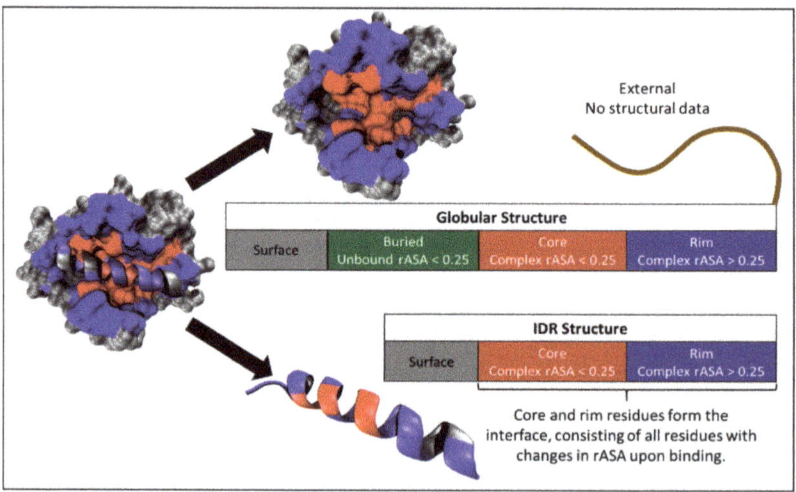

Figure 1. Structural regions analyzed in this study. The structural regions were defined based on solvent-accessible surface areas measured from protein complex structures. Residues with changes in relative solvent accessible surface area (rASA; see Methods) between the bound and unbound conformations were defined as core residues (red) if rASAs are smaller than 0.25 and rim residues (blue) if rASAs are greater than 0.25 in the bound structures. Buried residues are non-interface residues with rASAs smaller than 0.25 in the unbound structures, and the remainder are surface residues (gray). Because the full-length protein often contains regions without structural data coverage, these structurally undefined sequences were classified as external regions in our analyses.

2. Materials and Methods

2.1. Structural Data

The structural data of protein interactions consists of human proteins downloaded from the RCSB Protein Data Bank (PDB) in September 2018 (http://www.rcsb.org/). For each structure, the model of the biological unit was selected whenever available, and the first model was used when the PDB file contains multiple models. Complex structures that only consist of carbon-alpha coordinates or are too large for the computational software we utilized were removed. Protein interactions were analyzed pairwise by iterating through all pairs of protein chains in each PDB file, focusing only on human heteromeric interactions and removing pairs with no physical interaction, which was determined through calculating changes in solvent accessibility. FreeSASA was used for calculation of solvent accessible surface area (SASA) of the residues of each protein chain in their bound and unbound states [56], where the unbound state was the structure of each protein chain in isolation. Physically interacting protein chain pairs are those with a change in total SASA between their bound and unbound states.

For each interaction pair, relative solvent accessible surface area (rASA) of protein residues were calculated to categorize the residues into protein regions. The SASA of each residue of the protein structures was normalized by the SASA of the residue type "X" calculated in a Gly-X-Gly peptide in extended conformation [9]. The residues were placed into structural categories based on their rASA in their bound and unbound states [9]. Surface and buried regions consist of residues above and below 0.25 rASA in the isolated protein chain, respectively. All residues with a change in rASA between the bound and unbound states were defined as interface residues. The interaction interface consists of the rim residues, which have rASA > 0.25 in the bound state, and the core residues, which have rASA < 0.25 in the bound state. Subsequently, the categorized residues were mapped to UniProt sequences [57]. Supplementary Figure S1 provides an overview of the construction process of the interaction structure datasets, as well as the number of structures and proteins involved. The dataset of globular interactions encompasses all interaction pairs, which is justifiable since the dataset of IDR interactions is small in comparison.

2.2. Defining Intrinsically Disordered Regions (IDR) Interaction Datasets

IDR interactions were identified by mapping IDRs onto UniProt sequences and subsequently screening PDB complex structures for the IDRs. A curated dataset of IDRs was extracted from the MobiDB database in September 2018 [58]. The MobiDB database contains protein regions annotated as curated linear interacting peptides (LIPs), which consist of IDRs aggregated from multiple databases. For this study, entries from the ELM were excluded since they contain short linear motifs (SLiMs) that are found not only in disordered regions but also in globular regions [59].

For each UniProt sequence with IDRs defined by MobiDB, we iterated through all interaction pair structures to identify all instances of the IDRs. For an interaction pair structure to be labeled as an IDR interaction, one of the protein chains must overlap with an IDR sequence. A protein chain was labeled as an interacting IDR if more than 50% of the interface residues were within an IDR defined in MobiDB. Furthermore, protein chains with more than nine buried residues in their unbound states were excluded from interacting IDRs, thereby removing chains that potentially contain independently-folding regions. Once all the interacting IDR and IDR-partner interaction pair structures were defined, the remaining structures were excluded from the IDR interaction dataset.

2.3. Mapping Mutations to Globular and IDR Interaction Structural Data

SNVs in protein-coding regions that cause missense mutations and are associated with diseases were sourced from the SwissVar and COSMIC databases. The SwissVar SNV dataset consists of disease-related germline mutations [50]. The COSMIC SNV dataset consists of curated cancer mutations, excluding mutations annotated with genome-wide screens and single nucleotide polymorphisms [51]. The COSMIC database contains both cancer driver and passenger mutations, and the mutations may come from proteins that are labeled as oncoproteins or tumor suppressor as well as belonging to neither or both of those categories. Because of the functional differences between oncoproteins and tumor suppressors, we further divided the COSMIC-SNV-mapped proteins into those labeled exclusively as oncoproteins or tumor suppressors by using the datasets constructed by Brown et al. [14].

SNVs that cause missense mutations and are not associated with diseases, and thus are generally considered benign, were sourced from the gnomAD database [15]. The gnomAD SNV dataset consists of SNVs from the healthy human population as well as their frequencies, allowing their categorization into common and rare SNVs. We analyzed SNVs with a frequency between 0.1 and 10^{-6}. As the amount of common SNV data is very small, we used a comparatively relaxed threshold frequency of 0.001 to define our high-frequency SNV dataset, which provides a subset of SNVs with a greater fraction of benign mutations for analysis. For comparison, we also analyzed a set of rare SNVs that have frequencies between $5 * 10^{-6}$ and 10^{-6}.

We merged mutation data with structural data by mapping all missense mutations to the interaction pair structures through their shared UniProt sequences [57]. We subsequently iterated through all the interaction pair structures and merged all the structural and mutation data for each UniProt sequence. Merging the SwissVar SNV, COSMIC SNV, oncoprotein SNV, tumor suppressor SNV [14,51], and gnomAD SNV datasets with the globular, MobiDB interacting IDR [58], and IDR-partner datasets resulted in 15 combined datasets. For each of the 15 combined datasets, UniProt sequences lacking either structural data or mutations were removed. In case of overlap between multiple PDB structures, the residue structural label was decided by their priority from highest to lowest: core, rim, buried, surface, and unstructured (external region; see Figure 1). In other words, if a protein residue position was an interface core residue in one structure and a non-interface surface residue in another, the residue will be labeled as an interface core residue in the merged data. In the case of the IDR interaction dataset, the UniProt residues were also labeled as interacting IDR or IDR-partner. The tabulated residues and mutations for all datasets are presented in Supplementary Table S1.

2.4. Odds Ratio Calculations

We used odds ratios (ORs) to compare mutation enrichment between protein regions, as described previously by David and Sternberg [2]. OR values higher than one denote enrichment of missense mutations at the specified regions, while depletion results in values smaller than one. The probability of mutation (p) in region i was given by the number of mutated positions (m) in region i divided by the number of residues (r) in region i, i.e.,:

$$p_i = m_i / r_i \tag{1}$$

The odds ratio of mutations in region i over j is:

$$OR_{ij} = \frac{(p_i/(1-p_i))}{(p_j/(1-p_j))} \tag{2}$$

The standard error for the natural log of the odds ratio is [3]:

$$SE\_LOR_{ij} = \sqrt{\frac{1}{m_i} + \frac{1}{r_i - m_i} + \frac{1}{m_j} + \frac{1}{r_j - m_j}} \tag{3}$$

The standard error for the natural log of the odds ratio was used to estimate the standard error of the odds ratio [10]:

$$SE\_OR_{ij} \cong OR_{ij} * SE\_LOR_{ij} \tag{4}$$

The standard error of the odds ratio was used to define the error bars in the bar plots of ORs, which were generated using the ggplot2 module in R [60]. The p-values of ORs were calculated using the chi-square test in R and are reported in Supplementary Table S2. The enrichment of mutations at each protein region was determined using the full-length protein as the reference, i.e., region r_j is the total number of residues in a dataset. Therefore, an OR of the interacting IDR dataset would be calculated with p_j equal to the number of mutations in all proteins containing interacting IDRs divided by the total length of those proteins. Correspondingly, a p_j of the IDR-partner dataset would be calculated based on the subset of proteins containing IDR-partner structures.

3. Results

3.1. Disease-Associated Single Nucleotide Variants (SNVs) Are Enriched at IDR Interaction Interfaces

To pursue our goal of revealing whether IDR interfaces exhibit familiar trends in SNV enrichment/depletion that have been observed in globular interaction interfaces, we first repeated the enrichment analysis for globular interactions and set baselines for the comparisons with IDR interactions. We collected structures of heteromeric protein complexes from the Protein Data Bank (PDB) to assemble the globular dataset (see Methods for details). We divided residues in these complexes into structural regions based on their solvent accessibility in bound and unbound states (see Methods for details). Briefly, the surface and buried regions were defined as residues that are exposed and unexposed to solvent in the unbound state, respectively. The interface region was defined as residues that become more buried upon complex formation, i.e., residues that change in solvent exposure when comparing the complex to the separate protein chains (Figure 1). We further divided the interface region into core and rim, which are the central and peripheral sections of the interface, respectively, because of the differences in sequence and structural characteristics between the two regions [2,9].

We began our comparison of enrichments by analyzing the distribution of disease-associated missense mutations from the SwissVar database (SwissVar SNVs) in the globular dataset. After mapping the mutations to the structural regions, we calculated the enrichment/depletion of mutations at each structural region using odds ratios (ORs) [2], with OR > 1 indicating enrichment of mutations relative to the full sequence distribution (see Methods for details). Our globular interaction dataset shows the highest enrichment of SwissVar SNVs at the buried and interface core regions of proteins (buried OR = 1.9, p-value $\leq 10^{-99}$; core OR = 2.3, p-value $\leq 10^{-99}$; Figure 2). Buried residues of globular domains are typically more critical to the structure and stability of the protein, while core residues tend to contribute strongly to protein binding, so the substitution of these residues will more likely disrupt function. Thus, these enrichment patterns are consistent with the disease association of the mutations. These observations are also in agreement with previous studies of disease-associated missense mutations that reported the strongest mutation enrichment at the buried and core regions [1,2,61]. Although David et al. reported more significant enrichment at the buried region, this discrepancy could be explained by differences in rASA thresholds used in defining structural regions [2]. The rim region has an OR of 1.4 (p-value = 2.5 * 10^{-12}), which is much lower than the interface core region but still suggests stronger functional constraints than the non-interface surface region, which has an OR of 1.0 (Figure 2).

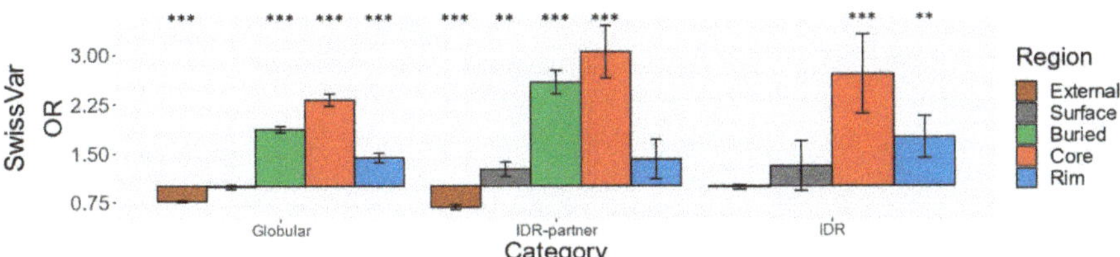

Figure 2. Odds ratios of SwissVar single nucleotide variants (SNVs). An odds ratio (OR) is calculated for each protein region using all residues in the dataset as the reference distribution. The bar graph plots the ORs (*Y*-axis) of each protein category and protein region (*X*-axis). Each OR is the odds of mutation in the specific region divided by the odds of the full-length parent proteins. The *Y*-axis is centered at one, and ORs > 1 show enrichment while ORs < 1 show depletion. Structural regions are color-coded (see Figure 1). Statistical significance is denoted by asterisks: * p-value < 0.05, ** p-value < 0.01, *** p-value < 0.001. ORs and p-values can be found in Supplementary Table S2.

Next, we repeated this analysis for IDR interactions. We identified these interactions, i.e., interactions between IDRs and folded IDR-partners, by mapping curated IDRs from the MobiDB database to PDB complex structures (see Methods for details). Notably, because PDB structures are often limited to crystallizable protein complexes, the IDRs in our datasets are generally regions that fold upon binding, such as MoRFs and peptide binding motifs (see Results Section 3.3 and Discussion). We then calculated ORs for the different protein regions, as it was done for the globular interactions, calculating the denominator odds using the number of mutations and sequence length of proteins that contain either IDR or IDR-partner structures (see Methods for details). IDR-partners in this dataset are independently folded domains while IDRs, by definition, are not, so we calculated the enrichments for

IDR-partners and interacting IDRs separately (Figure 2 middle and right). For the IDR-partner, the OR calculations reveal a similar picture of enrichment as for the globular interactions. Specifically, SwissVar diseases-associated mutations are found enriched at the buried parts as well as the interaction interface consisting of core and rim residues. The enrichment of SwissVar mutations at the IDR-partner interface core region as well as its buried parts are even more pronounced than at the globular interaction regions (buried OR = 2.6, *p*-value = 3.3 * 10^{-45}; core OR = 3.0 *p*-value = 7.3 * 10^{-19}). The IDRs themselves also show enrichment of SwissVar disease-associated mutations at interface locations (Figure 2 right). These mutations are found significantly enriched at both the core and the rim regions of the interface, but the enrichment is particularly pronounced at the interface core (core OR = 2.7, *p*-value = 4.3 * 10^{-6}; rim OR = 1.7, *p*-value = 2.4 * 10^{-3}). It needs to be noted that the IDRs lack buried residues because they predominantly interact by adopting secondary but not tertiary structures. This analysis of mutations from SwissVar clearly demonstrates that disease missense mutations are not only found enriched at the core of classical interfaces between folded domains but also at the core of interfaces between IDRs and their partners. This result may suggest that the interface core of interacting IDRs and IDR-partners have functional roles that are very susceptible to disruption by amino acid substitutions, maybe as susceptible as the core of globular interfaces.

Compared to germline SwissVar mutations, somatic cancer-associated mutations from the COSMIC database are known to have different enrichment patterns and mechanistic properties, which prompted us to analyze them independently. Past studies have shown a distinctively greater tendency for cancer-associated mutations to occur in protein surface and interface regions [6,11], in contrast to the disease-associated germline mutations that favor the buried region [2]. Using the same procedures for mapping mutations to structural data and evaluating mutation enrichment, we found that the COSMIC cancer-associated missense mutations also exhibit enrichment at functional regions of the globular and IDR interaction sets (Figure 3A), although to a lesser degree than SwissVar SNVs. Indeed, ORs closer to one, specifically in the globular interaction proteins and in the IDR-partners, indicate weaker enrichment patterns compared to SwissVar SNVs. This difference could be attributed, at least in part, to the presence of passenger mutations in the COSMIC SNV dataset, which are missense mutations identified in cancer tissue that do not contribute to tumor growth and are under weak negative or no selective pressure [62]; therefore, they are expected to have a more uniform distribution across protein regions. Notably, the enrichment at the globular buried region is significant but relatively weak (Table S2), which is consistent with previous observations [6,11]. Interestingly, the rim regions of both the globular interactions and the IDR-partners show significant enrichment levels equal to those of the core regions. This finding contrasts the observations for SwissVar SNVs (Figure 2) and is intriguing since rim residues tend to contribute less to binding affinity when compared to the core. Most importantly, the highest ORs are observed in the interacting IDR core and rim (core OR = 1.6, *p*-value = 9.6 * 10^{-5}; rim 1.5, *p*-value = 8.9 * 10^{-7}). This enrichment at interacting residues of the IDRs contrasts the known depletion of COSMIC SNVs within IDRs in general [41,63], emphasizing a strong association between cancer and interacting IDRs and, more specifically, their interface core residues.

Cancer development and progression are generally driven by the inactivation of tumor suppressors and the activation of oncoproteins. Therefore, selective pressures that act on tumor suppressors and oncoproteins in cancer cells may generate a distribution of missense mutations that reflects more closely the functional importance of the affected residues. While we mapped COSMIC missense mutations across many proteins for our analysis, only small subsets of these proteins are verified as tumor suppressors and oncoproteins that drive oncogenesis [13]. As we were interested in the mutation distribution difference between tumor suppressors and oncoproteins, we repeated the analysis while selecting only proteins that were labeled as tumor suppressors and oncoproteins, and we excluded proteins that were annotated in both categories to segregate them and study their differences, as it was done by Brown et al. [14]. As we expected, missense mutations in tumor suppressors show enrichment patterns reminiscent of the SwissVar SNVs, in both the globular interactions as well as IDR-partners (Figure 3B). Most prominent is the statistically significant enrichment of mutations at buried and interface core regions. This result is consistent with previous reports of cancer-associated missense mutation enrichment at the buried region of tumor suppressors [12] and an expected loss of function when mutations hit buried residues important for protein stability. In contrast, the enrichment patterns for globular and IDR-partner oncoproteins (Figure 3C) more resemble patterns observed in the full COSMIC SNV dataset (Figure 3A). Interestingly though, the core regions of IDR-partners that are oncoproteins are not statistically enriched in COSMIC missense mutations (core OR = 1.1, *p*-value = 0.4). Unfortunately, the numbers of cancer-associated missense mutations that map to interacting IDRs from tumor suppressors or oncoproteins are very small, too small for confident interpretation (i.e., all ORs are not statistically significant; see Table S2). Due to the limited data, which results in a lack of statistical significance, we can only speculate on the observed trends. Interacting IDR interface regions from tumor suppressors do not appear enriched in COSMIC missense mutations, which is consistent with the idea that mutations in an interacting IDR are less likely to lead to a loss of function of the protein compared to mutations in the buried regions of globular domains. By contrast, interacting IDR interface cores from oncoproteins appear enriched in cancer missense mutations, which mirrors the enrichment of cancer missense mutations observed in the IDR core of all analyzed proteins (Figure 3A) and implies that these IDR interactions hold functions in cancer-associated pathways.

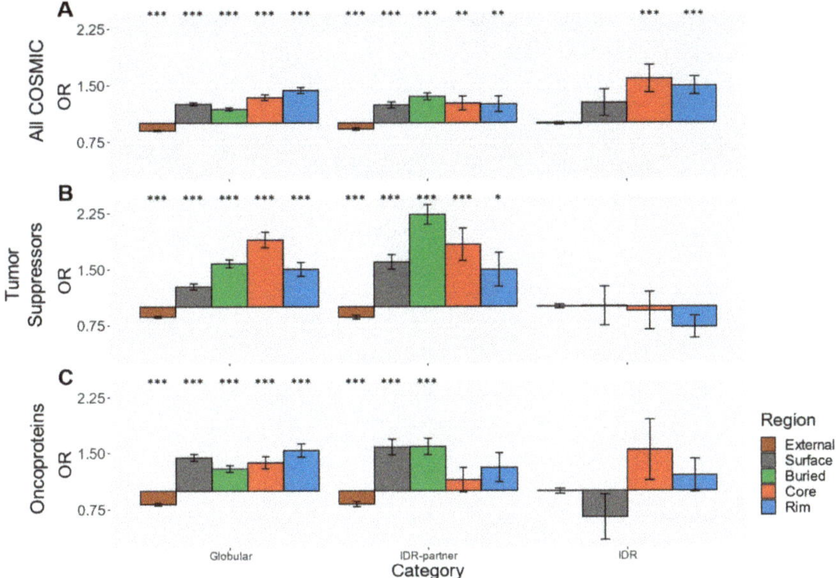

Figure 3. Odds ratios of COSMIC SNVs. (**A**) A bar graph of odds ratios of all COSMIC SNVs. The odds ratios of the subsets of proteins that were categorized as (**B**) tumor suppressors and (**C**) oncoproteins, respectively. See Figure 2 for details. *p*-values for all odds ratios can be found in Supplementary Table S2.

3.2. gnomAD SNVs Are Depleted at IDR Interaction Interfaces

Finally, we analyzed missense mutations from gnomAD to investigate how mutations present in the general population are distributed across structural regions in globular, interacting IDR and IDR-partner proteins. The mutations from gnomAD (gnomAD SNV) are observed in a population of healthy individuals, so these missense mutations are typically not directly associated with diseases. David and Sternberg previously studied non-disease-associated missense mutations annotated in the UniProt database and showed that these variants are depleted from functionally critical protein regions [2]. Specifically, they revealed enrichment at the rim and surface regions and depletion at the buried and interface core regions. The gnomAD SNV data is from large-scale genome sequencing projects, which also allows the study of rare missense mutations that were previously not detectable. Thus, in addition to analyzing all gnomAD SNVs, we also analyzed subsets of gnomAD SNVs with frequencies from $5 * 10^{-6}$ to 10^{-6} (rare SNVs) and from 0.1 to 0.001 (high-frequency SNVs). Studies have suggested that some mutations with very low frequencies can have deleterious effects [23,64], so the high-frequency SNVs should more accurately reflect the localization of benign mutations.

We first present results from the full gnomAD SNV dataset (frequency 0.1 to 10^{-6}) since it contains the largest number of mutations by far, thereby providing more reliable results. Results for the entire gnomAD SNV datasets were generated through the same procedures of mutation mapping and OR calculation. The ORs calculated for the globular dataset indicate that gnomAD SNVs causing missense mutations are significantly depleted at structured parts of proteins, especially at the functionally critical buried and interface core regions (Figure 4A). Compared to the ORs of the globular interaction set, the IDR-partner's ORs indicate more substantial depletions of gnomAD missense mutations from functional regions. Particularly depleted of gnomAD mutations is the core region of IDR-partners (OR = 0.5, *p*-value = $3.0 * 10^{-67}$). This finding is contrasted by the relatively high and significant ORs of the surface and rim regions of interacting IDRs. However, the interface core of interacting IDRs is also depleted of gnomAD missense mutations (OR = 0.8, *p*-value = $2.1 * 10^{-3}$), and interacting IDRs as a whole are not enriched in these SNVs (i.e., surface, buried, core, and rim combined; OR = 1.0; Figure 4A). Together, these findings highlight, again, the functional importance of the core residues in both IDR-partners and interacting IDRs.

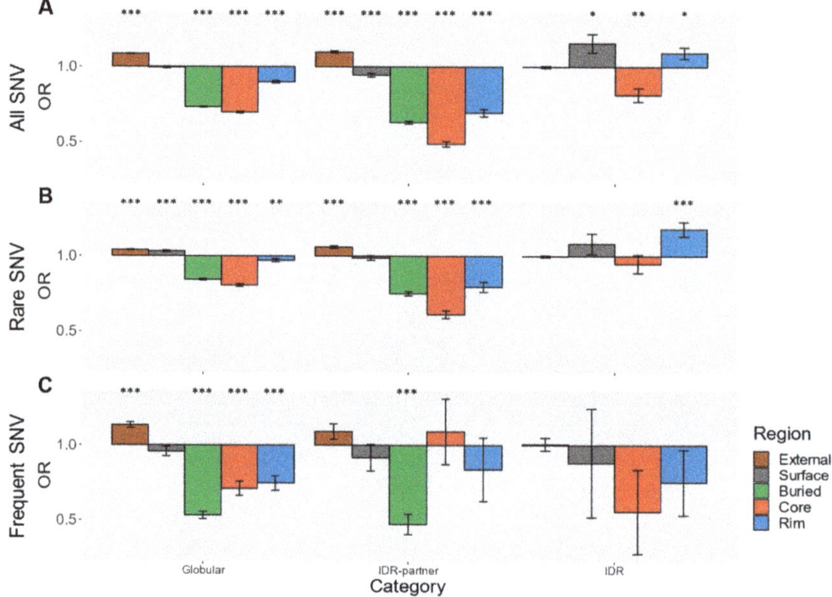

Figure 4. Odds ratios of gnomAD SNVs. (**A**) Odds ratios are calculated using gnomAD SNVs of frequencies between 0.1 to 10^{-6}. (**B**) Odds ratios are calculated using gnomAD SNVs of frequencies between $5 * 10^{-6}$ to 10^{-6}, i.e., rare SNVs. (**C**) Odds ratios are calculated using gnomAD SNVs of frequencies between 0.1 to 0.001, i.e., high-frequency SNVs. See Figure 2 for details.

When we isolated the rare SNVs (frequency $5 * 10^{-6}$ to 10^{-6}; Figure 4B), which contain variants that are generally seen only once in the available population sample, we still observed a significant depletion of the functional regions from missense mutations. This depletion is particularly pronounced for the IDR-partner core region (OR = 0.61, *p*-value = $2.8 * 10^{-27}$). However, the magnitude of the depletion is overall smaller than for all gnomAD SNVs. ORs closers to unity may be rationalized by the higher percentage of novel and deleterious SNVs among rare variants (see discussion). Interestingly, an exception to the weakening depletion pattern is the significant enrichment of rare SNVs at the interface rim of IDRs (OR = 1.2, *p*-value = 2.5 * 10^{-4}). Similar to the result for all gnomAD SNVs (Figure 4A), the missense mutation enrichment outside the core of the interacting IDR is consistent with the previous observations of higher mutation rates for IDRs in general [65,66].

The relatively subdued enrichment patterns of the rare SNV missense mutations are juxtaposed by the high-frequency SNV dataset. We investigated the high-frequency SNVs (frequency 0.1 to 0.001; Figure 4C) because these mutations are the most likely to be benign based on their recurrence in healthy individuals. In the globular interaction dataset, the high-frequency SNVs that cause missense mutations are generally more strongly depleted from the functional regions compared to the complete gnomAD SNVs dataset (Figure 4C), which is consistent with the expected greater proportion of benign mutations in the high-frequency SNVs. For both globular interactions and IDR-partners, the buried region is the most devoid of high-frequency SNVs (globular OR = 0.53, *p*-value = 4.9 * 10^{-46}; IDR-partner OR = 0.47, *p*-value = 7.7 * 10^{-8}). Interestingly, compared with all gnomAD SNVs, the ORs of the high-frequency SNVs for interface core regions do not decrease proportionately with the buried regions. The divergence of trends in the buried and core regions is most striking for the IDR-partners, which have a relatively large proportion of high-frequency SNVs in the core region (OR = 1.1, *p*-value = 0.7). In contrast, the high-frequency SNVs appear relatively depleted from the interacting IDR core and rim regions, but results are inconclusive due to the scarcity of high-frequency SNVs with structural data for interacting IDRs (core OR = 0.56, *p*-value = 0.2; rim OR = 0.75, *p*-value = 0.3).

3.3. Robustness of Datasets and Findings

As some of the datasets in our analysis are small, it is possible that our study was influenced by an overrepresentation of a few specific domains. Therefore, we searched our interaction sets to test for overrepresented Pfam domains [67]. Supplementary Figure S2 shows the number of proteins containing the 20 most frequent Pfam domains in each interaction set and mutation data analyzed. This analysis clearly shows that the majority of the highest-ranked domains have similar numbers of occurrences, with a few exceptions. In the globular dataset (Figure S2A–C), the protein kinase domain (Pfam: PF00069) stands out with an overall higher count. Among the globular proteins for which gnomAD missense mutations were mapped and analyzed (Figure S2A), 44 are observed to have this protein kinase domain. However, this number accounts for only 1.75% of the dataset (Figure S2A) due to its large size. In the much smaller IDR-partner and interacting IDR datasets (Figure S2D–I), ligand-binding domains of nuclear hormone receptors (Pfam: PF00104), PHD-finger domains (Pfam: PF00628) and core histone domains (Pfam: PF00125) stand out with high count numbers. The risk of bias is typically higher in these smaller datasets. For instance, the 11 ligand-binding domains of nuclear hormone receptors found among IDR-partners onto which SwissVar mutations were mapped (Figure S2F) make up nearly 12% of the dataset, potentially skewing the results of our enrichment analysis. Therefore, to test the robustness of our findings, we removed proteins containing the domains with overall high numbers of occurrences mentioned above and repeated our enrichment analysis. Essentially negligible changes are observed for the statistically significant ORs after the removal of these proteins (Figures S3–S5). Hence, the enrichment trends reported do not appear significantly biased by any Pfam domain overrepresentation.

The set of IDRs that we analyzed is likely enriched in those that fold upon binding, which includes peptide motifs and MoRFs [26,32,47], potentially leading to a bias against more dynamic forms of IDR interactions, namely fuzzy interactions [35,68]. To assess this concern, we compared the predicted level of intrinsic disorder and residue composition between our datasets and disordered regions from FuzDB, which is a database of fuzzy protein complexes [69]. We also compared both properties with Pfam domains to get the contrast with structured domains. First, we compared the datasets using scores from Disopred, a sequence-based predictor of disordered regions (Figure S6) [70]. The Disopred scores of the globular and IDR-partner sets are both very similar to the scores of the Pfam domains, all of which exhibit low levels of predicted disorder, which is expected from independently folding domains. On the other hand, the interacting IDRs show a distribution closer to that of FuzDB protein regions, although the overall level of predicted disorder is not as high. This difference is likely due to the presence of highly dynamic regions in FuzDB. This database includes not only polymorphic binding regions that sample multiple bound conformations but also flanking and clamping regions that are functionally important but are not the primary binding regions [69,71]. Next, we evaluated the residue composition of each dataset, which shows again that the globular and IDR-partner sets are closest to the Pfam set (Figure S7). Compared to the globular set, both FuzDB and interacting IDR sets are enriched in polar and charged residues, which are common in disordered regions. However, sequences in FuzDB have more polar residues, while those in the interacting IDR set have more charged residues. In summary, the interacting IDRs we analyzed are clearly distinguished from folded globular domains, but their sequence composition also differs slightly from disordered regions involved in fuzzy interactions. Overall, this analysis suggests that our mutation enrichment findings mainly pertain to IDR interactions that involve folding upon binding.

4. Discussion

IDR interactions are recognized not only for their critical role in cellular communication and regulation but also for their differences in molecular properties compared to the classical globular interactions [45,47,49,72]. It is reasonable to assume that the structural properties of protein interfaces will affect the susceptibility and tolerance of interface residues to missense mutations. In this study, we report evidence that IDR interactions are just as enriched in disease-associated mutations as globular interactions, suggesting that the interface residues from both categories of interactions are equally crucial for function. Most remarkable is the strong enrichment at the interface core of IDRs for disease mutations and its depletion in neutral ones. These trends are likely the consequence of the functional roles of the IDR interactions, which are often transient and specific interactions involved in signaling and regulation [31]. The abundance of some proteins with long IDRs is under tight cellular control [73], which would imply a high sensitivity to changes in binding affinity as well. Furthermore, some IDRs are involved in promiscuous interactions, both by flexibly binding multiple partners and by binding to globular proteins that have multiple partners (i.e., one-to-many and many-to-one interactions, respectively) [74,75]. Fewer neutral mutations and stronger evolutionary conservation were observed in residues that interact with multiple protein partners, which was postulated to be the result of additive constraints from multiple interactions [76].

The SwissVar SNV dataset enables the most direct interpretation due to the connection between these mutations and diseases. The enrichment patterns of SwissVar missense mutations indicate that both interacting IDRs and IDR-partner interfaces have residues that are critical for mediating interactions and, if mutated, lead to diseases. For IDR-partners, this interpretation is consistent with certain features we have previously found enriched at their interfaces, such as high rigidity, hydrophobicity, and conservation, which are all features associated with residues important for binding [49]. Concerning the interacting IDRs, which are inherently more dynamic, the enriched mutations are likely affecting highly conserved and often hydrophobic hotspot residues, which are key determinants of interaction affinity and are often part of conserved motifs, e.g., an SH3-binding motif [44,45,77]. Alternatively, disease-causing mutations in interacting IDRs may modulate the sampling of nascent, transient secondary structures in the unbound state. If these secondary structures are involved in binding and are present in the bound complex, changes in their sampling may alter the binding affinity [28,43]. This idea is exemplified by mutations in p53 that alter its residual helical structure and, consequently, change its affinity to MDM2 [31,78].

Similar to SwissVar SNVs, somatic cancer-associated mutations have also been found enriched at structural regions [63], but they are known to have a greater tendency to localize to the protein surface and interface regions [11]. These trends are reaffirmed in our COSMIC SNV globular interaction dataset, despite overall weaker enrichments. It is interesting to note that the rim of globular interactions is particularly enriched in somatic cancer-associated mutations. Preferential localization to the rim regions, which consists of polar and charged solvent-exposed residues, is only observed in the COSMIC dataset for the globular interactions. However, this observation is consistent with the previously reported tendency for cancer mutations to disrupt PPIs through substituting charged residues and perturbing the electrostatic component of binding affinities [11,79]. The most exciting finding of the analysis with the entire COSMIC SNV data is that the cores of interacting IDRs have the highest, statistically significant ORs. Many studies have demonstrated positive correlations between cancer and proteins harboring IDRs [41,44,53,80]. Furthermore, IDRs are enriched in sites of post-translational modifications such as phosphorylation, which are proposed to be prominent targets of cancer mutations [11,33,81]. However, a previous study had found globular domains more enriched in cancer-associated mutations than predicted interacting IDRs [41], and others have noted that cancer mutations are overrepresented within highly modular protein hubs, which incidentally tend to contain IDRs [41]. A good example of a protein with large segments of IDRs but has many more cancer-associated mutations within its globular domains is p53, which is the most frequently mutated protein in human cancers [41,82]. Therefore, the broadly-defined IDRs are likely depleted of cancer mutations compared to globular regions [41], but our results reveal the hidden enrichment within the more precise structurally-defined IDR interface regions, particularly their cores, which emphasizes the importance of detailed structural information in enrichment analyses.

Notably, only small subsets of the numerous proteins that are mutated in tumor cells are verified as oncoproteins and tumor suppressors that drive oncogenesis, which is why we further analyzed oncoproteins and tumor suppressors. Oncoproteins and tumor suppressors contribute to oncogenesis through diverging mechanisms, so, unsurprisingly, their mutation localization patterns are correspondingly different. Tumor suppressors are often deactivated through destabilizing and truncating mutations [12,13,83]. The localization patterns of COSMIC mutations in globular and IDR-partner tumor suppressors suggests that disruption of PPIs is also a deactivating mechanism. In contrast, the generally activating cancer-associated mutations in oncoproteins tend to be less destabilizing and more site-specific [79], which is reflected in the higher ORs in the protein surface and interface rim regions of the globular interaction set. Interestingly, the interface regions of IDR-partner oncoproteins have no statistically significant enrichment in cancer-associated mutations, which contrast the finding for globular oncoproteins. This finding suggests that some IDR-partner interfaces may be more robust to mutations, but this may also be the result of sparse data coverage. The lack of data also does not allow for an unambiguous interpretation of the mutation enrichments for interacting IDRs in tumor suppressors and oncoproteins. Overall, the analyses of the full COSMIC data reveal that IDR interaction interfaces are highly enriched in somatic cancer missense mutations, while those in tumor suppressors and oncoproteins exhibit intriguing differences compared to globular interfaces.

In contrast to the pathogenic disease-associated mutations, missense mutations that result from SNVs present in the general population (gnomAD) are assumed to be mostly benign, and as such, should be scarce in functionally critical regions. Buried and core regions in IDR-partners exhibit depletion levels of the entirety of gnomAD SNVs that surpasses the globular set, highlighting their functional importance. Interestingly, although the core region of interacting IDRs is also significantly depleted of gnomAD SNVs that cause missense mutations, the rim of interacting IDRs has a statistically significant enrichment of these variants, which suggests an overrepresentation of neutral and novel variants at this region. While interface rim residues also contribute to binding, there is a broader trend of enrichment of gnomAD SNVs within IDRs in general, which can be attributed to the weaker structural constraint compared to globular protein folds but is also proposed to be influenced by the higher mutation rate in the encoding genes [41,65]. In essence, the statistically robust results for all gnomAD SNVs that cause missense mutations reveal that the core regions of both IDR-partners and interacting IDRs are depleted of these variants but that the other areas of interacting IDRs are certainly more tolerant to these SNVs present in the general population.

Although individual gnomAD SNVs generally do not cause disease, one study suggested that 70% of rare mutations are mildly deleterious [23], and proteins enriched in rare mutations have been suggested to have stronger associations with diseases [84]. Consistent with this idea is our observation that rare SNVs have a much-subdued depletion pattern compared to the entire gnomAD dataset. In other words, we found a higher proportion of rare SNVs localizing to the functional regions of globular, IDR-partner, and IDR interaction structures (Figure 4B). In addition, novel mutations from rapid population growth likely contribute to the more uniform distribution of rare SNVs. By contrast, high-frequency SNVs are considered to be benign due to their common presence in the population and thus are more indicative of the tolerance to amino acid variation. Concordantly, we observed particularly strong depletion of high-frequency SNVs at the buried regions of globular proteins and IDR-partners, where substitutions would likely have the most damaging effects. Notably, the interface core of the globular interaction dataset is less depleted in high-frequency SNVs relative to the buried region. More strikingly, the IDR-partner core has a much higher OR for high-frequency SNVs. While the cores of the globular and IDR-partner sets contain many SwissVar disease-associated mutations, which are probably localizing to hotspot residues, we propose that some cores could at the same time accommodate mutations that result in disruptions mild enough to escape purifying selection and that this phenomenon is particularly relevant to the IDR-partner interfaces (see below). While most high-frequency SNVs likely arise from weak purifying selection, some may be driven by positive and balancing selection [20,23]. Indeed, sustaining a high frequency in the population is likely conditional on a positive molecular phenotype for some of these SNVs. Thus, the high proportion of such missense mutations within the IDR-partner core raises the possibility that some of these mutations could provide selective advantages or contribute to the genetic diversity in the population, which is important for evolutionary adaptation [19,85,86].

The combined observations from the different mutation categories reinforce the view that IDR-interactions are critical to human cellular functions and thus are susceptible to disease-causing missense mutations, but it also reveals a contrast between the interacting IDR and IDR-partners that needed to be addressed. While the cores of interacting IDRs demonstrate intolerance to mutations across the datasets, IDR-partners showed relatively weak depletion of high-frequency SNVs and weak enrichment of COSMIC SNVs within oncoproteins. The difference between IDR and IDR-partner interface residues is intriguing because their partnerships suggest shared functional constraints. One possible explanation for the observed difference is based on the participation of IDRs in promiscuous (i.e., one-to-many) interactions [74], which could mean more significant constraints for residues with multiple functions. However, the IDR-partners can also play the role of the promiscuous binder (i.e., many-to-one interactions) [74,87]. An alternative explanation is based on the molecular structure of the IDR interaction interfaces. We and others have previously noted that interface residues in the interacting IDR, especially the core residues, make greater individual contributions to binding than the interface residues in the IDR-partner [45,46,88]. Many IDRs interact using short motifs composed of relatively few residues. These few residues in the interacting IDR core, which are often hydrophobic and transition from being highly solvent-exposed in the unbound state to being buried upon binding, contribute more to interaction surfaces than residues on the IDR-partner side [46,47]. This difference between interacting IDR and IDR-partner interfaces may explain why both contain residues (probably hotspots) that, if mutated, lead to diseases, but that the relatively large IDR-partner interfaces also appear to accommodate other residues that may be more tolerant to variance. Proteins can evolve new functions through accumulating mutations, a process that is especially prominent in dynamic protein regions [65,85], so the IDR-partners' apparent higher tolerance for certain mutations may have a significant role in evolutionary adaption.

Nonetheless, we must mention some limitations of our study. While the use of experimentally determined structures provides crucial data for the determination of core, rim, surface, and buried regions, this approach may bias our findings to less dynamic complexes. While NMR experiments contributed a significant number of IDR complex structures, many structures were determined through X-ray crystallography experiments, which limits our dataset to IDRs that fold upon binding. Consequently, an increasingly recognized class of complexes that exhibit conformational heterogeneity in their bound state, known as fuzzy complexes, may be underrepresented [71]. Because a focal finding of this study is the importance of the core residues in interacting IDRs for function and disease, the method of identifying the core residues from IDR interaction structures is central to our investigation. Thus, fuzzy complexes in which key binding residues remain dynamic in their bound states may require a different approach to investigate mutation enrichment patterns in the future. It also needs to be stressed that some of the differences that we see between IDRs and IDR-partners are observed in sets with small numbers of data points. Therefore, additional analyses with larger datasets are required in the future to confirm the observed differences in SNV enrichments between IDRs and IDR-partners.

5. Conclusions

Investigating the enrichment of different categories of missense mutations within IDR interaction structures revealed several notable characteristics. Although limited in availability, IDR complex structures are crucial for precisely identifying the core residues of the interacting IDRs. While the categorization of PPI interface residues into core and rim is a well-established practice for globular proteins, the same is often not done for studying interacting IDRs. Once we identified the core residues, we more clearly observed that interacting IDR core residues are significantly enriched in SwissVar and COSMIC missense mutations as well as being depleted in gnomAD SNVs that cause missense mutations. These results suggest that interacting IDR core residues are highly intolerant to missense mutations, which support the view that the disruption of IDR interactions, and thus the cellular functions that they perform, is a common mechanism underlying many diseases. Interestingly, the trends that we observed suggest an asymmetry across the IDR interaction interface in the enrichment of certain missense mutation types. However, future analyses with more variant data will be required to confirm differences in variant enrichment between interacting IDRs and IDR-partners. In any case, the growing availability of protein structure and sequence data has enabled us to recognize important distinctions between globular and IDR-mediated interactions, and this trend continues to accelerate. Accounting for such differences will contribute to the understanding and prediction of the effects of missense mutations on disease susceptibility, which is a critical aspect of personalized medicine.

Supplementary Materials: The following are available online at http://www.mdpi.com/2218-273X/10/8/1097/s1, Figure S1: Flow chart of structural data processing procedure with sequence and structure tabulation. Figure S2: Pfam occurrences in mutation-mapped datasets. Figure S3: Odds ratios of SwissVar SNVs excluding selected frequent Pfam domains. Figure S4: Odds ratios of COSMIC SNVs excluding selected frequent Pfam domains. Figure S5: Odds ratios of gnomAD SNVs excluding selected frequent Pfam domains. Figure S6: Box plot of Disopred disorder prediction scores. Figure S7: Residue composition difference between the respective dataset and the globular dataset. Table S1: Tabulated residues and mutations used in odds ratios calculations. Table S2: Odds ratios and statistics.

Author Contributions: Conceptualization, E.T.C.W., J.M.B., J.G.; methodology, J.G., E.T.C.W.; validation, E.T.C.W., V.S.; formal analysis, E.T.C.W., V.S., M.G., N.M., E.R.K.; investigation, E.T.C.W., V.S., M.G.; resources, J.G.; data curation, E.T.C.W., V.S., M.G., N.M.; writing—original draft preparation, J.G., E.T.C.W.; writing—review and editing, J.G., E.T.C.W., E.R.K.; visualization, E.T.C.W.; supervision, J.G., E.T.C.W.; project administration, J.G.; funding acquisition, J.G. All authors have read and agreed to the published version of the manuscript.

Funding: This research was funded by Canadian Institutes of Health Research (CIHR); Natural Sciences and Engineering Research Council of Canada (NSERC).

Acknowledgments: We thank Dima Vavilov and Stephen MacDonald for IT support.

Conflicts of Interest: The authors declare no conflict of interest.

References

1. Sahni, N.; Yi, S.; Taipale, M.; Fuxman Bass, J.I.; Coulombe-Huntington, J.; Yang, F.; Peng, J.; Weile, J.; Karras, G.I.; Wang, Y.; et al. Widespread macromolecular interaction perturbations in human genetic disorders. *Cell* **2015**, *161*, 647–660. [CrossRef] [PubMed]
2. David, A.; Sternberg, M.J.E. The Contribution of Missense Mutations in Core and Rim Residues of Protein-Protein Interfaces to Human Disease. *J. Mol. Biol.* **2015**, *427*, 2886–2898. [CrossRef] [PubMed]
3. Savojardo, C.; Babbi, G.; Martelli, P.; Casadio, R. Functional and Structural Features of Disease-Related Protein Variants. *Int. J. Mol. Sci.* **2019**, *20*, 1530. [CrossRef] [PubMed]
4. Stehr, H.; Jang, S.-H.J.; Duarte, J.M.; Wierling, C.; Lehrach, H.; Lappe, M.; Lange, B.M.H. The structural impact of cancer-associated missense mutations in oncogenes and tumor suppressors. *Mol. Cancer* **2011**, *10*, 54. [CrossRef] [PubMed]
5. Wang, Z.; Moult, J. SNPs, protein structure, and disease. *Hum. Mutat.* **2001**, *17*, 263–270. [CrossRef] [PubMed]
6. David, A.; Razali, R.; Wass, M.N.; Sternberg, M.J.E. Protein-protein interaction sites are hot spots for disease-associated nonsynonymous SNPs. *Hum. Mutat.* **2012**, *33*, 359–363. [CrossRef] [PubMed]
7. Stefl, S.; Nishi, H.; Petukh, M.; Panchenko, A.R.; Alexov, E. Molecular mechanisms of disease-causing missense mutations. *J. Mol. Biol.* **2013**, *425*, 3919–3936. [CrossRef]
8. Chakrabarti, P.; Janin, J. Dissecting protein-protein recognition sites. *Proteins* **2002**, *47*, 334–343. [CrossRef]
9. Levy, E.D. A Simple Definition of Structural Regions in Proteins and Its Use in Analyzing Interface Evolution. *J. Mol. Biol.* **2010**, *403*, 660–670. [CrossRef]
10. Wang, X.; Wei, X.; Thijssen, B.; Das, J.; Lipkin, S.M.; Yu, H. Three-dimensional reconstruction of protein networks provides insight into human genetic disease. *Nat. Biotechnol.* **2012**, *30*, 159–164. [CrossRef]
11. Nishi, H.; Tyagi, M.; Teng, S.; Shoemaker, B.A.; Hashimoto, K.; Alexov, E.; Wuchty, S.; Panchenko, A.R. Cancer missense mutations alter binding properties of proteins and their interaction networks. *PLoS ONE* **2013**, *8*, e66273. [CrossRef] [PubMed]
12. Engin, H.B.; Kreisberg, J.F.; Carter, H. Structure-Based Analysis Reveals Cancer Missense Mutations Target Protein Interaction Interfaces. *PLoS ONE* **2016**, *11*, e0152929. [CrossRef] [PubMed]
13. Vogelstein, B.; Papadopoulos, N.; Velculescu, V.E.; Zhou, S.; Diaz, L.A.; Kinzler, K.W. Cancer genome landscapes. *Science* **2013**, *339*, 1546–1558. [CrossRef] [PubMed]
14. Brown, A.-L.; Li, M.; Goncearenco, A.; Panchenko, A.R. Finding driver mutations in cancer: Elucidating the role of background mutational processes. *PLoS Comput. Biol.* **2019**, *15*, e1006981. [CrossRef]
15. Karczewski, K.J.; Francioli, L.C.; Tiao, G.; Cummings, B.B.; Alföldi, J.; Wang, Q.; Collins, R.L.; Laricchia, K.M.; Ganna, A.; Birnbaum, D.P.; et al. The mutational constraint spectrum quantified from variation in 141,456 humans. *bioRxiv* **2020**. [CrossRef]
16. Sivley, R.M.; Dou, X.; Meiler, J.; Bush, W.S.; Capra, J.A. Comprehensive Analysis of Constraint on the Spatial Distribution of Missense Variants in Human Protein Structures. *Am. J. Hum. Genet.* **2018**, *102*, 415–426. [CrossRef]
17. Saint Pierre, A.; Génin, E. How important are rare variants in common disease? *Brief. Funct. Genom.* **2014**, *13*, 353–361. [CrossRef]
18. Choi, Y.; Sims, G.E.; Murphy, S.; Miller, J.R.; Chan, A.P. Predicting the functional effect of amino acid substitutions and indels. *PLoS ONE* **2012**, *7*, e46688. [CrossRef]
19. Lai, Y.-T.; Yeung, C.K.L.; Omland, K.E.; Pang, E.-L.; Hao, Y.; Liao, B.-Y.; Cao, H.-F.; Zhang, B.-W.; Yeh, C.-F.; Hung, C.-M.; et al. Standing genetic variation as the predominant source for adaptation of a songbird. *Proc. Natl. Acad. Sci. USA* **2019**, *116*, 2152–2157. [CrossRef]
20. Key, F.M.; Teixeira, J.C.; de Filippo, C.; André, A.M. Advantageous diversity maintained by balancing selection in humans. *Curr. Opin. Genet. Dev.* **2014**, *29*, 45–51. [CrossRef]
21. Yi, X.; Liang, Y.; Huerta-Sanchez, E.; Jin, X.; Cuo, Z.X.P.; Pool, J.E.; Xu, X.; Jiang, H.; Vinckenbosch, N.; Korneliussen, T.S.; et al. Sequencing of 50 Human Exomes Reveals Adaptation to High Altitude. *Science* **2010**, *329*, 75–78. [CrossRef] [PubMed]
22. Tennessen, J.A.; Bigham, A.W.; O'Connor, T.D.; Fu, W.; Kenny, E.E.; Gravel, S.; McGee, S.; Do, R.; Liu, X.; Jun, G.; et al. Evolution and functional impact of rare coding variation from deep sequencing of human exomes. *Science* **2012**, *337*, 64–69. [CrossRef]

23. Kryukov, G.V.; Pennacchio, L.A.; Sunyaev, S.R. Most rare missense alleles are deleterious in humans: Implications for complex disease and association studies. *Am. J. Hum. Genet.* **2007**, *80*, 727–739. [CrossRef] [PubMed]

24. Lek, M.; Karczewski, K.J.; Minikel, E.V.; Samocha, K.E.; Banks, E.; Fennell, T.; O'Donnell-Luria, A.H.; Ware, J.S.; Hill, A.J.; Cummings, B.B.; et al. Analysis of protein-coding genetic variation in 60,706 humans. *Nature* **2016**, *536*, 285–291. [CrossRef] [PubMed]

25. Marouli, E.; Graff, M.; Medina-Gomez, C.; Lo, K.S.; Wood, A.R.; Kjaer, T.R.; Fine, R.S.; Lu, Y.; Schurmann, C.; Highland, H.M.; et al. Rare and low-frequency coding variants alter human adult height. *Nature* **2017**, *542*, 186–190. [CrossRef]

26. Mohan, A.; Oldfield, C.J.; Radivojac, P.; Vacic, V.; Cortese, M.S.; Dunker, A.K.; Uversky, V.N. Analysis of molecular recognition features (MoRFs). *J. Mol. Biol.* **2006**, *362*, 1043–1059. [CrossRef]

27. Van der Lee, R.; Buljan, M.; Lang, B.; Weatheritt, R.J.; Daughdrill, G.W.; Dunker, A.K.; Fuxreiter, M.; Gough, J.; Gsponer, J.; Jones, D.T.; et al. Classification of intrinsically disordered regions and proteins. *Chem. Rev.* **2014**, *114*, 6589–6631. [CrossRef]

28. Tompa, P.; Davey, N.E.; Gibson, T.J.; Babu, M.M. A Million peptide motifs for the molecular biologist. *Mol. Cell* **2014**, *55*, 161–169. [CrossRef]

29. Seo, M.-H.; Kim, P.M. The present and the future of motif-mediated protein-protein interactions. *Curr. Opin. Struct. Biol.* **2018**, *50*, 162–170. [CrossRef]

30. Le Gall, T.; Romero, P.R.; Cortese, M.S.; Uversky, V.N.; Dunker, A.K. Intrinsic disorder in the Protein Data Bank. *J. Biomol. Struct. Dyn.* **2007**, *24*, 325–342. [CrossRef]

31. Wright, P.E.; Dyson, H.J. Intrinsically disordered proteins in cellular signalling and regulation. *Nat. Rev. Mol. Cell Biol.* **2015**, *16*, 18–29. [CrossRef] [PubMed]

32. Fuxreiter, M.; Simon, I.; Friedrich, P.; Tompa, P. Preformed structural elements feature in partner recognition by intrinsically unstructured proteins. *J. Mol. Biol.* **2004**, *338*, 1015–1026. [CrossRef] [PubMed]

33. Darling, A.L.; Uversky, V.N. Intrinsic disorder and posttranslational modifications: The darker side of the biological dark matter. *Front. Genet.* **2018**, *9*, 1–18. [CrossRef] [PubMed]

34. Pentony, M.M.; Jones, D.T. Modularity of intrinsic disorder in the human proteome. *Proteins Struct. Funct. Bioinf.* **2010**, *78*, 212–221. [CrossRef]

35. Tompa, P.; Fuxreiter, M. Fuzzy complexes: Polymorphism and structural disorder in protein-protein interactions. *Trends Biochem. Sci.* **2008**, *33*, 2–8. [CrossRef]

36. Uversky, V.N. Unusual biophysics of intrinsically disordered proteins. *Biochim. Biophys. Acta Rev. Cancer* **2013**, *1834*, 932–951. [CrossRef]

37. Tompa, P. The interplay between structure and function in intrinsically unstructured proteins. *FEBS Lett.* **2005**, *579*, 3346–3354. [CrossRef]

38. Mosca, R.; Pache, R.A.; Aloy, P. The role of structural disorder in the rewiring of protein interactions through evolution. *Mol. Cell. Proteom.* **2012**, *11*, M111.014969. [CrossRef]

39. Haynes, C.; Oldfield, C.J.; Ji, F.; Klitgord, N.; Cusick, M.E.; Radivojac, P.; Uversky, V.N.; Vidal, M.; Iakoucheva, L.M. Intrinsic disorder is a common feature of hub proteins from four eukaryotic interactomes. *PLoS Comput. Biol.* **2006**, *2*, e100. [CrossRef]

40. Hu, G.; Wu, Z.; Uversky, V.N.; Kurgan, L. Functional analysis of human hub proteins and their interactors involved in the intrinsic disorder-enriched interactions. *Int. J. Mol. Sci.* **2017**, *18*, 2761. [CrossRef]

41. Pajkos, M.; Mészáros, B.; Simon, I.; Dosztányi, Z. Is there a biological cost of protein disorder? Analysis of cancer-associated mutations. *Mol. Biosyst.* **2012**, *8*, 296–307. [CrossRef]

42. Brown, C.J.; Johnson, A.K.; Dunker, A.K.; Daughdrill, G.W. Evolution and disorder. *Curr. Opin. Struct. Biol.* **2011**, *21*, 441–446. [CrossRef] [PubMed]

43. Vacic, V.; Markwick, P.R.L.; Oldfield, C.J.; Zhao, X.; Haynes, C.; Uversky, V.N.; Iakoucheva, L.M. Disease-Associated Mutations Disrupt Functionally Important Regions of Intrinsic Protein Disorder. *PLoS Comput. Biol.* **2012**, *8*, e1002709. [CrossRef]

44. Uyar, B.; Weatheritt, R.J.; Dinkel, H.; Davey, N.E.; Gibson, T.J. Proteome-wide analysis of human disease mutations in short linear motifs: Neglected players in cancer? *Mol. BioSyst.* **2014**, *10*, 2626–2642. [CrossRef]

45. Mészáros, B.; Tompa, P.; Simon, I.; Dosztányi, Z. Molecular principles of the interactions of disordered proteins. *J. Mol. Biol.* **2007**, *372*, 549–561. [CrossRef] [PubMed]

46. Wong, E.T.C.; Na, D.; Gsponer, J. On the importance of polar interactions for complexes containing intrinsically disordered proteins. *PLoS Comput. Biol.* **2013**, *9*, e1003192. [CrossRef] [PubMed]

47. London, N.; Movshovitz-Attias, D.; Schueler-Furman, O. The Structural Basis of Peptide-Protein Binding Strategies. *Structure* **2010**, *18*, 188–199. [CrossRef]

48. Vacic, V.; Oldfield, C.J.; Mohan, A.; Radivojac, P.; Cortese, M.S.; Uversky, V.N.; Dunker, A.K. Characterization of molecular recognition features, MoRFs, and their binding partners. *J. Proteome Res.* **2007**, *6*, 2351–2366. [CrossRef]

49. Wong, E.T.C.; Gsponer, J. Predicting Protein–Protein Interfaces that Bind Intrinsically Disordered Protein Regions. *J. Mol. Biol.* **2019**, *431*. [CrossRef]

50. Mottaz, A.; David, F.P.A.; Veuthey, A.-L.; Yip, Y.L. Easy retrieval of single amino-acid polymorphisms and phenotype information using SwissVar. *Bioinformatics* **2010**, *26*, 851–852. [CrossRef]

51. Tate, J.G.; Bamford, S.; Jubb, H.C.; Sondka, Z.; Beare, D.M.; Bindal, N.; Boutselakis, H.; Cole, C.G.; Creatore, C.; Dawson, E.; et al. COSMIC: The Catalogue Of Somatic Mutations In Cancer. *Nucleic Acids Res.* **2019**, *47*, D941–D947. [CrossRef] [PubMed]

52. Wu, H.; Fuxreiter, M. The Structure and Dynamics of Higher-Order Assemblies: Amyloids, Signalosomes, and Granules. *Cell* **2016**, *165*, 1055–1066. [CrossRef] [PubMed]

53. Deiana, A.; Forcelloni, S.; Porrello, A.; Giansanti, A. Intrinsically disordered proteins and structured proteins with intrinsically disordered regions have different functional roles in the cell. *PLoS ONE* **2019**, *14*, e0217889. [CrossRef] [PubMed]

54. Anbo, H.; Sato, M.; Okoshi, A.; Fukuchi, S. Functional Segments on Intrinsically Disordered Regions in Disease-Related Proteins. *Biomolecules* **2019**, *9*, 88. [CrossRef]

55. Babu, M.M.; van der Lee, R.; de Groot, N.S.; Gsponer, J. Intrinsically disordered proteins: Regulation and disease. *Curr. Opin. Struct. Biol.* **2011**, *21*, 432–440. [CrossRef]

56. Mitternacht, S. FreeSASA: An open source C library for solvent accessible surface area calculations. *F1000Research* **2016**, *5*, 189. [CrossRef]

57. Magrane, M.; Consortium, U. UniProt Knowledgebase: A hub of integrated protein data. *Database* **2011**, *2011*, bar009. [CrossRef]

58. Piovesan, D.; Tabaro, F.; Paladin, L.; Necci, M.; Mičetić, I.; Camilloni, C.; Davey, N.; Dosztányi, Z.; Mészáros, B.; Monzon, A.M.; et al. MobiDB 3.0: More annotations for intrinsic disorder, conformational diversity and interactions in proteins. *Nucleic Acids Res.* **2018**, *46*, D471–D476. [CrossRef]

59. Gouw, M.; Michael, S.; Sámano-Sánchez, H.; Kumar, M.; Zeke, A.; Lang, B.; Bely, B.; Chemes, L.B.; Davey, N.E.; Deng, Z.; et al. The eukaryotic linear motif resource—2018 update. *Nucleic Acids Res.* **2018**, *46*, D428–D434. [CrossRef]

60. Wickham, H. *ggplot2: Elegant Graphics for Data Analysis*; Springer: New York, NY, USA, 2016; ISBN 978-3-319-24277-4.

61. Vitkup, D.; Sander, C.; Church, G.M. The amino-acid mutational spectrum of human genetic disease. *Genome Biol.* **2003**, *4*, R72. [CrossRef]

62. McFarland, C.D.; Korolev, K.S.; Kryukov, G.V.; Sunyaev, S.R.; Mirny, L.A. Impact of deleterious passenger mutations on cancer progression. *Proc. Natl. Acad. Sci. USA* **2013**, *110*, 2910–2915. [CrossRef]

63. Lu, H.-C.; Chung, S.S.; Fornili, A.; Fraternali, F. Anatomy of protein disorder, flexibility and disease-related mutations. *Front. Mol. Biosci.* **2015**, *2*, 47. [CrossRef] [PubMed]

64. Nishi, H.; Nakata, J.; Kinoshita, K. Distribution of single-nucleotide variants on protein-protein interaction sites and its relationship with minor allele frequency. *Protein Sci.* **2016**, *25*, 316–321. [CrossRef]

65. Forcelloni, S.; Giansanti, A. Evolutionary Forces and Codon Bias in Different Flavors of Intrinsic Disorder in the Human Proteome. *J. Mol. Evol.* **2020**, *88*, 164–178. [CrossRef] [PubMed]

66. Khan, T.; Douglas, G.M.; Patel, P.; Nguyen Ba, A.N.; Moses, A.M. Polymorphism Analysis Reveals Reduced Negative Selection and Elevated Rate of Insertions and Deletions in Intrinsically Disordered Protein Regions. *Genome Biol. Evol.* **2015**, *7*, 1815–1826. [CrossRef] [PubMed]

67. El-Gebali, S.; Mistry, J.; Bateman, A.; Eddy, S.R.; Luciani, A.; Potter, S.C.; Qureshi, M.; Richardson, L.J.; Salazar, G.A.; Smart, A.; et al. The Pfam protein families database in 2019. *Nucleic Acids Res.* **2019**, *47*, D427–D432. [CrossRef]

68. Fuxreiter, M. Fold or not to fold upon binding—Does it really matter? *Curr. Opin. Struct. Biol.* **2018**, *54*, 19–25. [CrossRef]

69. Miskei, M.; Antal, C.; Fuxreiter, M. FuzDB: Database of fuzzy complexes, a tool to develop stochastic structure-function relationships for protein complexes and higher-order assemblies. *Nucleic Acids Res.* **2017**, *45*, D228–D235. [CrossRef]

70. Jones, D.T.; Cozzetto, D. DISOPRED3: Precise disordered region predictions with annotated protein-binding activity. *Bioinformatics* **2015**, *31*, 857–863. [CrossRef]

71. Fuxreiter, M. Fuzziness in Protein Interactions-A Historical Perspective. *J. Mol. Biol.* **2018**, *430*, 2278–2287. [CrossRef]

72. Malhis, N.; Wong, E.T.C.; Nassar, R.; Gsponer, J. Computational identification of MoRFs in protein sequences using Hierarchical application of bayes rule. *PLoS ONE* **2015**, *10*, e0141603. [CrossRef] [PubMed]

73. Gsponer, J.; Futschik, M.E.; Teichmann, S.A.; Babu, M.M. Tight regulation of unstructured proteins: From transcript synthesis to protein degradation. *Science* **2008**, *322*, 1365–1368. [CrossRef] [PubMed]

74. Oldfield, C.J.; Meng, J.; Yang, J.Y.; Yang, M.Q.; Uversky, V.N.; Dunker, A.K. Flexible nets: Disorder and induced fit in the associations of p53 and 14-3-3 with their partners. *BMC Genom.* **2008**, *9*, S1. [CrossRef] [PubMed]

75. Dunker, A.K.; Cortese, M.S.; Romero, P.; Iakoucheva, L.M.; Uversky, V.N. Flexible nets. The roles of intrinsic disorder in protein interaction networks. *FEBS J.* **2005**, *272*, 5129–5148. [CrossRef]

76. Fornili, A.; Pandini, A.; Lu, H.-C.; Fraternali, F. Specialized Dynamical Properties of Promiscuous Residues Revealed by Simulated Conformational Ensembles. *J. Chem. Theory Comput.* **2013**, *9*, 5127–5147. [CrossRef]

77. Kurochkina, N.; Guha, U. SH3 domains: Modules of protein-protein interactions. *Biophys. Rev.* **2013**, *5*, 29–39. [CrossRef]

78. Yadahalli, S.; Li, J.; Lane, D.P.; Gosavi, S.; Verma, C.S. Characterizing the conformational landscape of MDM2-binding p53 peptides using Molecular Dynamics simulations. *Sci. Rep.* **2017**, *7*, 15600. [CrossRef]

79. Dincer, C.; Kaya, T.; Keskin, O.; Gursoy, A.; Tuncbag, N. 3D spatial organization and network-guided comparison of mutation profiles in Glioblastoma reveals similarities across patients. *PLoS Comput. Biol.* **2019**, *15*, e1006789. [CrossRef]

80. Meyer, K.; Kirchner, M.; Uyar, B.; Cheng, J.-Y.; Russo, G.; Hernandez-Miranda, L.R.; Szymborska, A.; Zauber, H.; Rudolph, I.-M.; Willnow, T.E.; et al. Mutations in Disordered Regions Can Cause Disease by Creating Dileucine Motifs. *Cell* **2018**, *175*, 239–253.e17. [CrossRef]

81. Reimand, J.; Bader, G.D. Systematic analysis of somatic mutations in phosphorylation signaling predicts novel cancer drivers. *Mol. Syst. Biol.* **2013**, *9*, 637. [CrossRef]

82. Baugh, E.H.; Ke, H.; Levine, A.J.; Bonneau, R.A.; Chan, C.S. Why are there hotspot mutations in the TP53 gene in human cancers? *Cell Death Differ.* **2018**, *25*, 154–160. [CrossRef] [PubMed]

83. Kamburov, A.; Lawrence, M.S.; Polak, P.; Leshchiner, I.; Lage, K.; Golub, T.R.; Lander, E.S.; Getz, G. Comprehensive assessment of cancer missense mutation clustering in protein structures. *Proc. Natl. Acad. Sci. USA* **2015**, *112*, E5486–E5495. [CrossRef] [PubMed]

84. Alhuzimi, E.; Leal, L.G.; Sternberg, M.J.E.; David, A. Properties of human genes guided by their enrichment in rare and common variants. *Hum. Mutat.* **2018**, *39*, 365–370. [CrossRef]

85. Tokuriki, N.; Tawfik, D.S. Protein dynamism and evolvability. *Science* **2009**, *324*, 203–207. [CrossRef] [PubMed]

86. Mahlich, Y.; Reeb, J.; Hecht, M.; Schelling, M.; De Beer, T.A.P.; Bromberg, Y.; Rost, B. Common sequence variants affect molecular function more than rare variants? *Sci. Rep.* **2017**, *7*, 1608. [CrossRef]

87. Kim, P.M.; Sboner, A.; Xia, Y.; Gerstein, M. The role of disorder in interaction networks: A structural analysis. *Mol. Syst. Biol.* **2008**, *4*, 179. [CrossRef]

88. London, N.; Raveh, B.; Schueler-Furman, O. Peptide docking and structure-based characterization of peptide binding: From knowledge to know-how. *Curr. Opin. Struct. Biol.* **2013**, *23*, 894–902. [CrossRef]

 *biomolecules*

Article

Ancient Evolutionary Origin of Intrinsically Disordered Cancer Risk Regions

Mátyás Pajkos [1], András Zeke [2] and Zsuzsanna Dosztányi [1],*

[1] Department of Biochemistry, ELTE Eötvös Loránd University, Pázmány Péter stny 1/c, H-1117 Budapest, Hungary; matyaspajkos@caesar.elte.hu

[2] Research Centre for Natural Sciences, Magyar tudósok körútja 2, H-1117 Budapest, Hungary; zeke.andras@ttk.mta.hu

* Correspondence: dosztanyi@caesar.elte.hu

Received: 21 June 2020; Accepted: 20 July 2020; Published: 28 July 2020

Abstract: Cancer is a heterogeneous genetic disease that alters the proper functioning of proteins involved in key regulatory processes such as cell cycle, DNA repair, survival, or apoptosis. Mutations often accumulate in hot-spots regions, highlighting critical functional modules within these proteins that need to be altered, amplified, or abolished for tumor formation. Recent evidence suggests that these mutational hotspots can correspond not only to globular domains, but also to intrinsically disordered regions (IDRs), which play a significant role in a subset of cancer types. IDRs have distinct functional properties that originate from their inherent flexibility. Generally, they correspond to more recent evolutionary inventions and show larger sequence variations across species. In this work, we analyzed the evolutionary origin of disordered regions that are specifically targeted in cancer. Surprisingly, the majority of these disordered cancer risk regions showed remarkable conservation with ancient evolutionary origin, stemming from the earliest multicellular animals or even beyond. Nevertheless, we encountered several examples where the mutated region emerged at a later stage compared with the origin of the gene family. We also showed the cancer risk regions become quickly fixated after their emergence, but evolution continues to tinker with their genes with novel regulatory elements introduced even at the level of humans. Our concise analysis provides a much clearer picture of the emergence of key regulatory elements in proteins and highlights the importance of taking into account the modular organisation of proteins for the analyses of evolutionary origin.

Keywords: intrinsically disordered regions; linear motifs; gene duplications; de novo; evolutionary origin

---

1. Introduction

Most human genes are thought to have an extensive and very deep evolutionary history. In line with the thought "Nature is a tinkerer, not an inventor" [1], major human gene families date back to the earliest Eukaryotic evolutionary events, or even beyond. The very oldest layers of human genes encode metabolically, structurally, or otherwise essential proteins that typically go back to unicellular evolutionary stages. Mutations to this core biochemical apparatus can prove disruptive to all aspects of cellular life, and indeed, there are known mutational targets associated with genome stability and cancer. In contrast to these "caretaker" genes, a more novel set of genes have emerged at the transition to a multicellular stage. These "gatekeeper" proteins are involved in cell-to-cell communication, especially in early embryonic development and tissue regeneration. Gatekeeper genes that control cell division are among the best known cancer-associated oncogenes and tumor suppressors [2].

In order to establish the evolutionary origins of cancer genes, Domazet-Loso and Tautz carried out a systematic analysis based on phylostratigraphic tracking [3]. By correlating the evolutionary origin of genes with particular macroevolutionary transitions, they found that a major peak connected to the emergence of cancer genes corresponds to the level where multicellular animals have emerged. However, many cancer genes have a more ancient origin and can be traced back to unicellular organisms. These trends seem to apply to the appearance of disease genes [4] and novel genes in general as well [5]. These studies were based on the evolutionary history of the founder domains. However, new genes can also be generated by duplication either in whole or from part of existing genes, when the duplicate copy of a gene becomes associated with a different phenotype to its paralogous partner. This mechanism can also influence the emergence of disease genes [5].

By taking advantage of the flux of cancer genome data, several new proteins have been identified to play a direct role in driving tumorigenesis during recent years [6]. One of the key signatures of cancer drivers is the presence of mutation hotspot regions, where many different patients might show a similarly recurrent pattern of mutations [7]. These hotspots are typically located within well-folded, structured domains. However, many cancer associated proteins have a complex modular architecture, incorporating not only globular domains, but also intrinsically disordered segments, which can also be sites of cancer mutations. In our recent work, we systematically collected disordered regions that are directly targeted by cancer mutations and analyzed their basic functional and system level properties. [8]. While only a relatively small subset of such disordered cancer drivers was identified, their mutations can be the main driver event in certain cancer types. These disordered regions can function in a variety of ways including post-transcriptional modification sites (PTMs), linear motifs, linkers, and larger sized functional modules typically involved in binding to macromolecular complexes. These disordered cancer drivers have a characteristic functional repertoire and increased interaction potential, and their perturbation can give rise to all ten hallmarks of cancer independently of ordered drivers [8].

In general, owing to the lack of structural constraints, disordered segments show more evolutionary variability [9]. In particular, linear motifs can easily emerge to a previously non-functional region of protein sequence by only a few mutations, or disappear as easily, leaving little trace after millions or billions of years [10]. However, elements fulfilling a critical regulatory function might linger on for a longer time. So far, the evolutionary origin of intrinsically disordered regions that have a critical function proven by a human disease association has not been analyzed.

In the current study, we studied the evolutionary origin of disordered cancer risk regions. For this, we used a dataset of cancer driving proteins in which cancer mutations specifically targeted intrinsically disordered regions [8]. We retrieved phylogeny data from the ENSEMBL Compara database. Using a novel conservation and phylogenetic-based strategy, we determined the evolutionary origin not only at the gene level, but also at the region level. In addition, we also investigated the emergence mechanism of disordered cancer risk regions and how evolutionary constraints, selection, and gene duplications events influenced the fate of these examples. Finally, we presented interesting case studies that demonstrate the ancient evolutionary origin of these examples and the continuing evolution of their genes built around the critical conserved functional module.

2. Materials and Methods

2.1. Dataset

We used a subset of the previously identified disordered cancer risk regions [8]. These regions were identified based on genetic variations collected from the COSMIC database [11] using the method that located specific regions that are enriched in cancer mutations [7]. Disorder status of these regions was verified based on experimental data collected from dedicated databases and from the literature when available, or based on consensus disorder prediction methods [8]. Mapping was not feasible for CDKN2A isoform (Tumor suppressor ARF), because it was not present in the ENSEMBL database we used in our study), hence this protein was excluded from the further analyses. Proteins in which both disordered and ordered cancer regions were identified were filtered out in order to be able to focus clearly on the disordered regions. Regions that were primarily mutated by in-frame insertion and deletion and contained less than 15 missense mutations were also excluded because of our conservation calculation method (see below). Finally, histone proteins were merged, keeping the single entry of HIST1H3B. Ultimately, we obtained a list of 36 disordered cancer risk regions of 32 proteins APC (Adenomatous polyposis coli protein): 1284–1537, ASXL1 (Polycomb group protein ASXL1): 1102–1107, BCL2 (Apoptosis regulator Bcl-2): 2–80, CALR(Calreticulin): 358–384, CARD11 (Caspase recruitment domain-containing protein 11): 111–134; 207–266; 337–436, CBL (E3 ubiquitin-protein ligase CBL): 365–374, CCND3 (G1/S-specific cyclin-D3): 278–290, CD79B (B-cell antigen receptor complex-associated protein beta chain): 191–199, CEBPA (CCAAT/enhancer-binding protein alpha): 293–327, CSF1R (Macrophage colony-stimulating factor 1 receptor): 969–969, CTNNB1 (Catenin beta-1): 32–45, EIF1AX (Eukaryotic translation initiation factor 1A, X-chromosomal): 4–15, EPAS1 (Endothelial PAS domain-containing protein 1): 529–539, ESR1 (Estrogen receptor): 303–303, FOXA1 (Hepatocyte nuclear factor 3-alpha): 248–268, FOXL2 (Forkhead box protein L2): 134–134, FOXO1 (Forkhead box protein O1): 19–26, HIST1H3B (Histone H3.1): 28–28, ID3 (DNA-binding protein inhibitor ID-3): 48–70, MED12 (Mediator of RNA polymerase II transcription subunit 12): 44–44, MLH1 (DNA mismatch repair protein Mlh1): 379–385, MYC (Myc proto-oncogene protein): 57–60, MYCN(N-myc proto-oncogene protein): 44–44, MYOD1(Myoblast determination protein 1): 122–122, NFE2L2 (Nuclear factor erythroid 2-related factor 2): 20–38; 75–82, PAX5 (Paired box protein Pax-5): 75–80, RPS15 (40S ribosomal protein S15): 129–145, SETBP 1 (SET-binding protein): 858–880, SMARCB1 (SWI/SNF-related matrix-associated actin-dependent regulator of chromatin subfamily B member 1): 368–381, SRSF2 (Serine/arginine-rich splicing factor 2): 95–95, USP8 (Ubiquitin carboxyl-terminal hydrolase 8): 713–736, VHL (von Hippel-Lindau disease tumor suppressor): 54–136; 144–193.

2.2. Evolutionary Framework

In this work, we calculated the evolutionary origin of cancer risk regions within our dataset of disordered proteins. Our approach focused on the age of orthologous gene families, instead of focusing on the evolutionary origin of founder domains. Assignment of age of human gene families (origin) was carried out using the ENSEMBL genome browser database. To identify the origin of individual human gene families, we fetched the phylogenies and analysed the evolutionary supertrees built by the pipeline of the ENSEMBL Compara multi-species comparisons project [12,13]. The used release (99) of the project contained 282 reference species including 277 vertebrata, 4 eumetazoa, and 1 opisthokonta (*S. cerevisiae*) species. Note that, in these phylogenies, the most ancient node can be the ancestor of yeast. The origin of the gene family was identified by taking the taxonomy level of the most ancient node of the phylogenetic supertrees. Taxonomy levels were broken into major nested age categories (mammals, vertebrates, eumetazoa, opisthokonta), similarly to previous studies [14].

To define the evolutionary origin of regions, we built a customized pipeline that included collecting and mapping mutations from COSMIC database to ENSEMBL entries, constructing multiple sequence alignments of protein families, and mapping the cancer regions among orthologs and paralogs. According to the ENSEMBL supertrees, protein sequences of human paralogs (including the cancer gene) and their orthologs were queried from the database using the Rest API function. Then, multiple sequence alignments of the corresponding sequences were created with MAFFT (default settings) [15]. On the basis of the sequence alignments, cancer regions were mapped onto the sequences. In the mapping step, cancer regions were considered as functional units (linear motifs, linkers, disordered domains) and borders of the regions were defined according to this. When the highly mutated regions covered only a single residue, it was extended to cover the known functional linear motif or using its sequence neighbourhood. On this basis, the subset of paralogs, in which the mapped cancer region was found to be conserved, was identified.

Next, the set of sequences containing regions that showed evolutionary similarity to the mutated regions were identified among the collected orthologs and paralogs. Conservation of the regions among paralogs was evaluated relying on two strategies, by calculating the similarity of mutated positions in the cancer risk regions (see below) and based on HMM profiles. This consideration was taken into account in order to reduce the chance of false conservation interpretation arising from the difficulty of aligning disordered proteins. The HMM profiles were built from conserved cancer regions of vertebrate model organisms using the HMMER (version 3.3) method [16]. The identified region hits were manually checked to minimize the chance of false positives or negatives. Next, we identified the evolutionarily most distant relative in which the cancer region was declared to be conserved. As a result, the origin of the region could differ from the origin of the orthologous gene family, when paralogue sequences that contained the conserved motif had a more ancient origin. Basically, we treated the cancer risk regions as the founder of the family. The taxonomy level of this ortholog was defined as the level in which the cancer region emerged in the common ancestor of this ortholog and *H. sapiens*.

2.3. Region Conservation

Within the identified cancer risk region, some of the positions could be more heavily mutated and are likely to be more critical for the function of this region. We took this into account when calculating the region conservation. Mutations for each position collected from the COSMIC database were mapped to the corresponding ENSEMBL human entry. On the basis of the sequence alignment corresponding to the cancer risk regions, we identified the positions that were similar to the reference sequence. Two positions were considered similar when the substitution score was non-negative according to the BLOSUM62 substitution matrix. A given cancer region was considered to be conserved between homologs, when the conserved residues carried more than 50% of missense mutations.

2.4. Positive Selection: Selectome and McDonald and Kreitman (MK) Test Results

For each entry in our dataset, we collected information about positive selection using the Selectome database (current version 6) [17]. This database contains collected sites of positive selection detected on a single branch of the phylogeny using the systematic branch-site test of the CODEML algorithm from the PAML [18] phylogenetic package version 4b. The ratio of non-synonymous and synonymous substitutions (ω) can be interpreted as a measurement of selective pressure indicating purifying (ω values < 1), neutral (ω values = 1), or positive (ω values > 1) selection. In our work, positions under positive selection that have a posterior probability higher than 0.9 were extracted from the database and mapped onto our gene set.

However, the branch-site model generally cannot detect species-specific positive selection. Potential cases of human-specific positive selection may be detected effectively by comparing divergence to polymorphism data, as in the McDonald and Kreitman (MK) test. Human-specific positive selection detected by MK test previously calculated [19] was mapped onto our dataset of disordered cancer genes.

3. Results

3.1. Evolutionary Origin of Genes and Regions

Altogether, we collected 36 cancer risk regions of 32 disordered proteins and investigated the evolutionary origin at the level of genes and regions. The age estimation of disordered cancer genes was obtained using the last common ancestor of descendants using the ENSEMBL supertrees, which includes phylogeny of gene families returning not only individual gene history, but also relationships of ancient paralogs and their history (see Material and Methods). Using this strategy instead of analysing the evolution of individual genes or simply the emergence of the founder domain, we could define the origin of regions more precisely, even the ancient ones, without introducing any bias of overprediction of origins. However, some ambiguity still remained and was manually checked (Supplementary Materials 1). The genes were traced back to opisthokonta (in accordance with the ENSEMBL database) and divided into four major phylostratigraphic groups, which are associated with the emergence of unicellular, multicellular organisms, vertebrates, and mammals.

Previous results identified the level of eumetazoa as the main age for the emergence of cancer associated proteins [3]. We observed a similar trend in the case of disordered cancer proteins.

Specifically, we found that 21 disordered cancer proteins, the majority of cases, have emerged at the level of eumetazoa (Figure 1). Fourteen cases were found to be even more ancient and could be traced back to single cell organisms, at least to opisthokonta. The only protein that emerged more recently, at the level of vertebrates, was CD79B, the B-cell antigen receptor complex-associated protein β chain. Its appearance is in agreement with the birth of many immune receptors [20] and is assumed to be driven by the insertion of transposable elements.

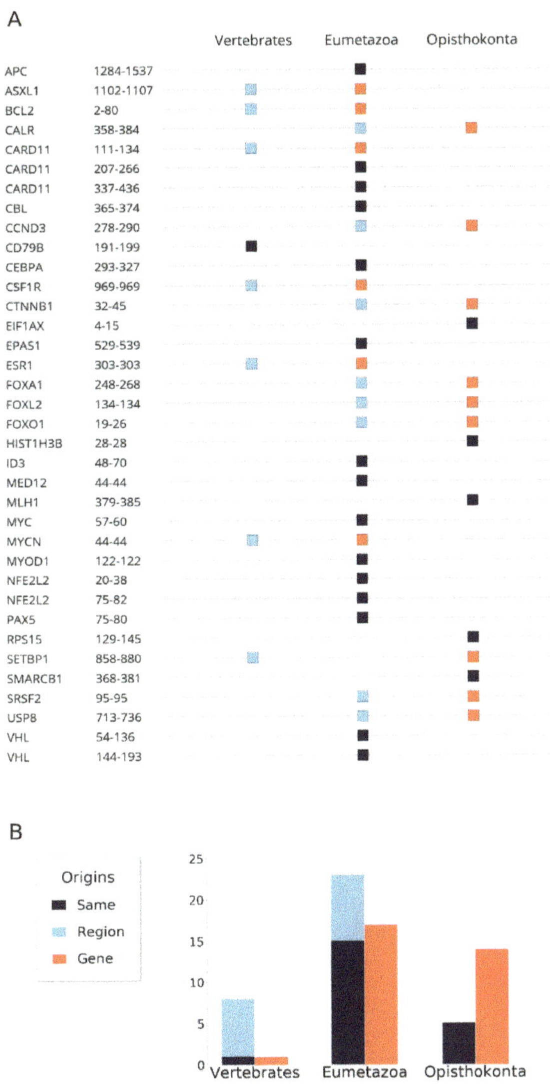

Figure 1. Conservation-based evolutionary origin of disordered cancer regions and genes. (**A**) The orange and sky blue squares represent the origin of genes and regions, respectively. Gunmetal squares indicate the same evolutionary origin at both region and gene levels. (**B**) Summary barchart of origins in the three gene-age categories.

In around half of the cases (21), the emergence of the mutated region was the same as the emergence of the protein (Figure 1). Strikingly, these included five cases (EIF1AX, HIST1H3B, MLH1, RPS15, SMARCB1) where not only the gene/protein, but also the region primarily mutated in human cancers were very ancient and could be traced back to unicellular organisms. Fifteen regions with Eumetazoa and one with Vertebrata origin could be traced back to the same level as their corresponding gene. However, in several cases, the emergence of the region was a more recent event compared with the emergence of the gene. Of these, eight regions emerged at the Eumetazoa and seven at the Vertebrate level. In general, there was only one level difference between the emergence of the gene and the region at this resolution. The only exception was SETBP1. In this case, the region itself emerged at the vertebrate level. However, the gene could be traced back to opisthokonta level, although the eumetazoa origin cannot be completely ruled out (see Supplementary Materials 1). Overall, many of the disordered regions were more recent evolutionary inventions compared with the origin of their genes, and date back to the common ancestors of eumetazoans or vertebrates. Nevertheless, the ancestors of all of the regions were already present from the vertebrate level.

3.2. Position Conservation

Overall, these results point to the ancient evolutionary origin of disordered regions involved in cancer, not only at the gene level, but also at the region level. To take a closer look, we also calculated the conservation of individual positions within the regions based both in terms of homologous substitutions and identity. The results show that these residues are highly conserved even compared with the conservation of the whole sequence (Figure 2). Here, 86% of the regions have more than 0.8 average conservation value even based on identities (Figure 2A). Among the cases with the four lowest values, the conservation of VHL, CALR, and APC, which all correspond to relatively longer segments, was still relatively high. The only outlier was BCL2. In this case, the mutations are distributed along the N-terminal, encompassing the highly conserved BH4 motif, as well as the linker region between the BH4 and C-terminal part, which is conserved only in mammals (Figure S1).

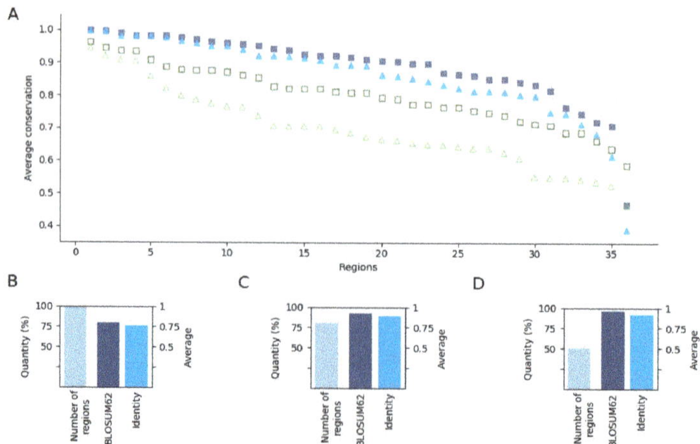

Figure 2. Representation of average conservation values. (**A**) Sorted conservation values for each region having positions with at least one mutation and for the whole protein. Squares (dark blue—region, green—whole sequence) and triangles (light blue—regions, green—full sequence) represent BLOSUM62 and identity based conservation values, respectively. The outlier at the very end of the sequence corresponds to the region of BCL2. (**B–D**) The number of regions and average conservation value of regions having positions with at least 1, 15, and 25 mutations, respectively. The conservation values are based on BLOSUM62 and identity, and the number of regions are colored by dark, medium, and sky blue, respectively.

Next, we investigated how this average value is altered when only the highly mutated positions are considered. We repeated that analysis for positions that had at least 15 and 25 missense mutations, which slightly decreased the number of regions considered. The remaining 28 and 17 regions with positions having at least 15 and 25 mutations had 0.93, 0.89, 0.96, and 0.92 average conservation values based on substitutions and identity, respectively (Figure 2C,D). This reflects a very clear trend with positions with a higher number of cancer mutations showing higher evolutionary conservation.

We also collected sites of potential positive selection mapped onto our genes based on the Selectome database [17], which provides information on likely molecular selection both at the level of the evolutionary branch and the sequence position based on the ratio of non-synonymous and synonymous substitutions (ω). According to these results, positive selection affected only three genes on the human lineage in our dataset, CALR, CTNNB1, and VHL. All of these selections could be mapped onto the vertebrates division with multiple positions (see Material and Methods) (Table 1).

Table 1. Positive selection within disordered cancer genes. Positions within cancer risk regions are colored blue. The numbers in brackets are the posterior probability of positive selection for each position.

Gene	Positions under Positive Selection Referring to the Human Protein Sequence
CALR	83(0.971), 155(0.971), 177(0.990), 267(0.995), 307(0.994), 336(0.991), 360(0.999)
CTNNB1	121(0.999), 206(0.993), 250(0.998), 287(0.991), 411(0.998), 433(0.993), 525(0.997), 552(0.998), 556(0.916)
VHL	127(0.957), 132(0.942), 141(0.923), 171(0.947), 183(0.963), 185(0.920)

However, these positions showed limited overlap with the mutated regions. In the case of CTNNB1, none of the positions under selection overlapped with the cancer mutated region. In the case of CALR, there was only a single position under selection within the cancer risk region, but it was not directly targeted by cancer mutations. In the case of VHL, six positions were detected with selective pressure and five of them were situated within the significantly mutated region. However, none of them corresponded to a highly mutated residue.

Taking advantage of an earlier analysis [19], we also analyzed if there was any human specific positive selection. As the ω based approach can not be used without uncertainty to identify human-specific positive selection, this work relied on the McDonald and Kreitman (MK) test, which compares the divergence to polymorphism data using closely related species, such as human and chimp. There was only a single entry in our database, ESR1, that showed human specific evolutionary changes (see case studies).

3.3. Contribution of Duplications to the Emergence of Disease Risk Regions

Gene duplications often drive the appearance of a novel function through the process called neofunctionalization. In these cases, after a duplication event, one copy may acquire a novel, beneficial function that becomes preserved by natural selection. Here, we have evaluated whether the emergence of disordered cancer regions corresponds to such neofunctionalization events. For this analysis, we collected paralog sequences and evaluated if there were regions present in these sequences that showed clear evolutionary similarity to the cancer mutated region.

The evolutionary history of many genes is quite complex and can involve multiple duplication events. We focused on the level where the cancer regions emerged and distinguished the following scenarios based on the relationship between the duplication and the presence of the region among the paralogs. The first scenario corresponds to duplication induced neofunctionalization. In this case, an ancient cancer region emerged directly after a given gene duplication and became preserved in only one of the branches that appeared after the duplication (Figure 3A). There are two basic scenarios in which the duplication cannot be directly linked with the emergence of the regions. One possible scenario is when both branches contain the region, which indicates that the region must have emerged before the duplication (Figure 3B). The other possible scenario is when the region emerged at a later evolutionary stage after a duplication, and duplication cannot be directly linked to neofunctionalization (Figure 3B).

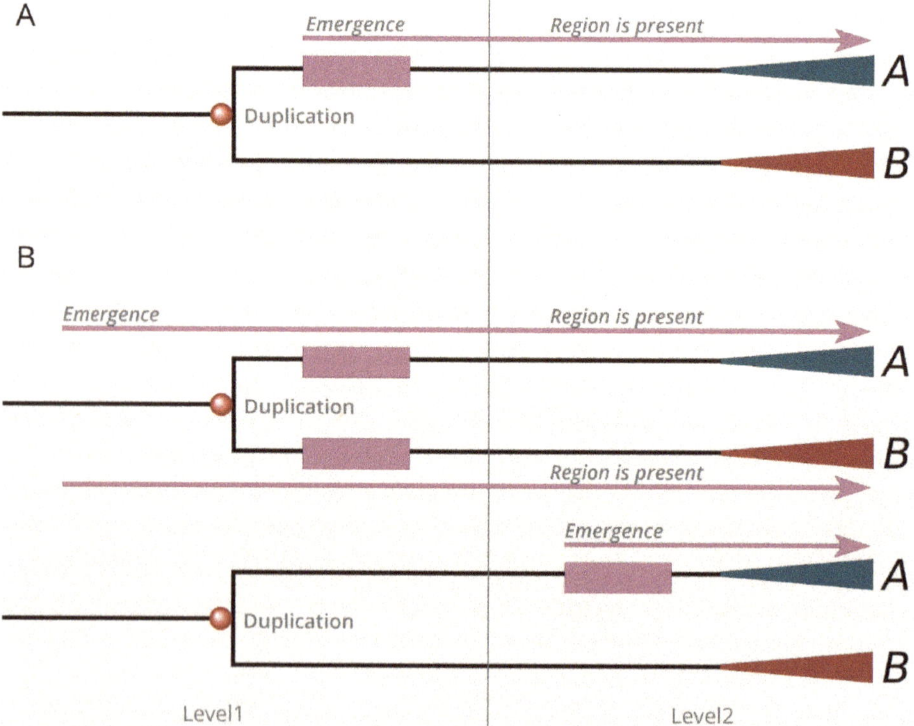

Figure 3. The mechanisms of emergence of regions by neofunctionalization and de novo. (**A**) Demonstration of the model of duplication induced (neofunctionalization) cancer region emergence. (**B**) Depiction of the two sub-scenarios of the de novo region emergence. Mallow boxes and arrows explain the evolution of the region. Red and green triangles symbolize the further evolution of paralogs after gene duplications.

Surprisingly, the duplication induced neofunctionalization was much less common than we expected, with only seven cases showing this behaviour. One example for this scenario is presented by the β-catenin family, where the degron motif [21] based cancer risk region that emerged after duplication is present only on the branch of β-catenin and junctional plakoglobin (JUP). In contrast, we found that 23 regions have evolved by de novo emergence, which seemed to be the dominant mechanisms for the emergence of the analyzed cancer mutated disordered regions (Figure 4A). For example, ID3 underwent multiple duplications, but all paralogs contain the cancer risk region, which indicates that the region emerged prior to the duplication. Another example is ESR1, in which case the paralogs were born at the level of eumetazoa; however, this event is not directly linked to the emergence of the cancer region, which appeared only at the level of the ancient vertebrates. In addition, there were two singletons in our dataset, RPS15 and SMARCB1, which did not have any detectable paralogs. In the cases of ASXL1, CCND3, SETBP1, and the first region of CARD11, the evolutionary scenarios could not be unambiguously established. These six examples formed the "Other" group.

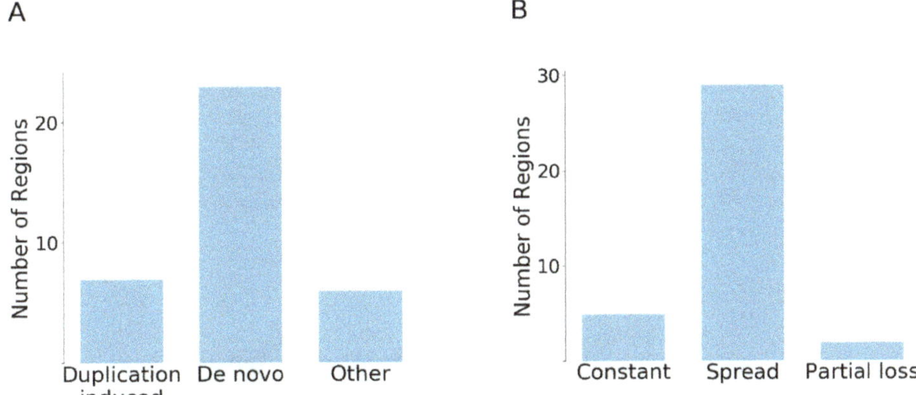

Figure 4. Categorization of emergence scenarios and evolutionary fates of cancer regions. (**A**) The number of regions that have emerged by duplication or de novo. Six regions were not categorized (Other). (**B**) Classification of cancer regions in terms of their evolutionary fate after emergence.

We also analyzed if additional duplication events occurred after the emergence of regions and whether the novel paralogues retained the regions. There are basically three scenarios that can occur: (i) the region is preserved without any further duplications; (ii) the region spreads and becomes preserved in all of the novel duplicates; (iii) partial loss scenario, that is, the region is preserved in some duplicates, but is lost in others. Our results show that the most common evolutionary fate is the second one (Figure 4B). In 29 cases, at least one duplication that inherited the region can be observed after the emergence of the cancer region. In contrast, only five regions were not duplicated. Some ancient cases, such as MLH1 and USP8, are also included among the non-duplicated ones, which means that the reason for the lack of duplications is not the short evolutionary time. The partial loss scenario was observed in only two cases, in the case of VHL and NFE2L2. For instance, in the case of VHL, there was a relatively recent gene duplication at the level of mammals. While the N-terminal segment is present on both paralogs (VHL and VHLL), the C-terminal segment is only present in VHL, but was lost from VHLL. In a similar fashion, NFE2L2 underwent a more recent gene duplication at the level of vertebrates, but the newly emerged paralog did not retain the two linear motifs that are primarily targeted by cancer mutations.

3.4. Case Studies

3.4.1. MLH1

One of the most ancient examples in our dataset corresponds to MLH1 (MutL Homolog 1), an essential protein in DNA mismatch repair (MMR). As one of the classic examples of a caretaker function, mutations of MLH1 can lead to cancer by increasing the rate of single-base substitutions and frameshift mutations [22]. Several positions of MLH1 are mutated in people with Lynch syndrome, also known as hereditary nonpolyposis colorectal cancer (HNPCC). However, according to the COSMIC database of somatic cancer mutations, the most common mutation of MLH1 is V384D. Mutational studies of V384D using yeast assays and in vitro MMR assay did not indicate a strong phenotype, but still showed a limited decrease of MMR activity [23]. However, it was shown that the (mostly germline) V384D variant is clearly associated with increased colorectal cancer susceptibility [24], and it is highly prevalent in HER2-positive luminal B breast cancer [25].

MLH1 is an ancient protein that is present from bacteria to humans. It has a highly conserved domain organization that involves ordered N- and C-terminal domains connected by a disordered linker [26] (Figure 5). This underlines the functional importance not only of the structured domains, but also of the connecting disordered region. In our previous work, we identified the region from 379 to 385 to be significantly mutated [7], which is located within the disordered segment. Recently, it was shown that the linker can regulate both DNA interactions and enzymatic activities of neighboring structured domains [27]. In agreement with the linker function, both the composition and length of this intrinsically disordered region (IDR) are critical for efficient MMR. Overall, most of the linker shows relatively low sequence conservation, however, the identified cancer risk region is highly conserved from across all eukaryotic sequences (Figure 5), in an island-like manner. Although the exact function of this region is not known, the strong evolutionary conservation indicates a highly important function, not yet explored in detail.

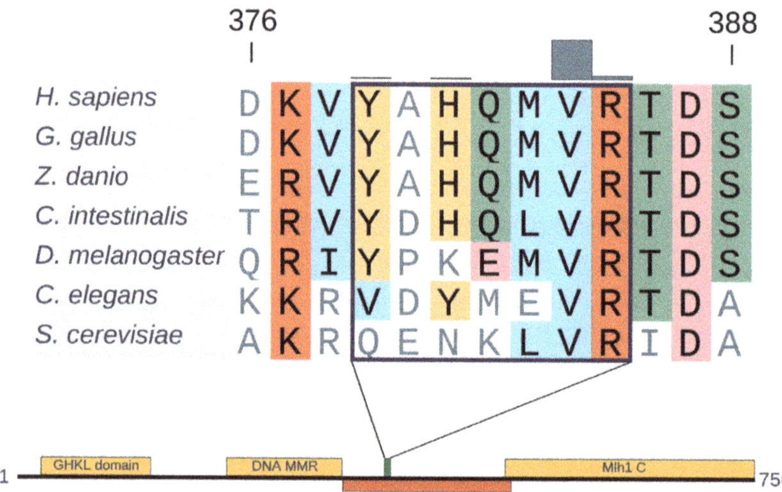

Figure 5. Alignment of MLH1 orthologs generated with MAFFT [15] and domain structure of human MLH1. The segment of the alignment represents the cancer region (highlighted by a rectangle) with the missense mutation distribution depicted by gray bars. Domains are depicted by yellow, disordered regions by red boxes, while the green box indicates the cancer risk region.

3.4.2. VHL

VHL, the Von Hippel-Lindau disease tumor suppressor protein possesses an E3 ligase activity. It plays a key role in cellular oxygen sensing by targeting hypoxia-inducible factors for ubiquitylation and proteasomal degradation. To carry out its function, VHL forms a complex with elongin B, elongin C, and cullin-2 and the RING finger protein RBX1 [28,29]. VHL has an α-domain (also known as the VHL-box, residues 155 to 192) that forms the principal contacts with elongin C, and a larger β-domain (residues 63 to 154) that directly binds the proline hydroxylated substrate, HIF1α. The positions mutated across various types of cancers cover a large part of the protein, including both the α and β domains. While these regions form a well-defined structure in complex with elongin B, elongin C, and cullin-2, they are disordered in isolation and rapidly degraded [30].

The VHL gene emerged de novo at the level of Eumetazoa together with HIFα and PHD, the other key components of the hypoxia regulatory pathway. However, more recently, the gene underwent various evolutionary events. The VHL gene showed slightly higher evolutionary variations compared with other cancer risk regions (Figure 2). Some positions, including K171, showed signs of positive selection at the level of Sarcopterygii, which might implicate the occurrence of an important evolutionary event. It was shown that the SUMO E3 ligase PIASy interacts with VHL and induces VHL SUMOylation on lysine residue 171 [31]. VHL also undergoes ubiquitination on K171 (and K196), which is blocked by PIASy. In the proposed model of the dynamic regulation of VHL, the interaction of VHL with PIASy results in VHL nuclear localization, SUMOylation, and stability for blocking ubiquitylation of VHL. Meanwhile, PIASy dissociation with VHL or attenuation of VHL SUMOylation facilitates VHL nuclear export, ubiquitylation, and instability. This dynamic process of VHL with reversible modification acts in concert to inhibit HIF1α [32].

A novel acidic repeat region appeared at the N-terminal region of the protein at the level of Sarcopterygii, and this region underwent further repeat expansion in the lineage leading up to humans (Figure 6). These GxEEx repeats are generally thought to confer additional regulation to the long isoform of VHL (translated from the first methionine), with a number of putative (USP7) or experimentally detected (p14ARF) interactors [33]. Although poorly studied, this repetitive region also seems to harbour casein kinase 2 (CK2) phosphorylation as well as proteolytic cleavage sites, regulating VHL half-life (consistent with a deubiquitinase, such as USP7 binding role) [34]. As a result of a recent gene duplication, the human genome even encodes a VHL-like protein (VHLL), which has lost the C-terminal segment including the α domain. Consequently, VHLL cannot nucleate the multiprotein E3 ubiquitin ligase complex. Instead, it was suggested that VHLL functions as a dominant-negative VHL to serve as a protector of HIF1α [35]. This example demonstrates that, while the basic cancer risk region remains largely unchanged during evolution, additional regulatory mechanisms can emerge to further fine-tune the function of the protein.

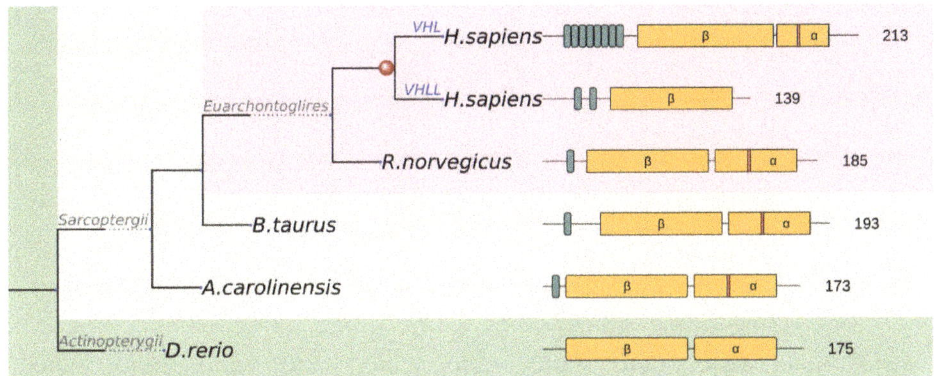

Figure 6. Schematic representation of the evolutionary scenario of the VHL family and the functional units of the members. Repeat units in varying numbers and the α and β core domains are depicted by green and yellow boxes, respectively. Red stripe in the α domain of human VHL indicates K171 identified to emerge by positive selection on the Sarcopterygii branch (mapped K171 to other Sarcopterygii are also indicated by red stripes).

3.4.3. ESR1

Estrogen receptor 1 (ESR1) is a member of the nuclear hormone receptor family with eumetazoan origin. The most common mutation in both primary and tamoxifen therapy associated samples corresponds to a single mutation (K303R). This single site emerged more recently (Figure 7) and is located in a rather complex switch region adjacent to the ligand-binding domain (Figure S2). The highly mutated K303 of ESR1 (more than 200 K303R missense mutations are seen in COSMIC) is a part of a motif-based molecular switch region involving several mutually exclusive PTMs. At positions 302, 303, and 305, methylation by SET7/9, acetylation by p300, and phosphorylation by PKA or PAK1 were observed in previous studies, respectively [36–40]. Our results show that this region is conserved only in Sarcopterygii, which indicates a relatively young evolutionary origin of the switching mechanism. However, while the methylation and acetylation sites are well conserved, the phosphorylation motif appears to be specific only to H. sapiens. We came to this conclusion because R300 and K302 as well as L306 are required for the protein kinase A (PKA) phosphorylation consensus and the oncogenic mutation K303R is expected to turn this region into an even better PKA substrate [41,42]. Curiously, these residues are not found in any other mammal, supposing species specific adaptive changes.

Comparison of substitutions and polymorphic sites is a powerful approach to identify specific changes in a pair of closely related species, like H. sapiens and chimpanzee. Relying on this approach, 198 of 9785 analyzed genes were identified to show human-specific changes including ESR1 [19]. In ESR1, there are three more changes besides R300 and K306 (L44, Q502, S559) between H. sapiens and chimp that are also thought to be adaptive substitutions according to the MK test. Phosphorylation of S559 was experimentally identified, suggesting this residue is also a H. sapiens specific PTM [43,44], but there is no specific data in the literature about the biological function of L44 and Q502. Yet, we know that phosphorylation of S305 allows the increase of estrogen sensitivity by external stimuli other than steroids, and permits ESR1 activity even when the canonical estrogen effect is completely blocked by tamoxifen [40,42]. In mice, ESR1 activity is essential for the estrogen effect and normal estrous episodes [45,46]. Although we lack information, we theorize that this human-specific signaling crosstalk might somehow be connected to the continuous menstrual cycle of H. sapiens (quite unusual among mammals), or some other human-specific reproductive adaptation.

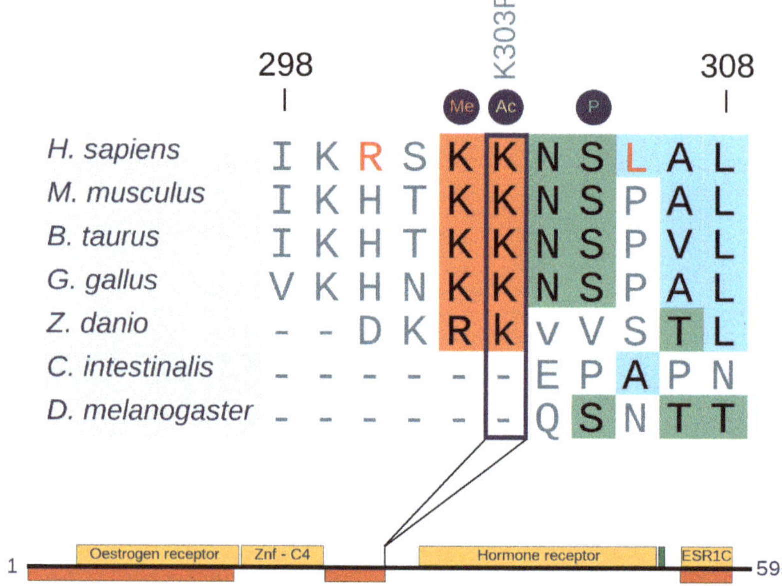

Figure 7. Insertion-free sequence alignment of estrogen receptor 1 (ESR1) orthologs and domain structure of human ESR1. The alignment generated with MAFFT [15] represents the cancer region with sites of post-translational modifications. Borders of non-depicted insertion of zebrafish are indicated by lower case letters. The highly mutated position (K303R) is highlighted by a rectangle. PTM sites are indicated by circles above the alignment. *H. sapiens* specific changes are colored in red. Domains are depicted in yellow, disordered regions are depicted by red boxes, while the green boxes indicate the cancer risk regions.

4. Discussion

In our study, we aimed to estimate the evolutionary origin of disordered regions that are specifically targeted in cancer. Intrinsically disordered protein regions play essential roles in a wide-range of biological processes and can function as linear motifs, linkers, or other intrinsically disordered domain-sized segments [47]. They are integral parts of many cancer associated proteins and, in a smaller number of cases, they can also be the direct targets of cancer driving mutations. In general,

IDRs are believed to be of more recent evolutionary origin, and exhibit higher rates of evolutionary variations compared with that of folded globular domains [9]. However, this is not what we see in the case of disordered cancer genes. Instead, we observed that cancer-targeted disordered regions are extremely conserved with deep evolutionary origins, which underlines their critical function. The two main ages for emergence of disordered cancer genes can be linked to unicellular organisms and the emergence of multicellularity, in agreement with the result of phylostratigraphic tracking of cancer genes in general [3].

One of the most unexpected findings of our study is the examples of disordered cancer genes that can be traced back to unicellular organisms. Mechanistically, the group of cancer genes that emerged in unicellular organisms were suggested to play a caretaker role and contribute to tumorigenesis by increasing mutation rates and genome instability. In contrast, cancer genes that emerged at the level of multicellularity were suggested to typically have a gatekeeper function and promote tumour progression directly by changing cell differentiation, growth, and death rates [48]. MLH1 is one of the best characterized examples of a gene with a caretaker function [49]. It is involved in mismatch repair (MMR) of DNA bases that have been misincorporated during DNA replication. Thus, disruptive mutations of MLH1 greatly increase the rate of point mutations in genes and underline various inherited forms of cancer. However, the most commonly seen alterations in patients are located in the flexible internal linker. Mutational studies indicate that this highly conserved segment might not be directly involved in MMR, but likely has an important, currently uncharacterized function. The other ancient examples are also involved in basic cellular processes, however, they are associated with a broader set of functions. HIST1H3B, SMARCB1, and SETBP1 are involved in epigenetic regulation and their mutations can alter gene expression patterns [50,51]. Mutations of EIF1AX and RPS15 are likely to perturb translation events [52,53]. However, SRSF2, which is responsible for orchestrating splicing events, can also have a global influence on cellular states [54]. Therefore, the caretaker function is also a subject of evolution and some of its components emerged as a result of more recent evolutionary events.

A clear novelty of our approach is to focus at the origin of sub-gene elements; that is, regulatory regions, modules, and domains, instead of full genes. The genes can be built around founder genes that have an extremely ancient origin, but their biological function and regulation can change fundamentally during subsequent evolution. In several cases, the origin of the cancer mutated region was substantially more recent than the origin of the gene. Nevertheless, after their emergence, disordered cancer regions were fixated rapidly and showed little variations afterwards. However, their evolution at the gene level was not set in stone and there are several indications that this process continues indefinitely. In several cases, the cancer genes underwent gene duplications, further regulatory regions were added, or fine-tuned by changing some of the less critical positions. We highlighted a fascinating case when such an event occurred when our species, *H. sapiens*, separated from its primate relatives.

In general, the rate of gene duplications is very high (0.01 per gene per million years) over evolution, which provides the source of emergence of evolutionary novelties [55]. According to the general view, paralogs go through a brief period of relaxed selection directly after duplications—this time ensures the acquisition of novelties—and subsequently experience strong purifying selection, preserving the newly developed function. However, our results showed that only a few disordered cancer regions have emerged in a duplication induced manner and the vast majority of disordered cancer regions emerged de novo, independent of duplications. The evolution of disordered regions is better described by the ex-nihilo motif theory, which is based on the rapid disappearance and emergence of linear motifs by the change of only a few residues within a given disordered protein segment [10]. This evolutionary phenomenon is commonly observed in the case of linear motifs, for example, in the case of NFE2L2. This protein carries a pair of crucial linear motifs that have emerged in the ancient eumetazoa, but are not preserved in the most recent duplicates. In an evolutionary biology aspect, our results suggest that the evolution of functional novelties in the case of disordered region mediated functions requires a more complex model.

Exploring the evolutionary origin of cancer genes is an important step to understand how this disease can emerge. This knowledge can also have important implications of how their regulatory networks are disrupted during tumorigenesis and can be incorporated into developing improved treatment options [56]. In this work, we focused on a subset of cancer genes that belong to the class of intrinsic disordered proteins, which rely on their inherent flexibility to carry out their important functions. While the selected examples represent only a small subset of cancer genes, they are highly relevant for several specific cancer types [8]. In general, disordered proteins are evolutionarily more variable compared with globular proteins, however, the disordered cancer risk regions showed remarkable conservation with ancient evolutionary origin, highlighting their importance in core biological processes. Nevertheless, we found several examples where the region specifically targeted by cancer mutations emerged at a later stage compared with the origin of the gene family. Our results highlight the importance of taking into account the complex modular architecture of cancer genes in order to get a more complete understanding of their evolutionary origin.

Supplementary Materials: The following are available online at http://www.mdpi.com/2218-273X/10/8/1115/s1, Figure S1: Sequence alignment of BCL2 cancer region. Figure S2: Schematic representation of the evolutionary scenario and functional units of the ESR1 and ESR2 proteins. Supplementary Materials 1: Evolutionary origins of selected cases.

Author Contributions: Conceptualization, M.P., A.Z. and Z.D.; Data curation, M.P.; Formal analysis, M.P.; Funding acquisition, Z.D.; Investigation, M.P., A.Z. and Z.D.; Methodology, M.P.; Supervision, Z.D.; Visualization, M.P.; Writing—original draft, M.P., A.Z. and Z.D. All authors have read and agreed to the published version of the manuscript.

Funding: This research was funded by the "FIEK" grant from the National Research, Development, and Innovation Office (FIEK16-1-2016-0005) and the ELTE Thematic Excellence Programme (ED-18-1-2019-003) supported by the Hungarian Ministry for Innovation and Technology.

Conflicts of Interest: The authors declare no conflict of interest.

References

1. Jacob, F. Evolution and tinkering. *Science* **1977**, *196*, 1161–1166. [CrossRef] [PubMed]
2. Kinzler, K.W.; Vogelstein, B. Cancer-susceptibility genes. Gatekeepers and caretakers. *Nature* **1997**, *386*, 761–763. [CrossRef] [PubMed]
3. Domazet-Lošo, T.; Tautz, D. Phylostratigraphic tracking of cancer genes suggests a link to the emergence of multicellularity in metazoa. *BMC Biol.* **2010**, *8*, 66. [CrossRef]
4. Domazet-Loso, T.; Tautz, D. An ancient evolutionary origin of genes associated with human genetic diseases. *Mol. Biol. Evol.* **2008**, *25*, 2699–2707. [CrossRef] [PubMed]
5. Dickerson, J.E.; Robertson, D.L. On the origins of mendelian disease genes in man: The impact of gene duplication. *Mol. Biol. Evol.* **2012**, *29*, 2284. [CrossRef]
6. Bailey, M.H.; Tokheim, C.; Porta-Pardo, E.; Sengupta, S.; Bertrand, D.; Weerasinghe, A.; Colaprico, A.; Wendl, M.C.; Kim, J.; Reardon, B.; et al. Comprehensive characterization of cancer driver genes and mutations. *Cell* **2018**, *173*, 371–385.e18. [CrossRef]
7. Mészáros, B.; Zeke, A.; Reményi, A.; Simon, I.; Dosztányi, Z. Systematic analysis of somatic mutations driving cancer: Uncovering functional protein regions in disease development. *Biol. Direct* **2016**, *11*, 23. [CrossRef]
8. Mészáros, B.; Hajdu-Soltész, B.; Zeke, A.; Dosztányi, Z. Intrinsically disordered protein mutations can drive cancer and their targeted interference extends therapeutic options. *Bioinform. bioRxiv* **2020**, 2443. [CrossRef]
9. Brown, C.J.; Johnson, A.K.; Dunker, A.K.; Daughdrill, G.W. Evolution and disorder. *Curr. Opin. Struct. Biol.* **2011**, *21*, 441–446. [CrossRef]
10. Davey, N.E.; Cyert, M.S.; Moses, A.M. Short linear motifs—Ex nihilo evolution of protein regulation. *Cell Commun. Signal.* **2015**, *13*, 43. [CrossRef]
11. Sondka, Z.; Bamford, S.; Cole, C.G.; Ward, S.A.; Dunham, I.; Forbes, S.A. The COSMIC cancer gene census: Describing genetic dysfunction across all human cancers. *Nat. Rev. Cancer* **2018**, *18*, 696–705. [CrossRef] [PubMed]

12. Flicek, P.; Amode, M.R.; Barrell, D.; Beal, K.; Brent, S.; Carvalho-Silva, D.; Clapham, P.; Coates, G.; Fairley, S.; Fitzgerald, S.; et al. Ensembl 2012. *Nucleic Acids Res.* **2011**, *40*, D84–D90. [CrossRef]

13. Herrero, J.; Muffato, M.; Beal, K.; Fitzgerald, S.; Gordon, L.; Pignatelli, M.; Vilella, A.J.; Searle, S.M.J.; Amode, R.; Brent, S.; et al. Ensembl comparative genomics resources. *Database* **2016**, *2016*, bav096. [CrossRef] [PubMed]

14. Liebeskind, B.J.; McWhite, C.D.; Marcotte, E.M. Towards consensus gene ages. *Genome Biol. Evol.* **2016**, *8*, 1812–1823. [CrossRef] [PubMed]

15. Katoh, K.; Standley, D.M. MAFFT multiple sequence alignment software version 7: Improvements in performance and usability. *Mol. Biol. Evol.* **2013**, *30*, 772–780. [CrossRef]

16. Eddy, S.R. Accelerated Profile HMM Searches. *PLoS Comput. Biol.* **2011**, *7*, e1002195. [CrossRef]

17. Moretti, S.; Laurenczy, B.; Gharib, W.; Castella, B.; Kuzniar, A.; Schabauer, H.; Studer, R.A.; Valle, M.; Salamin, N.; Stockinger, H.; et al. Selectome update: Quality control and computational improvements to a database of positive selection. *Nucleic Acids Res.* **2013**, *42*, D917–D921. [CrossRef]

18. Yang, Z. PAML 4: Phylogenetic Analysis by Maximum Likelihood. *Mol. Biol. Evol.* **2007**, *24*, 1586–1591. [CrossRef]

19. Gayà-Vidal, M.; Alba, M.M. Uncovering adaptive evolution in the human lineage. *BMC Genom.* **2014**, *15*, 1–12. [CrossRef]

20. Berg, T.K.V.D.; Yoder, J.A.; Litman, G. On the origins of adaptive immunity: Innate immune receptors join the tale. *Trends Immunol.* **2004**, *25*, 11–16. [CrossRef]

21. Wu, G.; Xu, G.; Schulman, B.A.; Jeffrey, P.D.; Harper, J.W.; Pavletich, N.P. Structure of a beta-TrCP1-Skp1-beta-catenin complex: Destruction motif binding and lysine specificity of the SCF(beta-TrCP1) ubiquitin ligase. *Mol. Cell* **2003**, *11*, 1445–1456. [CrossRef]

22. Shcherbakova, P.V.; Kunkel, T.A. Mutator Phenotypes conferred by MLH1 Overexpression and by Heterozygosity for mlh1 Mutations. *Mol. Cell. Biol.* **1999**, *19*, 3177–3183. [CrossRef] [PubMed]

23. Takahashi, M.; Shimodaira, H.; Andreutti-Zaugg, C.; Iggo, R.; Kolodner, R.D.; Ishioka, C. Functional analysis of human MLH1 variants using yeast and in vitro mismatch repair Assays. *Cancer Res.* **2007**, *67*, 4595–4604. [CrossRef] [PubMed]

24. Akagi; Ohsawa, T.; Sahara, T.; Muramatsu, S.; Nishimura, Y.; Yathuoka, T.; Tanaka, Y.; Yamaguchi, K.; Ishida, H. Colorectal cancer susceptibility associated with the hMLH1 V384D variant. *Mol. Med. Rep.* **2009**, *2*, 887–891. [CrossRef]

25. Lee, S.E.; Lee, H.S.; Kim, K.-Y.; Park, J.-H.; Roh, H.; Park, H.Y.; Kim, W.S. High prevalence of the MLH1 V384D germline mutation in patients with HER2-positive luminal B breast cancer. *Sci. Rep.* **2019**, *9*, 10966. [CrossRef] [PubMed]

26. Gueneau, E.; Dhérin, C.; Legrand, P.; Tellier-Lebègue, C.; Gilquin, B.; Bonnesoeur, P.; Londino, F.; Quemener, C.; Le Du, M.-H.; A Márquez, J.; et al. Structure of the MutLα C-terminal domain reveals how Mlh1 contributes to Pms1 endonuclease site. *Nat. Struct. Mol. Biol.* **2013**, *20*, 461–468. [CrossRef]

27. Kim, Y.; Furman, C.M.; Manhart, C.M.; Alani, E.; Finkelstein, I.J. Intrinsically disordered regions regulate both catalytic and non-catalytic activities of the MutLα mismatch repair complex. *Nucleic Acids Res.* **2018**, *47*, 1823–1835. [CrossRef]

28. Kamura, T.; Maenaka, K.; Kotoshiba, S.; Matsumoto, M.; Kohda, D.; Conaway, R.C.; Conaway, J.W.; Nakayama, K.I. VHL-box and SOCS-box domains determine binding specificity for Cul2-Rbx1 and Cul5-Rbx2 modules of ubiquitin ligases. *Genes Dev.* **2004**, *18*, 3055–3065. [CrossRef]

29. Cardote, T.A.; Gadd, M.S.; Ciulli, A. Crystal structure of the Cul2-Rbx1-EloBC-VHL Ubiquitin Ligase complex. *Structure* **2017**, *25*, 901–911.e3. [CrossRef]

30. Sutovsky, H.; Gazit, E. The von Hippel-Lindau tumor suppressor protein is a molten globule under native conditions. *J. Biol. Chem.* **2004**, *279*, 17190–17196. [CrossRef]

31. Cai, Q.; Verma, S.C.; Kumar, P.; Ma, M.; Robertson, E.S. Hypoxia Inactivates the VHL tumor suppressor through PIASy-Mediated SUMO modification. *PLoS ONE* **2010**, *5*, e9720. [CrossRef] [PubMed]

32. Cai, Q.; Robertson, E.S. Ubiquitin/SUMO modification regulates VHL protein stability and nucleocytoplasmic localization. *PLoS ONE* **2010**, *5*, e12636. [CrossRef] [PubMed]

33. Minervini, G.; Mazzotta, G.; Masiero, A.; Sartori, E.; Corrà, S.; Potenza, E.; Costa, R.; Tosatto, S.C.E. Isoform-specific interactions of the von Hippel-Lindau tumor suppressor protein. *Sci. Rep.* **2015**, *5*, 12605. [CrossRef] [PubMed]

34. German, P.; Bai, S.; Liu, X.-D.; Sun, M.; Zhou, L.; Kalra, S.; Zhang, X.; Minelli, R.; Scott, K.L.; Mills, G.B.; et al. Phosphorylation-dependent cleavage regulates von Hippel Lindau proteostasis and function. *Oncogene* **2016**, *35*, 4973–4980. [CrossRef] [PubMed]

35. Qi, H.; Gervais, M.L.; Li, W.; DeCaprio, J.A.; Challis, J.R.G.; Ohh, M. Molecular cloning and characterization of the von Hippel-Lindau-like protein. *Mol. Cancer Res.* **2004**, *2*, 43–52.

36. Dhayalan, A.; Kudithipudi, S.; Rathert, P.; Jeltsch, A. Specificity analysis-based identification of new methylation targets of the SET7/9 protein lysine methyltransferase. *Chem. Biol.* **2011**, *18*, 111–120. [CrossRef]

37. Wang, C.; Fu, M.; Angeletti, R.H.; Siconolfi-Baez, L.; Reutens, A.T.; Albanese, C.; Lisanti, M.P.; Katzenellenbogen, B.S.; Kato, S.; Hopp, T.; et al. Direct Acetylation of the Estrogen receptor α hinge region by p300 regulates transactivation and hormone sensitivity. *J. Biol. Chem.* **2001**, *276*, 18375–18383. [CrossRef]

38. Wang, R.-A.; Mazumdar, A.; Vadlamudi, R.K.; Kumar, R. P21-activated kinase-1 phosphorylates and transactivates estrogen receptor-α and promotes hyperplasia in mammary epithelium. *EMBO J.* **2002**, *21*, 5437–5447. [CrossRef]

39. Michalides, R.; Griekspoor, A.; Balkenende, A.; Verwoerd, D.; Janssen, L.; Jalink, K.; Floore, A.; Velds, A.; vant Veer, L.; Neefjes, J. Tamoxifen resistance by a conformational arrest of the estrogen receptor α after PKA activation in breast cancer. *Cancer Cell* **2004**, *5*, 597–605. [CrossRef]

40. Cui, Y.; Zhang, M.; Pestell, R.; Curran, E.M.; Welshons, W.V.; Fuqua, S.A.W. Phosphorylation of estrogen receptor α blocks its Acetylation and regulates estrogen sensitivity. *Cancer Res.* **2004**, *64*, 9199–9208. [CrossRef]

41. Rust, H.L.; Thompson, P.R. Kinase consensus sequences: A breeding ground for crosstalk. *ACS Chem. Biol.* **2011**, *6*, 881–892. [CrossRef] [PubMed]

42. De Leeuw, R.; Flach, K.; Toaldo, C.B.; Alexi, X.; Canisius, S.; Neefjes, J.; Michalides, R.; Zwart, W. PKA phosphorylation redirects ERα to promoters of a unique gene set to induce tamoxifen resistance. *Oncogene* **2012**, *32*, 3543–3551. [CrossRef] [PubMed]

43. Atsriku, C.; Britton, D.J.; Held, J.M.; Schilling, B.; Scott, G.K.; Gibson, B.W.; Benz, C.C.; Baldwin, M.A. Systematic mapping of posttranslational modifications in human estrogen receptor-α with emphasis on novel phosphorylation sites. *Mol. Cell. Proteom.* **2008**, *8*, 467–480. [CrossRef] [PubMed]

44. Williams, C.C.; Basu, A.; El-Gharbawy, A.; Carrier, L.; Smith, C.L.; Rowan, B.G. Identification of four novel phosphorylation sites in estrogen receptor α: Impact on receptor-dependent gene expression and phosphorylation by protein kinase CK2. *BMC Biochem.* **2009**, *10*, 36. [CrossRef]

45. Walker, V.R.; Korach, K. Estrogen receptor knockout mice as a model for endocrine research. *ILAR J.* **2004**, *45*, 455–461. [CrossRef] [PubMed]

46. Porteous, R.; Herbison, A.E. Genetic deletion of Esr1 in the mouse preoptic area disrupts the LH surge and estrous cyclicity. *Endocrinology* **2019**, *160*, 1821–1829. [CrossRef]

47. Van Der Lee, R.; Buljan, M.; Lang, B.; Weatheritt, R.J.; Daughdrill, G.W.; Dunker, A.K.; Fuxreiter, M.; Gough, J.; Gsponer, J.; Jones, D.T.; et al. Classification of intrinsically disordered regions and proteins. *Chem. Rev.* **2014**, *114*, 6589–6631. [CrossRef] [PubMed]

48. Lengauer, C.; Kinzler, K.W.; Vogelstein, B. Genetic instabilities in human cancers. *Nature* **1998**, *396*, 643–649. [CrossRef]

49. Ellison, A.R.; Lofing, J.; Bitter, G.A. Human MutL homolog (MLH1) function in DNA mismatch repair: A prospective screen for missense mutations in the ATPase domain. *Nucleic Acids Res.* **2004**, *32*, 5321–5338. [CrossRef]

50. Duchatel, R.J.; Jackson, E.R.; Alvaro, F.; Nixon, B.; Hondermarck, H.; Dun, M.D. Signal transduction in diffuse intrinsic Pontine Glioma. *Proteomics* **2019**, *19*, e1800479. [CrossRef]

51. Piazza, R.; Magistroni, V.; Redaelli, S.; Mauri, M.; Massimino, L.; Sessa, A.; Peronaci, M.; Lalowski, M.M.; Soliymani, R.; Mezzatesta, C.; et al. SETBP1 induces transcription of a network of development genes by acting as an epigenetic hub. *Nat. Commun.* **2018**, *9*, 2192. [CrossRef] [PubMed]

52. Martin-Marcos, P.; Zhou, F.; Karunasiri, C.; Zhang, F.; Dong, J.; Nanda, J.; Kulkarni, S.D.; Sen, N.D.; Tamame, M.; Zeschnigk, M.; et al. eIF1A residues implicated in cancer stabilize translation preinitiation complexes and favor suboptimal initiation sites in yeast. *eLife* **2017**, *6*, e31250. [CrossRef] [PubMed]

53. Bretones, G.; Álvarez, M.G.; Arango, J.R.; Rodríguez, D.; Nadeu, F.; Prado, M.A.; Valdés-Mas, R.; Puente, D.A.; Paulo, J.A.; Delgado, J.; et al. Altered patterns of global protein synthesis and translational fidelity in RPS15-mutated chronic lymphocytic leukemia. *Blood* **2018**, *132*, 2375–2388. [CrossRef]

54. Masaki, S.; Ikeda, S.; Hata, A.; Shiozawa, Y.; Kon, A.; Ogawa, S.; Suzuki, K.; Hakuno, F.; Takahashi, S.-I.; Kataoka, N. Myelodysplastic syndrome-associated SRSF2 mutations cause splicing changes by altering binding motif sequences. *Front. Genet.* **2019**, *10*, 338. [CrossRef] [PubMed]

55. Assis, R.; Bachtrog, D. Rapid divergence and diversification of mammalian duplicate gene functions. *BMC Evol. Biol.* **2015**, *15*, 138. [CrossRef]

56. Trigos, A.S.; Pearson, R.B.; Papenfuss, A.T.; Goode, D. How the evolution of multicellularity set the stage for cancer. *Br. J. Cancer* **2018**, *118*, 145–152. [CrossRef] [PubMed]

Article

A Phosphorylation-Induced Switch in the Nuclear Localization Sequence of the Intrinsically Disordered NUPR1 Hampers Binding to Importin

José L. Neira [1,2,*], Bruno Rizzuti [3], Ana Jiménez-Alesanco [2], Martina Palomino-Schätzlein [4], Olga Abián [2,5,6,7,8], Adrián Velázquez-Campoy [2,5,6,7,9] and Juan L. Iovanna [10,*]

[1] Instituto de Biología Molecular y Celular, Universidad Miguel Hernández, 03202 Elche, Spain
[2] Instituto de Biocomputación y Física de Sistemas Complejos (BIFI), Joint Units IQFR-CSIC-BIFI, and GBsC-CSIC-BIFI, Universidad de Zaragoza, 50009 Zaragoza, Spain; ajimenez@bifi.es (A.J.-A.); oabifra@unizar.es (O.A.); adrianvc@unizar.es (A.V.-C.)
[3] CNR-NANOTEC, Licryl-UOS Cosenza and CEMIF.Cal, Department of Physics, University of Calabria, Via P. Bucci, Cubo 31 C, 87036 Arcavacata di Rende, Cosenza, Italy; bruno.rizzuti@cnr.it
[4] Centro de Investigación Príncipe Felipe, 41930 Valencia, Spain; martina@tinet.org
[5] Instituto de Investigación Sanitaria Aragón (IIS Aragón), 50009 Zaragoza, Spain
[6] Centro de Investigación Biomédica en Red en el Área Temática de Enfermedades Hepáticas y Digestivas (CIBERehd), 28029 Madrid, Spain
[7] Departamento de Bioquímica y Biología Molecular y Celular, Universidad de Zaragoza, 50009 Zaragoza, Spain
[8] Instituto Aragonés de Ciencias de la Salud (IACS), 50009 Zaragoza, Spain
[9] Fundación ARAID, Gobierno de Aragón, 50009 Zaragoza, Spain
[10] Centre de Recherche en Cancérologie de Marseille (CRCM), INSERM U1068, CNRS UMR 7258, Aix-Marseille Université and Institut Paoli-Calmettes, Parc Scientifique et Technologique de Luminy, 163 Avenue de Luminy, 13288 Marseille, France
[*] Correspondence: jlneira@umh.es (J.L.N.); juan.iovanna@inserm.fr (J.L.I.); Tel.: +34-96-6658475 (J.L.N.); +33-(0)4-9182-8803 (J.L.I.)

Received: 23 August 2020; Accepted: 9 September 2020; Published: 11 September 2020

Abstract: Several carrier proteins are involved in protein transport from the cytoplasm to the nucleus in eukaryotic cells. One of those is importin α, of which there are several human isoforms; among them, importin $\alpha3$ (Imp$\alpha3$) has a high flexibility. The protein NUPR1, a nuclear protein involved in the cell-stress response and cell cycle regulation, is an intrinsically disordered protein (IDP) that has a nuclear localization sequence (NLS) to allow for nuclear translocation. NUPR1 does localize through the whole cell. In this work, we studied the affinity of the isolated wild-type NLS region (residues 54–74) of NUPR1 towards Imp$\alpha3$ and several mutants of the NLS region by using several biophysical techniques and molecular docking approaches. The NLS region of NUPR1 interacted with Imp$\alpha3$, opening the way to model the nuclear translocation of disordered proteins. All the isolated NLS peptides were disordered. They bound to Imp$\alpha3$ with low micromolar affinity (1.7–27 μM). Binding was hampered by removal of either Lys65 or Lys69 residues, indicating that positive charges were important; furthermore, binding decreased when Thr68 was phosphorylated. The peptide phosphorylated at Thr68, as well as four phospho-mimetic peptides (all containing the Thr68Glu mutation), showed the presence of a sequential NN($i,i + 1$) nuclear Overhauser effect (NOE) in the 2D-^{1}H-NMR (two-dimensional–proton NMR) spectra, indicating the presence of turn-like conformations. Thus, the phosphorylation of Thr68 modulates the binding of NUPR1 to Imp$\alpha3$ by a conformational, entropy-driven switch from a random-coil conformation to a turn-like structure.

Keywords: circular dichroism; flexibility; fluorescence; importin; intrinsically disordered protein; isothermal titration calorimetry (ITC); molecular docking; nuclear magnetic resonance (NMR); nuclear protein 1 (NPR1); peptide

---

1. Introduction

Active nuclear translocation happens through importins (also known as karyopherins), together with other proteins such as the GTPase Ran and nucleoporins [1–3]. The classical nuclear import pathway is started by recognition of a nuclear localization sequence (NLS) in the cargo by importin α [4]. The complex cargo importin α binds to importin β; then, this complex goes through the nuclear pore complex (NPC). The GTPase Ran dissociates the ternary complex within the nucleus by interacting with importin β, and both importins α and β are recycled back to the cytoplasm [4]. The human genome encodes seven isoforms of importin α, with three subtypes [4–6]. These isoforms have a role in cell differentiation, gene regulation [5,7], and even in viral infections, because some viral proteins are recognized by specific importins [8].

Importin α is a modular protein built of α-helix repeat armadillo (ARM) units [1,4]. It has two domains: (i) a N-terminal importin β-binding (IBB) domain, approximately 60-residues-long, which is used for binding to importin β before transport through the NPC, and (ii) a C-terminal NLS-binding motif formed by ten ARM units [9]. Structures of several truncated importin α, without the IBB domain [8,9], have shown that the cargo NLS region binds in a disordered conformation. This interaction occurs at a concave site of the elongated structure, involving ARM motifs 2 to 4 (major site) or 6 to 8 (minor site) for the shortest classical monopartite NLSs or both sets of ARM motifs for the largest bipartite NLS regions. When importin β is not present, the IBB domain, which mimics an NLS region, occupies the ARM motifs involved in NLS recognition [9]. This intramolecular interaction has an autoinhibitory role, and it is thought to be relevant in cargo dissociation in the nucleoplasmic side [9].

Intrinsically disordered proteins (IDPs) do not have a unique stable conformation, resulting in a dynamic conformational ensemble that is reflected in a high structural flexibility. They are involved in cell cycle control, signaling, molecular recognition, replication, and transcription processes [10–13]. The discovery of IDPs has shown that protein biological activity is possible even without a well-defined structure [12–14] but, rather, with an extreme structural flexibility. However, IDPs may have a propensity to adopt structures at the local level; this acquisition of local order can be achieved by, among other factors, post-translational modifications [14]. Such modifications, in turn, can widen their biological functions [11,15]. NUPR1 (UniProtKB O60356) is an 82-residue-long (8 kDa), highly basic, monomeric IDP that is overexpressed during the acute phase of pancreatitis [16,17] and in almost any, if not all, cancer tissues [18]. Its exact functions are unknown, but NUPR1 is a key element in the cell-stress response and cell-cycle regulation [18,19]. Moreover, NUPR1 intervenes in apoptosis through the formation of a complex with the oncoprotein ProTα [20] and in DNA repair [21,22]. In the interactions with all these partners and other synthetic molecules, NUPR1 uses two hotspots around residues Ala33 and Thr68 [22–24]. In addition, NUPR1 has a bipartite NLS region around Thr68, which is fully functional [25]. Thus, even though NUPR1 is a relatively small protein, it might require the assistance of the importin system for nuclear translocation due to its unfolded nature and its large radius of gyration, which would be closer to the limit of free diffusion through the NPC. In addition, NUPR1 might require the presence of importins to avoid undesired interactions with other macromolecules in the cytoplasm, due to its basic nature [26].

In this work, we have studied the interaction of human importin α3 (Impα3), also called KPNA4, and that of its truncated species, without the IBB domain (ΔImpα3), to either NUPR1 or peptides encompassing its NLS (NLS-NUPR1). We have chosen Impα3 as a target for NUPR1 because of its larger flexibility when compared with other importins, as concluded by the structural factors from the X-ray data, which confers in it a greater ability to interact with cargos, having a higher variety of conformations [8]. From an experimental pint of view, Impα3 can be also easily expressed and purified for in vitro structural studies [8]. Interestingly, it has also been shown to be crucial in pain pathways [27]. In addition, by studying both importin species (with and without the IBB), we were interested in finding out whether the absence of the IBB domain affected the binding of NLS-NUPR1. The NLS-NUPR1 peptides had mutations at: (i) the two lysines in the sequence (Lys65 and Lys69), which are important for nuclear translocation, according to in vivo studies [25], and (ii) Thr68, where we have either introduced phospho-threonine or, alternatively, we have designed phospho-mimetic mutations (with a glutamic residue). We have used several spectroscopic and biophysical techniques—namely, steady-state fluorescence, circular dichroism (CD), nuclear magnetic resonance (NMR), isothermal titration calorimetry (ITC), and molecular docking—to address the binding of the peptides to both importins. Our results indicate that the isolated wild-type (wt) NLS-NUPR1, as well as the mutants, were monomeric and disordered in the solution. The wt NLS-NUPR1 peptide bound to both importins, and the affinity was larger for ΔImpα3 (0.95 μM versus 1.7 μM for Impα3), indicating that the IBB region must have an inhibitory effect; this result is in agreement with other binding studies involving intact, well-folded protein cargos [9], but to the best of our knowledge, this is the first time tested with an IDP. The binding of NLS-NUPR1 peptides to both importins was hampered by removal of either Lys65 or Lys69, and it was almost abolished when Thr68 was phosphorylated or when the phospho-mimetics were assayed. Interestingly enough, the phosphorylated peptide at Thr68 and the four phospho-mimetics showed the presence of turn-like conformations, which were not observed in the wt NLS-NUPR1 peptide or in the Lys65Ala or Lys69Ala mutants. We concluded that the phosphorylation of Thr68 modulates the binding of NUPR1 to importin by a conformational switch from a random-coil to a turn-like conformation.

2. Materials and Methods

2.1. Materials

Isopropyl-β-D-1-tiogalactopyranoside and ampicillin were obtained from Apollo Scientific (Stockport, UK). Imidazole, kanamycin, Trizma base, and His-Select HF nickel resin were from Sigma-Aldrich (Madrid, Spain). Protein marker (PAGEmark Tricolor) and Triton X-100 were from VWR (Barcelona, Spain). Amicon centrifugal devices were from Millipore (Barcelona, Spain), and they had a cut-off molecular weight of 30 or 50 kDa. The rest of the materials were of analytical grade. Water was deionized and purified on a Millipore system.

2.2. Protein Expression and Purification

The His-tagged ΔImpα3 (residues 64-521) was obtained from BL21 (DE3) cells as described [8]. The DNA of the codon-optimized, intact Impα3 with a His-tag at the N terminus was synthesized by NZYtech (Lisbon, Portugal) and cloned into the pHTP1 vector (with kanamycin resistance). Expression and purification of Impα3 were carried out as those for ΔImpα3 in the same *Escherichia coli* strain. Concentration of both species was determined from their six tyrosines and six tryptophans [28].

2.3. Design and Synthesis of the Peptides

The peptides were synthesized by NZYtech with a purity of 95%. The peptides comprised the NLS region of NUPR1 (Table 1); peptides were named with the accompanying name within parenthesis for each sequence, as reported in Table 1. All peptides were acetylated and amidated at the N and C termini, respectively, to avoid fraying effects. As the wt NLS had no tyrosine, we introduced one at the N terminus to allow for absorbance measurements [28]. We synthesized eight peptides with different mutations, with the following rationale: (i) we studied the importance of positions Lys65 and Lys69 in the binding to both importins by mutating the two positions to alanine, (ii) we mutated Thr68 to the glutamic T68E peptide to have a phosphomimic at this position, (iii) we combined this mutation at

Thr68 with either of the other two as double mutants, as well as to both in a triple mutant, and (iv) we designed the phosphorylated peptide at position Thr68 (pT68 peptide) to study the effects of this single post-translational modification.

Table 1. Hydrodynamic properties of the nuclear localization sequence (NLS) NUPR1 peptides.

Peptide [a]	D (cm^2 s^{-1}) × 10^6 (R_h, Å) [b]	R_h, Å [c]
YT^{54}NRPSPGGHERKLVTKLQNSE (wt)	1.85 ± 0.04 (11 ± 1)	13 ± 3
YTNRPSPGGHER**A**LVTKLQNSE (K65A)	1.94 ± 0.08 (11 ± 1)	13 ± 3
YTNRPSPGGHERKLVT**A**LQNSE (K69A)	1.79 ± 0.06 (12 ± 2)	13 ± 3
YTNRPSPGGHERKLV**E**KLQNSE (T68E)	2.17 ± 0.06 (10 ± 1)	13 ± 3
YTNRPSPGGHER**A**LV**E**KLQNSE (K65AT68E)	1.76 ± 0.06 (12 ± 1)	13 ± 3
YTNRPSPGGHERKLV**EA**LQNSE (T68EK69A)	1.87 ± 0.08 (11 ± 1)	13 ± 3
YTNRPSPGGHER**A**LV**EA**LQNSE (K65AT68EK69A)	2.4 ± 0.2 (9 ± 2)	13 ± 3
YTNRPSPGGHERKLV**pT**KLQNSE (pT68)	1.89 ± 0.08 (11 ± 1)	13 ± 3

[a] Mutations with respect to the wild-type sequence are indicated in bold. The last peptide has a phospho-threonine at position 68 (indicated with a "pT"). [b] The R_h was determined from the translational diffusion coefficient of dioxane (R_h = 2.12 Å) added to each sample. [c] Calculated from the scale law: $R_h = (0.027 ± 0.01)$ MW$^{(0.50 ± 0.01)}$ [29], where MW is the molecular weight of the peptide. D: translational diffusion coefficient.

2.4. Fluorescence

2.4.1. Steady-State Fluorescence

Fluorescence spectra were collected on a Cary Varian spectrofluorometer (Agilent, Santa Clara, CA, USA) with a Peltier unit. The samples were prepared the day before and left overnight at 278 K; before experiments, samples were left for 1 h at 298 K. A 1-cm-pathlength quartz cell (Hellma, Kruibeke, Belgium) was used. Concentrations of the peptides were 10 μM and that of importins was 4 μM. Samples containing the isolated peptide, the isolated importin, and the mixture of both, at those concentrations, were prepared for each peptide and each importin. Experiments were acquired at pH 7.0 in 50-mM phosphate buffer.

Protein samples were excited at 280 and 295 nm (although the samples of the isolated peptides did not show any fluorescence at the latter value). The other experimental parameters and the buffers used have been described elsewhere [30]. Appropriate blank corrections were made in all spectra.

2.4.2. Thermal Denaturations

Thermal denaturations were performed at 60 K/h with an average time of 1 s for all samples. Thermal scans were collected at 315, 330, and 350 nm after excitation at 280 or 295 nm from 298 to 358 K. The rest of the experimental set-up was the same as described above. Thermal denaturations for both importins were irreversible, as well as that of the complexes with any peptide. The apparent thermal denaturation midpoint was estimated from a two-state equilibrium equation as described [30].

2.5. CD

Far-ultraviolet (UV) CD spectra were collected on a Jasco J810 spectropolarimeter (Jasco, Tokyo, Japan) with a thermostated cell holder and interfaced with a Peltier unit at 298 K. The instrument was periodically calibrated with (+)-10-camphorsulphonic acid. A path length cell of 0.1 cm was used (Hellma, Kruibeke, Belgium). All spectra were corrected by subtracting the corresponding baseline. The concentration of each polypeptide was the same used in the fluorescence experiments. The buffer was the same used in the fluorescence experiments.

2.5.1. Far-Ultraviolet (UV) Spectra

Isothermal wavelength spectra of each sample were acquired with six scans at a scan speed of 50 nm/min, with a response time of 2 s and a bandwidth of 1 nm. The samples were prepared the day before and left overnight at 278 K to allow for equilibration. Before starting the experiments, the samples were further left for 1 h at 298 K.

2.5.2. Thermal Denaturations

The experiments were performed at 60 K/h and a response time of 8 s. Thermal scans were collected by following the changes in ellipticity at 222 nm from 298 to 343 K. The rest of the experimental set-up was the same as reported in the steady-state experiments. Thermal denaturations were not reversible for any of the samples, as shown by: (i) comparison of the spectra before and after the heating and (ii) changes in the voltage of the instrument detector [31]. The apparent thermal denaturation midpoint of the samples was estimated as described [30].

2.6. ITC

The experimental set-up and data processing of ITC experiments has been described previously [32]. Calorimetric titrations, performed in an Auto-iTC200 calorimeter (MicroCal, Malvern-Panalytical, Malvern, UK) consisted of series of 19 2-μL injections, with 150 s time spacing and a 750-rpm stirring speed. Impα3 or ΔImpα3 (at 10–20 μM) was loaded into the calorimetric cell and NLS-NUPR1 peptides in the syringe (150–300 μM); all solutions were prepared in buffer Tris 50 mM, pH 8. The temperature for all the experiments was 298 K. The experiments were analyzed by applying a model considering a single ligand binding site (1:1 stoichiometry) implemented in Origin 7.0 (OriginLab, Northampton, MA, USA). The binding affinity (association constant) and the binding enthalpy were estimated through a least-squares nonlinear regression data analysis, from which the Gibbs energy and the entropic contribution to the binding were calculated using well-known thermodynamic relationships. Since the binding stoichiometry is constrained by the model, the parameter n provides a fraction of the active or binding competent protein. Experiments for each peptide and importin species were performed, at least, in duplicates.

2.7. NMR

The NMR experiments were acquired at 283 K on a Bruker 500 MHz Advance III spectrometer (Bruker GmbH, Karlsruhe, Germany) equipped with a triple-resonance probe and z-pulse field gradients. Temperature of the probe was calibrated with methanol [33]. All experiments were carried out at pH 7.2, 50-mM deuterated Tris buffer (not corrected for isotope effects). The spectra were calibrated with TSP ((trimethylsilyl)-2,2,3,3-tetradeuteropropionic acid) by considering pH-dependent changes of its chemical shifts [33].

2.7.1. 1D-^{1}H-NMR (One-Dimensional Proton NMR) Spectra

In all cases, 128 scans were acquired with 16 K acquisition points and using concentrations of 1.0–1.2 mM. Homonuclear 1D-^{1}H-NMR spectra were processed with Bruker TopSpin 3.1 (Bruker GmbH, Karlsruhe, Germany) after zero-filling and apodization with an exponential window.

2.7.2. Translational NMR Diffusion Ordered Spectroscopy (DOSY)

Concentrations of peptides in all DOSY experiments were 120 μM, and 128 scans, where the gradient strength was varied, were acquired for each curve. Translational self-diffusion measurements were performed with the pulsed gradient spin-echo sequence in the presence of 100% D_2O. Experimental details have been described elsewhere [30]. Briefly, the gradient strength was varied in sixteen linear steps between 2% and 95% of the total power of the gradient coil. The gradient strength was previously calibrated by using the value of the translational diffusion coefficient, D, for the residual proton water line in a sample containing 100% D_2O in a 5-mm tube [34]. In our experiments for each peptide, the duration of the gradient was 2.25 ms, the time between the two pulse gradients in the pulse sequence was set to 200 ms, and the recovery delay between the bipolar gradients was set to

100 µs. The methyl groups with signals between 1.0 and 0.80 ppm were used for integration. Fitting of the exponential curves obtained from experimental data was carried out with KaleidaGraph (Synergy Software, Version 3.5), as described [30]. A final concentration of 1% of dioxane, which was assumed to have a hydrodynamic radius $R_h = 2.12$ Å [34], was added to the solutions of each of the peptides to have a comparison for estimating their sizes.

2.7.3. 2D-^{1}H-NMR Spectra

Two-dimensional spectra in each dimension were acquired in phase-sensitive mode by using the time-proportional phase incrementation technique (TPPI) and a spectral width of 7801.69 Hz [35]; the final concentration was the same used in the 1D-^{1}H-NMR experiments. Standard total correlation spectroscopy (TOCSY) (with a mixing time of 80 ms) [36] and nuclear Overhauser effect spectroscopy (NOESY) experiments (with a mixing time of 250 ms) [37] were performed with a data matrix size of $4K \times 512$. The DIPSI (decoupling in the presence of scalar interactions) spin-lock sequence [38] was used in the TOCSY experiments with 1 s of relaxation time. Typically, 64 scans were acquired per increment in the first dimension, and the residual water signal was removed by using the WATERGATE sequence [39]. NOESY spectra were collected with 96 scans per increment in the first dimension, with the residual water signal removed again by the WATERGATE sequence and 1 s of relaxation time. Data were zero-filled and resolution-enhanced in each dimension, with a square sine-bell window function optimized in each spectrum, baseline-corrected, and processed with Bruker TopSpin 3.1. The ^{1}H resonances were assigned by standard sequential assignment processes [40]. The chemical shift values of H_α protons in random-coil regions were obtained from tabulated data, corrected by neighboring residue effects [41,42] and taking into account the phosphorylation of Thr68 [43,44] for the corresponding peptide.

2.8. Molecular Docking

Molecular docking was performed using AutoDock Vina (Version 1.1.2) [45], largely following a protocol we have previously described for screening NUPR1 sequence fragments [24]. The structure of ΔImpα3 was modeled on the basis of the Protein Data Bank (PDB) entry 5 × 8N [46], which reports the X-ray structure of monomeric Impα1 bound to the NLS of the Epstein-Barr virus EBNA-LP protein. The search volume was centered on the macromolecule and had the size 50 Å × 90 Å × 90 Å, which was sufficient to carry out a blind search on the whole protein surface.

The peptides used in our experiments encompassed residues 53–74 of NUPR1, with a number of rotatable dihedral angles ranging from 85 to 91. Their conformational space was too large to be reasonably treated by molecular docking; therefore, we followed a two-fold approach [47] that consisted in reducing the number of degrees of freedom and using a longer search protocol. The number of rotatable dihedrals was halved by considering the reduced sequence that encompasses residues 63–71 of NUPR1, and therefore, it includes only the core region of the NLS. These shorter peptide sequences were capped with an acetyl and N-methyl group at the two main chain endings, to mimic the fact that they are internal portions of the sequence of the protein, as well as of their full-length parent peptides. An extensive search was performed with very high exhaustiveness, 16 times larger than the recommended default value [48].

3. Results

3.1. The Isolated wt NLS-NUPR1 and Its Mutants Were Monomeric and Disordered in Aqueous Solution

We first determined the conformational propensities of isolated peptides by using CD and NMR. We did not use fluorescence to characterize their conformational features, because the peptides only have a single tyrosine at their N terminus, whose maximum wavelength (~308 nm) does not change under different environments in solutions [49]. The CD spectra of isolated peptides did show an intense minimum at ~200 nm (Figure S1), indicating that they were mainly in a random-coil conformation.

This was further confirmed by 1D-^{1}H-NMR spectra, which showed, for all the peptides, a clustering of the signals of all the amide protons between 8.0 and 8.5 ppm (Figure S2) and grouping of the methyl protons between 0.8 and 1.0 ppm, which is a feature of disordered polypeptide chains [40].

The peptides were monomeric, as concluded from the values of D measured by the DOSYs and the calculation of the estimated R_h from a random-coil polypeptide according to an exponential law [29] (Table 1).

To further confirm the disordered nature of the peptides, we also carried out homonuclear 2D-^{1}H-NMR experiments (Tables S1–S8). For all peptides, NOEs between the H_α protons of Arg56 or Ser58 and the H_δ of the two following residues (Pro57 and Pro59, respectively) were always observed (Figure 1); these findings suggest that the Arg56-Pro57 and Ser58-Pro59 peptide bonds predominantly adopted a trans-conformation in all the peptides (other minor signals were not observed). Two lines of evidence confirmed the disordered nature of the peptides (further pinpointing the findings from far-UV CD (Figure S1) and the 1D-^{1}H-NMR spectra (Figure S2)). First, the sequence-corrected conformational shifts ($\Delta\delta$) of H_α protons [40–44] were within the commonly accepted range for random-coil peptides ($\Delta\delta \leq 0.1$ ppm) (Tables S1–S8). It is interesting to note at this stage that, in the phosphorylated Thr68 of the pT68 peptide, the signals from the H_β protons were downfield shifted when compared to those of the wt peptide (4.58 versus 4.15 ppm, respectively), as well as the chemical shift of the amide proton: 8.62 versus 8.33, respectively (Tables S1 and S3), as it has been reported to occur for phosphorylated threonines [43,44], thus confirming the phosphorylation of this particular threonine and not of the other one in the sequence, Thr54. Second, in any of the peptides, no long- or medium-range NOEs were generally detected but, rather, only strong sequential ones (αN(i,i + 1)) (Figure 1). Only in the pThr68 peptide and in the four phospho-mimics (T68E, K65AT68E, T68EK69A, and K65AT68EK69A peptides), we observed a weak NOE (NN(i,i + 1)) between the amide protons of Val67 and Thr68 (Figure S3). This NOE, although weak when compared with the intensity of sequential αN(i,i + 1) NOEs, is a fingerprint signature of turn-like conformations [40].

Although there are some isolated short peptides that are partially structured (such as the isolated Ribonuclease S peptide [33,40]), our findings by CD and NMR indicate that the isolated NLS-NUPR1 peptides were mainly disordered in aqueous solution when isolated.

3.2. The NLS-NUPR1 Peptides Bound to Both Impα3 and ΔImpα3

In the present work, we measured the affinity of intact NUPR1 for ΔImpα3, obtaining a value for the dissociation constant of 0.4 μM (Figure S4), and we have previously measured the affinity of intact NUPR1 for Impα3, and a value of 1.4 μM has been obtained (shown in Figure S4; for a comparison, [50]). Furthermore, we tried to dissect the affinity of the NLS region of NUPR1 for Impα3 by using a "divide and conquer" approach with the peptides comprising the region. The interaction between full-length NUPR1 and its mutants with Impα3 and ΔImpα3 was the focus of this study, but instead, we employed NLS peptides to elucidate the binding mechanism to Impα3. The reason behind such an approach relies in the fact that we have observed that, very often, mutations at any place of the polypeptide length of NUPR1 result in a poor expression of the corresponding mutant, and mutations in some positions lead to no expression at all [24].

First, we decided to investigate a possible interaction between the NLS NUPR1 peptides and Impα3 in vitro by using fluorescence and CD. As a representative example, we describe our findings for the wt peptide. We observed changes in the fluorescence spectrum of this peptide after excitation at 280 nm (whereas there were no changes at 295 nm); that is, the additional spectrum obtained from the spectra of isolated wt peptide and either Impα3 or ΔImpα3 was different to those of their respective complexes (Figure 2A). These results indicate that tyrosine residues of at least one of the biomolecules (peptides with either Impα3 or ΔImpα3) were mainly involved in the binding. The changes were small for Impα3, and there were no changes for ΔImpα3; furthermore, thermal denaturations followed by fluorescence did not show a variation in the apparent thermal denaturation midpoint for both Impα3 and ΔImpα3 (Figure S5). On the other hand, the comparison of the additional spectrum and that of the complex obtained by far-UV CD did show differences (both for Impα3 and ΔImpα3), indicating that there were changes in the secondary structure of at least one of the macromolecules upon binding (Figure 2B); however, there were no differences in the determined thermal denaturation midpoint for isolated Impα3 (or ΔImpα3) and that of the complex (Figure S5). It is important to note that the far-UV CD region is sensitive to elements of secondary structures (α-helix and β-sheet); however, local structural elements and nonregular structures might also be present, which could be masked by the presence of long disordered regions. The above results indicate that there was binding between the wt peptide and both importins, but the binding did not induce large changes in the structures of both macromolecules.

The situation was slightly different in the case of the Impα3 and ΔImpα3 complexes with the other mutant peptides. As an example, we described our results with the K65A peptide, and the findings for the other peptides were basically similar to those described here. Where the far-UV CD spectra of the addition and that of the complex with both importins also showed small differences (Figure S6A,B), the fluorescence spectra did not have modifications (either by excitation at 280 or 295 nm) (Figure S6C,D). In general, for the mutant peptides, the changes were smaller than for the wt peptides.

The above experiments were sufficient to conclude that the NLS-NUPR1 peptides interacted with Impα3 or ΔImpα3, but we also carried out ITC experiments to measure the binding affinity. The results (Table 2 and Figure 3) indicate that: (i) the highest affinity towards either Impα3 or ΔImpα3 was that observed for the wt peptide, (ii) the affinity for most of the peptides was higher for binding to ΔImpα3 (the only exceptions were the T68EK69A and pT68 peptides), (iii) removal of Lys65 or Lys69 residues decreased the affinity (and the variations in affinity were higher for ΔImpα3 than for Impα3), and (iv) the phosphorylation or mutation to Glu (phospho-mimics) of Thr68 decreased the affinity by almost one order of magnitude when compared to the other mutations for both importin species. Therefore, the ITC findings mirrored the results obtained by fluorescence: there were lesser structural changes (as reported by fluorescence) in the binding of the peptide mutants than for the wt one, and the affinity of the former peptides for importins was lower (Table 2).

Taking together all these findings, we conclude that the isolated region of NUPR1 comprising its NLS was capable of binding to Impα3 and that this binding was strongly modulated by the phosphorylation state of Thr68 and the charges at positions Lys65 and Lys69.

Biomolecules 2020, 10, 1313

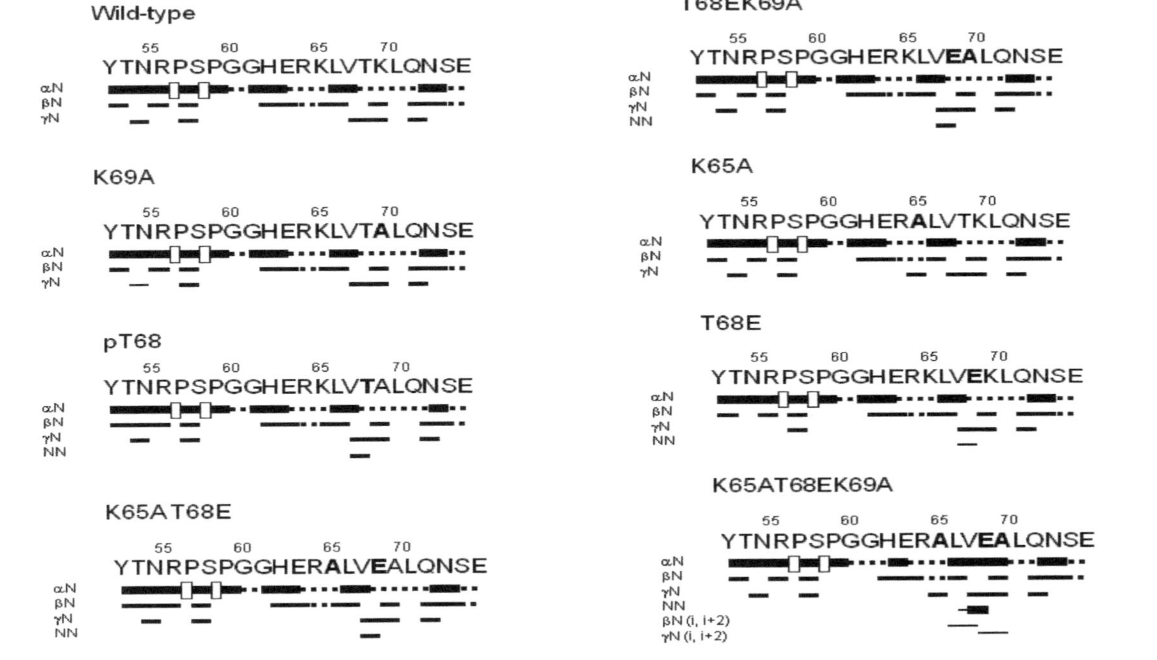

Figure 1. NMR structural characterization of the nuclear localization sequence (NLS) NUPR1 peptides. Nuclear Overhauser effects (NOEs) are classified into strong, medium, or weak, as represented by the height of the bar underneath the sequence; the signal intensity was judged by visual inspection from the nuclear Overhauser effect spectroscopy (NOESY) experiments. The symbols αN, βN, γN, and NN correspond to the sequential contacts (that is, for instance, the NN corresponds to the NN $(i, i+1)$ contacts). The corresponding H_α NOEs with the H_δ of the following proline residues are indicated by an open bar in the row corresponding to the αN contacts. The dotted lines indicate NOE contacts that could not be unambiguously assigned due to signal overlap. The numbering of the residues corresponds to that of the whole sequence of NUPR1.

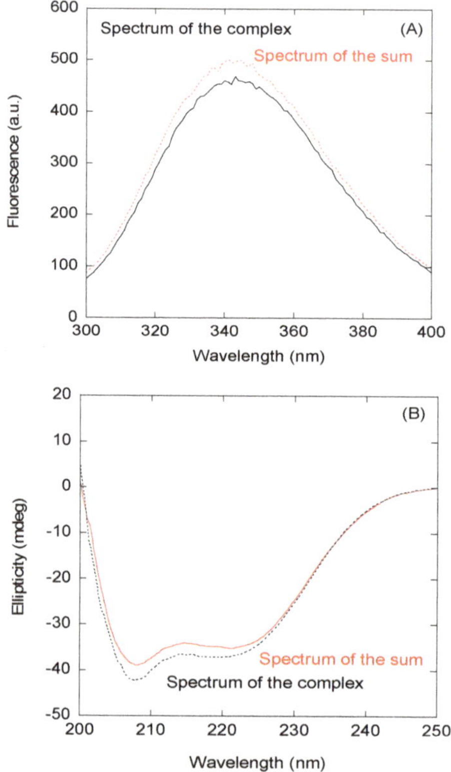

Figure 2. Binding of the wild-type (wt) peptide to importin α3 (Impα3) monitored by spectroscopic techniques: (**A**) Fluorescence spectrum obtained by excitation at 280 nm of the complex between Impα3 and the wt peptide and the addition spectrum obtained by the sum of the spectra of both isolated macromolecules. (**B**) Far-UV CD (ultraviolet circular dichroism) spectrum of the complex between the Impα3 and wt peptides and the additional spectrum obtained by the sum of the spectra of both isolated macromolecules.

3.3. Binding Regions in the Docking of NUPR1 Peptides to Importins

Since we have shown that there was binding between the peptides and both importins, and we have identified the most important residues for attaining such binding, we performed molecular docking to determine details on the location and binding energy of the NUPR1 peptides on the surface of Impα3. When applied to our case, the docking techniques possess three caveats that are worth mentioning explicitly. First, even in the case of our relatively short peptides, the number of degrees of freedom to be considered is too large to be computationally tractable. This number was halved by considering reduced sequences (nine amino acids, corresponding to residues 63–71 of NUPR1), which included all the mutation sites plus at least two more residues at each end. Second, it is impossible with this technique to discriminate differences in the binding between Impα3 and ΔImpα3, and therefore, only the latter protein structure was considered. Third, molecular docking does not take into account the dynamics of a protein-ligand complex, which could also contribute to the binding. Keeping in mind these limitations, the protein surface was blindly explored by considering a volume that included the whole structure and using a high exhaustiveness of search that is equivalent to running multiple (>10) distinct simulations.

Figure 4 summarizes the predictions obtained in our docking calculations. In particular, Figure 4A illustrates the energetically most favorable poses obtained for the wild-type (capped) sequence ERKLVTKLQ mapped on the surface of importin. The best eight poses are reported for clarity and to obtain a more direct comparison with the cluster of the single best pose for each of the eight different peptides (see below, Figure 4D). The results clearly show that the most favorable binding modes cluster into a single location that consists of the major NLS-binding site, located on ARM repeats 2–4. As shown in Figure 4B, the best structure found for our peptide sequence overlaps quite remarkably with that of the NLS of the Epstein-Barr virus EBNA-LP protein, whose structure has been previously determined in crystallography [46]. A number of different amino acids participate in the binding, including some key tryptophan residues (see the details in Figure 4C) that are known to play an important role in the formation of the importin-cargo complex. The binding energy in the docking for the most favorable conformation was −7.2 kcal/mol, indicating a moderate affinity in the low micromolar range. Compared to the experimental values found for the whole wild-type sequence YTNRPSPGGHERALVTKLQNSE (−7.87 and −8.22 kcal/mol for Impα3 and ΔImpα3, respectively; see Table 2), this finding indicates that the reduced docked sequence provides the major contribution to the binding-free energy of the full-length peptide.

Figure 4D shows the best docking poses obtained for the seven mutant sequences compared to the wt one, which is also reported. Again, in this case, all the most favorable binding modes (and, more generally, even the first ten docking poses for each peptide species) clustered in the same location correspond to the major NLS-binding site. This observation suggests that the mutations do not modify essentially the binding location of the peptides but only their affinity towards importin. The calculated binding energies ranged from −5.6 to −6.6 kcal/mol, indicating that any of the explored mutations reduced the binding affinity with respect to the wt sequence, in agreement with our experimental results (Table 2). We observed a poor correlation between the computational and experimental rankings of the mutated peptides in terms of affinity towards the protein, although this could reasonably be explained, because the experimental binding energies are, in most cases, very close to each other (Table 2). This finding did not let us push too far the interpretation of our results in terms of the molecular details that assist the binding. Nevertheless, the contribution of the protein tryptophan residues to the binding still seemed to be, in all cases, an important determinant (even though we did not observe changes in the fluorescence spectra (either by excitation at 280 or 295 nm) when binding for some of the mutant peptide sequences was explored, Figure S6).

To sum up, a number of important conclusions can be drawn from the docking results reported: (i) all the sequences investigated interacted with the same region of importin; (ii) this region matched unambiguously with the major NLS-binding site of the protein; (iii) the ligand with the highest binding affinity corresponded to the wt sequence of NUPR1 (in agreement with the experimental results from ITC; Table 2); (iv) the major contribution to the binding energy of the parent peptides (i.e., those used in this work) was due to such a restricted sequence portion, which includes only nine residues (and this region includes Lys65, Thr68, and Lys69); (v) this essential sequence fragment corresponded to the core region of the predicted NLS of NUPR1; (vi) the binding region roughly mapped around Thr68 (where the residue name and number refers to wild-type, intact NUPR1 numbering), which therefore appears to be a key amino acid; and (vii) the most favorable predicted structure for the NLS region of wild-type NUPR1 essentially overlapped with the conformation of the NLS of a different protein (the Epstein-Barr virus EBNA-LP protein) determined in crystallography.

Table 2. Thermodynamic parameters at 298 K in the binding reaction of NLS NUPR1 peptides to the two importin species [a].

Peptide	Impα3				ΔImpα3			
	K_d (μM)	ΔH (kcal/mol)	$-T\Delta S$ (kcal/mol)	n	K_d (μM)	ΔH (kcal/mol)	$-T\Delta S$ (kcal/mol)	n
wt	1.7	0.8	−8.7	0.9	0.95	−3.7	−4.5	1.0
K65A	3.9	−2.8	−4.6	1.4	2.7	−10.2	2.6	1.4
K69A	11	−10.8	4.0	1.3	7.6	−21.3	14.3	1.4
T68E	22	−11.1	4.7	(1)	12	−17.5	10.8	(1)
K65AT68E	21	−7.8	1.4	(1)	14	−17.9	11.3	(1)
T68EK69A	17	−7.5	1.0	(1)	17	−21.2	14.7	(1)
K65AT68EK69A	27	−16.3	9.1	(1)	24	−28.5	22.2	(1)
pT68	27	−14.8	3.6	(1)	29	−28.2	22.0	(1)

[a] Relative error in K_d (dissociation constant) is 30%, absolute errors in ΔH (enthalpy) and $-T\Delta S$ (entropy) are 0.5 and 0.7 kcal/mol, respectively, and absolute error in n (the stoichiometry) is 0.2. The parenthesis in n values indicate that this parameter had to be fixed in order to get convergence in the fit due to low affinity.

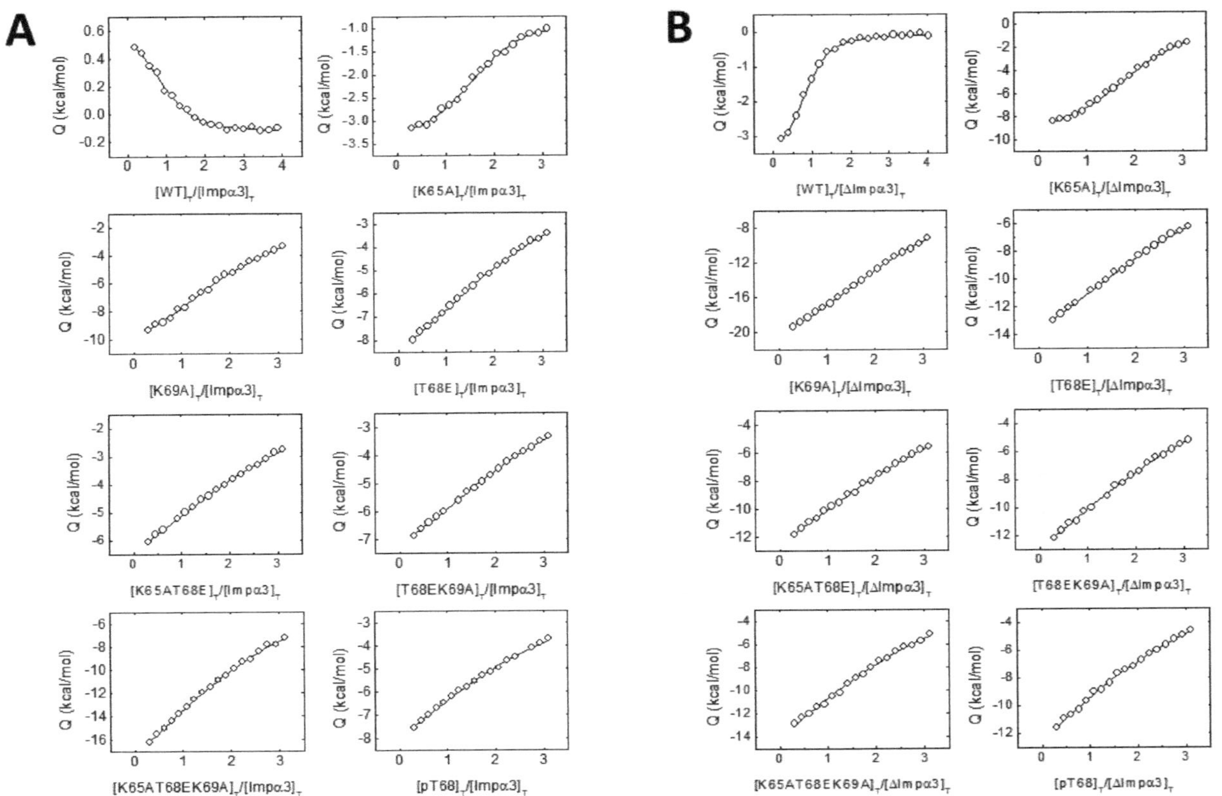

Figure 3. The interaction of the wt and mutant NLS NUPR1 peptides with both importins as measured by isothermal titration calorimetry (ITC). Interaction isotherms (ligand normalized heat effect per injection as a function of the ligand:protein molar ratio) with Impα3 (**A**) and ΔImpα3 (**B**) are shown. Binding parameters were estimated by a nonlinear least-squares regression data analysis of the interaction isotherms applying a single ligand binding site model, implemented in Origin 7.0.

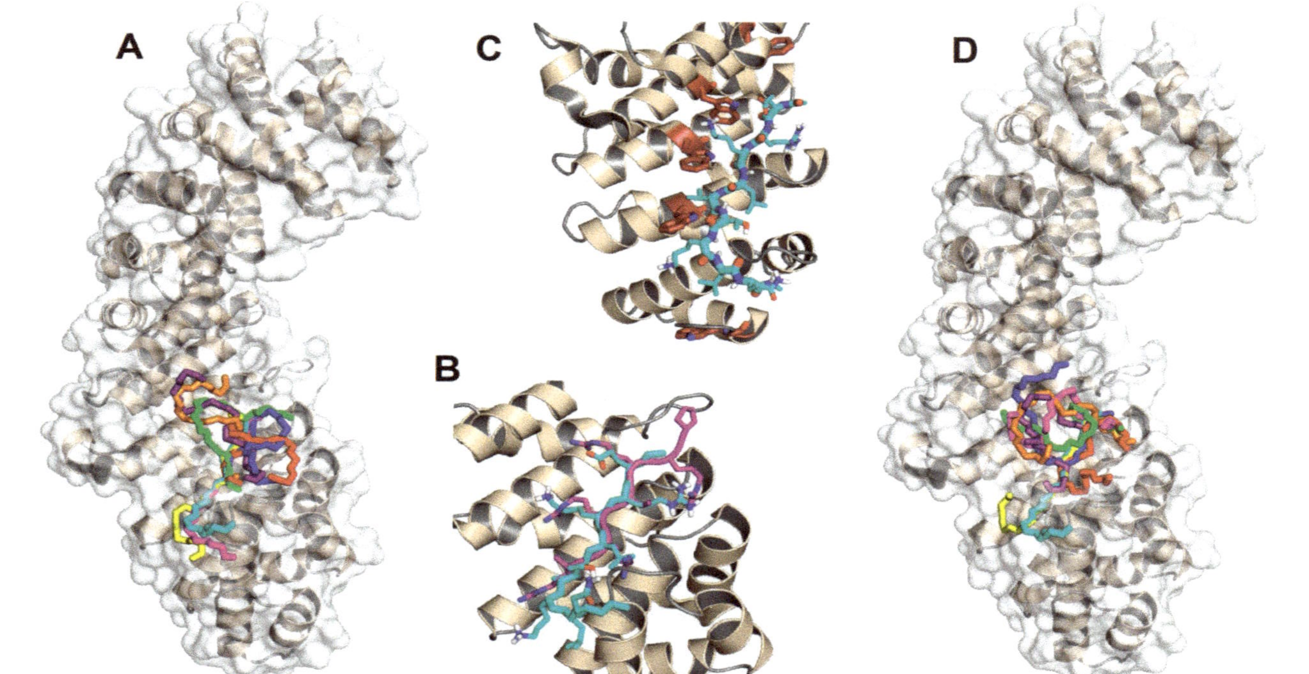

Figure 4. Predicted docking poses for the NLS of NUPR1 on importin. (**A**) Backbone (–N–C$^\alpha$–C– atoms) representation of the best eight docking poses on ΔImportin for the wt sequence ERKLVTKLQ (the N terminus is on the top), which constitutes the core region for the NLS of NUPR1. (**B**) Most favorable binding pose for the same sequence (cyan), compared to the crystallographic conformation [46] of the NLS of the Epstein-Barr virus EBNA-LP protein (purple). For clarity, atoms are shown in standard colors only in the side chains of the two peptides, and the main-chain O and H atoms are omitted; apolar H atoms are not present. (**C**) Trp residues (brown) in the major NLS-binding site of importin play a key role in the binding of the most favorable conformation of the NLS of wild-type NUPR1. The view is slightly rotated with respect to previous representations to evidence the tryptophan side chains. (**D**) Most favorable docking poses for the eight peptide sequences: wild type (cyan), K65A (magenta), K69A (yellow), T68E (blue), K65AT68E (red), T68EK69A (green), K65AT68EK69A (orange), and pT68 (violet). PyMol was used for all displays.

4. Discussion

4.1. Molecular Mechanisms for Impα3 Recognition of NUPR1: The Influence of Lys65 and Lys69

In this work, we tried first to find out whether the theoretically predicted NLS region of NUPR1 was capable of binding in isolation to Impα3. Second, we tried to elucidate, for the first time, the molecular bases behind the binding of an NLS region of an IDP to an importin. Our results indicate that the isolated, wild-type NLS region of NUPR1 interacted with the intact Impα3 and ΔImpα3, with an affinity similar to that for intact NUPR1 (1.4 μM, Figure S4), and within the same range measured for the affinities to natural partners of NUPR1 [22,24,51] and synthetic molecules [23,52]. Furthermore, our results also address the molecular importance of IBB in the binding of cargos to importins.

As it happens for the intact NUPR1 (whose dissociation constants are 1.4 μM for Impα3 and 0.44 μM for ΔImpα3 (Figure S4)), the wt peptide bound to ΔImpα3 with a two-fold larger affinity (0.95 μM) than that for Impα3 (Table 2) (1.7 μM). These findings allow us to draw several conclusions. First, the presence of the IBB region (which contains a large quantity of lysine amino acids) exerts an autoinhibitory effect, and the domain hampers the entrance of the NLS peptide into the major NLS-binding region of Impα3, as it has been suggested in other studies with well-folded proteins [9]. However, this is the first time such a hypothesis is tested in an IDP. Modulation of the assembly complex formation between importins and their cargos has been attributed to the IBB domain [4]; this domain has been found to be involved even in the formation of a homodimeric species between importins [53], with a reduced ability to bind cargos. Second, although the affinities of the wt peptide for both importins were smaller than those for intact NUPR1, many of the interactions implicated in the binding to importin could be ascribed to a region comprised within the wt peptide, as concluded from the similarities among the dissociation constants (0.44 (intact NUPR1) and 0.95 μM (wt peptide) for ΔImpα3 and 1.4 (intact NUPR1) and 1.7 μM (wt peptide) for Impα3). Third, given the similarities among the affinity constants for Impα3 of the wt peptide and NUPR1, the peptide could be used as a lead compound to design an inhibitor of its nuclear translocation.

We have previously shown in vivo that a mutant of NUPR1 at positions Lys65, Lys69, Lys76, and Lys77 is present through the whole cell, whereas the wild-type NUPR1 species is localized exclusively into the nucleus [25]. In this work, we have found that the mutation Lys65Ala decreased two-fold the affinity for Impα3, and the mutation Lys69Ala decreased six-fold the affinity. Thus, the decrease in the affinity was larger with the removal of Lys69, probably indicating that this residue makes more contacts with importin, as pinpointed by our docking models. In fact, we observed in the simulation that both lysine residues were involved in hydrophobic and polar contacts (the latter with their NH_3^+ moieties), with residues of importin α. The removal of the long side chains would disfavor those contacts, thus decreasing the affinity (Table 2). The importance of lysines is key in determining the binding to importins of other well-folded proteins through their disordered NLS regions, as shown by several structural studies [3,8,54,55]. It could be thought that our study does not provide new mechanistic insight into the function of importins, because the results obtained with an IDP pinpoint, for the first time, the importance of positive charges (as it happens in folded proteins) in the binding of their cargos; however, to the best of our knowledge, this is the first reported case where the importance of such residues is addressed in vitro for the NLS of an IDP, and our results acquire more relevance considering recent findings, where it has been suggested that IDPs do not require the presence of importins to be translocated into the nucleus, although demonstrated mostly for acidic proteins [56]. Then, our results indicated that IDPs require the help of importins to be translocated into the nucleus, and it seems that the rules governing such processes are similar to those observed in well-folded proteins.

The same decrease in affinity was observed for the K65A and K69A peptides towards ΔImpα3, but, compared to the wt peptide, the variation was larger than that observed for both mutants with Impα3 (Table 2). Furthermore, as it happens with the intact importin, the decrease in affinity was larger for the K69A peptide. These findings indicate that, although the IBB region maintains its independence within the whole Impα3 in terms of conformation, its removal may either alter the structure of some regions of the ARM repeats involved in the major NLS-binding site (which relies on hydrophobic contacts to anchor the cargo, therefore altering its docking) or, alternatively, IBB removal may change the whole protein dynamics and its stability.

4.2. Molecular Mechanisms for Impα3 Recognition of NUPR1: The Influence of Thr68 and Its Phosphorylation-Triggered Conformational Switch

Apart from the importance of the two lysines of NUPR1 in the binding to importins, we also wanted to address the importance of Thr68. It is well-established that Thr68 is a key residue in the binding of NUPR1 to any partner, either natural or synthetic [22–24,57]; in fact, together with Ala33, it constitutes one of the two hotspot regions of NUPR1. We decided to address such a question by following two approaches: (i) we mutated Thr68 to Glu to have the phospho-mimics, and (ii) we synthesized a peptide with the phosphorylated Thr (pT68 peptide).

Among all the mutants explored, the peptide with the smallest affinity for Impα3 or ΔImpα3 (~30 μM for both species) was the pThr68 peptide (Table 2). Phosphorylation affects the binding probably by inhibiting long-range electrostatic contacts with both importins. Where the affinity of the wt peptide for ΔImpα3 was larger, the changes due to the addition of the phosphate group in the pThr68 peptide were even larger, further pinpointing subtle structural changes in the major NLS-binding region upon removal of the IBB. Around a third of the eukaryotic proteins can be phosphorylated, and the majority of those phosphorylation sites belong to intrinsically disordered regions because of their accessibility to kinases [14]. Phosphorylation is a key regulatory mechanism in translation, transcription, and other processes.

The phospho-mimetic peptide of NLS-NUPR1, the T68E peptide, also showed a smaller affinity for both importins than the wt one (22 μM for Impα3 and 12 μM for ΔImpα3, Table 2), but the decrease was not as large as that in the pT68 peptide (27 μM for Impα3 and 29 μM for ΔImpα3, Table 2), indicating that the phospho-mimics did not cause the same effect as phosphorylation. Phosphorylation at Thr68 replaces the neutral OH (hydroxyl) group with a tetrahedral PO_4^{2-} (phosphoryl group) with two negative charges, which modifies the electrostatic, chemical, and steric properties of the threonine environment. The double-negative charge of the PO_4^{2-} and its large surrounding hydration shell make the situation chemically different from the Glu phospho-mimic, which has a smaller hydration shell and a single negative charge. Differences among the affinities of phospho-mimics and phosphorylated threonines for a well-folded protein have been also observed in the affinities measured in other protein systems [58], as well as in other IDPs [59].

The values of the affinity constants of the K65AT68E and T68EK69A peptides were similar to that of the T68E peptide (Table 2). This finding indicates that: (i) the effect of Thr68Glu in the binding to importins surpassed those caused by substitutions of the single lysines (and then, Thr68 must have a greater importance in the interaction), and (ii) the effect of removing a lysine when the threonine is phospho-mimicked is not additive for the double mutants, probably because the remnant lysine establishes electrostatic interactions with the glutamic residue. However, the accumulation of the three mutations (in the K65AT68EK69A peptide) led to a large decrease of the affinity constant (Table 2), further highlighting the influence of electrostatic effects between the lysines and the phospho-mimics in the bindings with the two importins. Other studies of phosphorylation of threonines in IDPs indicate that the proximity of arginines can stabilize the charge of the phosphoryl moiety and the stabilization of turn-like structures [60]. We suggest that, in the case of NUPR1, lysines, instead of arginines, would play the role of stabilizing the conformation.

Interestingly enough, the peptides containing the phospho-mimic mutation (T68E) or the phosphorylated Thr68 (pT68 peptide) did show an NN(i,i + 1) NOE (Figure 1) between Val67 and phosphorylated Thr68 (or Glu68). We did not observe such a NOE in the intact NUPR1 when we assigned it [22]. It could be thought that the absence of such a NOE in the wt peptide may be due to the fact that the chemical shifts of the amide protons of those residues (Val67 and Thr68) were similar (Table S1), and then, the NOE could not be observed because of its proximity to the spectrum diagonal. However, the chemical shifts of amides of both residues in the K65A peptide were different enough (8.25 and 8.35 ppm for Val67 and Thr68, respectively; Table S6) to allow for its detection, and nevertheless, we did not observe any NOE (Figure S3). Thus, the presence of such a NOE, although it is weak in intensity, indicates that, upon phosphorylation, the two residues populated a turn-like conformation [40]; the presence of this turn is further supported by the observation of $\beta N(i,i + 2)$ and $\gamma N(i,i + 2)$ NOEs and an additional $NN(i,i + 1)$ contact for the K65AT68EK69A peptide involving residues Leu66–Thr68 (Figure 1), due to the large, intrinsic propensity of alanine to populate helix-like conformations [61,62]. As the affinity of the peptides for both importins decreased when Thr68 was phosphorylated or was phospho-mimicked (Table 2), we can conclude that the decrease in the affinity of peptides upon phosphorylation was structurally related to a conformational switch around Thr68, as a consequence of the introduced negative charge, shifting the population at equilibrium from a random-coil conformation to a turn-like one. The decrease in affinity for both importins may be related to the reduction in entropy of the polypeptide chain upon acquisition of the turn-like conformation and a concomitant conformational energetic penalty for the binding. Interestingly enough, two decades ago, we showed by using FTIR (Fourier transform infra-red spectroscopy) and CD that the unspecific phosphorylation of the serines and threonines in NUPR1 led to a higher population of α-helix- and/or turn-like conformations in the intact protein [21]; at the moment, however, we do not have any evidence for the biological importance of the particular phosphorylation of Thr68 in vivo. Nonetheless, we have recently shown that the mutation of Thr68 to Gln hampers the formation of several complexes of NUPR1 with other proteins involved in SUMOylation processes [50]. Our previous result is confirmed in this work by our new findings obtained with Thr68. Phosphorylation, as well as other post-translational modifications, can affect protein conformations: (i) on a local scale—for instance by affecting the population of *cis* proline isomers [63], (ii) determining a change of entropy of the conformational ensemble [64], (iii) modulating the binding to other macromolecules and triggering phase separation [65], (iv) in an allosteric manner, by affecting distant residues from the phosphorylation site [66], and, (iv) causing a conformational change [67,68]. Conformational switching affecting a threonine in several IDPs has been described [59,69,70]. For instance, the phosphorylation of Thr51 in the IDP prostate-associated gene protein increases the population of transient turn-like populations [70]; the difference with our results is that the turn-like structures in NUPR1 were stabilized in a much shorter polypeptide region, although we cannot rule out that phosphorylation at other sites of NUPR1 could help in stabilizing this conformation. On the other hand, the p27 protein, which modulates the mammalian cell cycle by the inhibition of cyclin-dependent kinases, contains some disordered regions, and the phosphorylation of residue Thr157 in breast cancer cells prevents its interaction with the nuclear import machinery, leading to the accumulation of this protein in the cytoplasm, whereas it is normally found in the nucleus [69]; however, no indication on the particular structure acquired upon phosphorylation at Thr157 has been provided. Finally, it is important to note that recent theoretical molecular dynamic simulations have shown that the binding of importin α to heterochromatin protein 1 α is modulated by phosphorylation at residues in its importin-binding region [71].

Thr68 is, together with the polypeptide patch around Ala33, the hotspot region of NUPR1, involved in binding to its natural partners [21,22,24,51] and to other synthetic molecules and macromolecules [23,52,57]. We have previously observed that the mutation of Thr68 to glutamine hampers the binding to those other molecules [24,51]. Such a mutation will probably cause a shift of the ensemble population from a random-coil towards turn-like conformations, and it is the adoption of such a local fold that hampers bindings to those other natural partners or synthetic molecules. Moreover, as the affinity of NUPR1 to its partners is basically the same in all cases described to date [21,22,24,51,52,57], its binding features can also be modulated by phosphorylation at Thr68 at least partially, since the region around Ala33 is also involved in the binding. In addition, since this threonine is also associated with the binding of drugs strongly effective against pancreatic cancer in mice [23,32], we hypothesize that the molecular effects of such drugs could be the induction of a stable fold (turn-like) by this polypeptide region, besides competitive steric hindrance, preventing binding to other natural partners of NUPR1, and hampering the protein cascades where it is involved.

5. Conclusions

We have described the interaction between the NLS region of NUPR1, a nuclear intrinsically disordered protein involved in cancer, and Impα3 by using a series of peptides comprising that polypeptide patch. Binding to Impα3 is modulated by the charges of Lys64 and Lys69 but, most importantly, by phosphorylation at Thr68, which constitutes an entropy-driven conformational switch, shifting the population of the dynamic ensemble towards a turn-like conformation. As Thr68 is also a hotspot for NUPR1 interactions, these results open the venue to modulating the binding to its partners by targeting this residue. Furthermore, it also suggests a possible mechanism for the action of drugs targeting NUPR1, which also bind through Thr68.

Supplementary Materials: The following are available online at http://www.mdpi.com/2218-273X/10/9/1313/s1, Figure S1: Structural features of NLS-NUPR1 peptides as monitored by far-UV CD, Figure S2: Structural features of NLS-NUPR1 peptides as monitored by 1D-^{1}H-NMR, Figure S3: The amide region of 2D-^{1}H-NOESY spectra of NLS-NUPR1 peptides, Figure S4: Interaction between (left) Impα3 and (right) ΔImp α 3 with full-length NUPR1 as observed by ITC, Figure S5: Thermal denaturation of the complexes followed by spectroscopic techniques, Figure S6: Interaction between Imp α3 and ΔImp α3 with K65A peptide measured by different spectroscopic techniques, Table S1: Chemical shifts (δ, ppm from TSP) of wt peptide in aqueous solution (pH 7.2, 283K), Table S2: Chemical shifts (δ, ppm from TSP) of K69A peptide in aqueous solution (pH 7.2, 283 K), Table S3: Chemical shifts (δ, ppm from TSP) of pT68 peptide in aqueous solution (pH 7.2, 283 K), Table S4: Chemical shifts (δ, ppm from TSP) of T68EK69A peptide in aqueous solution (pH 7.2, 283 K), Table S5: Chemical shifts (δ, ppm from TSP) of K65AT68E peptide in aqueous solution (pH 7.2, 283 K), Table S6: Chemical shifts (δ, ppm from TSP) of K65A peptide in aqueous solution (pH 7.2, 283 K), Table S7: Chemical shifts (δ, ppm from TSP) of T68E peptide in aqueous solution (pH 7.2, 283 K), Table S8: Chemical shifts (δ, ppm from TSP) of K65AT68EK69A peptide in aqueous solution (pH 7.2, 283 K).

Author Contributions: Conceptualization, J.L.N., B.R., A.V.-C., O.A., and J.L.I.; methodology, J.L.N., B.R., A.V.-C., O.A., and J.L.I.; investigation, J.L.N., B.R., A.V.-C., M.P.-S., and A.J.-A.; data analysis, J.L.N., B.R., A.V.-C., and A.J.-A.; writing—original draft preparation, J.L.N., B.R., and A.V.-C.; writing—review and editing, J.L.N., B.R., J.L.I., M.P.-S., A.J.-A., O.A., and A.V.-C.; and funding acquisition, J.L.N., A.V.-C., O.A., and J.L.I. All authors have read and agreed to the published version of the manuscript.

Funding: This research was funded by the Spanish Ministry of Economy and Competitiveness and European ERDF Funds (MCIU/AEI/FEDER, EU) (RTI2018-097991-B-I00 to J.L.N. and BFU2016-78232-P to A.V.C.); La Ligue Contre le Cancer, INCa, Canceropole PACA, and INSERM to J.L.I.; Miguel Servet Program from Instituto de Salud Carlos III (CPII13/00017 to OA); Fondo de Investigaciones Sanitarias from Instituto de Salud Carlos III and European Union (ERDF/ESF, "Investing in your future") (PI15/00663 and PI18/00349 to O.A.); Diputación General de Aragón (Protein Targets and Bioactive Compounds Group E45_17R to A.V.C. and Digestive Pathology Group B25_17R to O.A.); and the Centro de Investigación Biomédica en Red en Enfermedades Hepáticas y Digestivas (CIBERehd). The NMR equipment used in this work was funded by the Generalitat Valenciana and cofinanced with ERDF funds (OP ERDF of Comunitat Valenciana 2014-2020).

Acknowledgments: We thank J. K. Forwood (Charles Sturt University, Waga Waga, Autralia) for the kind gift of the ΔImpα3 vector. B.R. acknowledges the kind hospitality and use of computational resources in the European Magnetic Resonance Center (CERM), Sesto Fiorentino (Florence), Italy. We thank the three anonymous reviewers for their helpful comments and suggestions.

References

1. Stewart, M. Molecular mechanism of the nuclear protein import cycle. *Nat. Rev. Mol. Cell Boil.* **2007**, *8*, 195–208. [CrossRef]

2. Bednenko, J.; Cingolari, G.; Gerace, L. Nucleo-cytoplasmic transport navigating the channel. *Traffic* **2003**, *4*, 127–135. [CrossRef] [PubMed]

3. Cingolani, G.; Bednenko, J.; Gillespie, M.T.; Gerace, L. Molecular basis for the recognition of a non-classical nuclear localization signal by importin beta. *Mol. Cell* **2002**, *10*, 1345–1353. [CrossRef]

4. Goldfarb, D.S.; Corbett, A.H.; Mason, D.A.; Harreman, M.T.; Adam, S.A. Importin α: A multipurpose nuclear-transport receptor. *Trends Cell Boil.* **2004**, *14*, 505–514. [CrossRef] [PubMed]

5. Pumroy, R.A.; Cingolani, G. Diversification of importin-α isoforms in cellular trafficking and disease states. *Biochem. J.* **2015**, *466*, 13–28. [CrossRef]

6. Mason, D.A.; Stage, D.E.; Goldfarb, D. Evolution of the metazoan-specific importin α gene family. *J. Mol. Evol.* **2009**, *68*, 351–365. [CrossRef]

7. Miyamoto, Y.; Loveland, K.L.; Yoneda, Y. Nuclear importin α and its physiological importance. *Commun. Integr. Boil.* **2012**, *5*, 220–222. [CrossRef]

8. Smith, K.M.; Tsimbalyuk, S.; Edwards, M.R.; Cross, E.M.; Batra, J.; Da Costa, T.P.S.; Aragão, D.; Basler, C.; Forwood, J. Structural basis for importin alpha 3 specificity of W proteins in Hendra and Nipah viruses. *Nat. Commun.* **2018**, *9*, 3703. [CrossRef]

9. Kobe, B. Autoinhibition by an internal nuclear localization signal revealed by the crystal structure of mammalian importin α. *Nat. Struct. Biol.* **1999**, *6*, 388–397. [CrossRef]

10. Berlow, R.B.; Dyson, H.J.; Wright, P.E. Expanding the Paradigm: Intrinsically Disordered Proteins and Allosteric Regulation. *J. Mol. Boil.* **2018**, *430*, 2309–2320. [CrossRef]

11. Xie, H.M.; Vucetic, S.; Iakoucheva, L.M.; Oldfield, C.J.; Dunker, A.K.; Uversky, V.N.; Obradovic†, Z. Functional Anthology of Intrinsic Disorder. 1. Biological Processes and Functions of Proteins with Long Disordered Regions. *J. Proteome Res.* **2007**, *6*, 1882–1898. [CrossRef] [PubMed]

12. Babu, M.M.; Van Der Lee, R.; De Groot, N.S.; Gsponer, J. Intrinsically disordered proteins: Regulation and disease. *Curr. Opin. Struct. Boil.* **2011**, *21*, 432–440. [CrossRef] [PubMed]

13. Gsponer, J.; Futschik, M.E.; Teichmann, S.A.; Babu, M.M. Tight Regulation of Unstructured Proteins: From Transcript Synthesis to Protein Degradation. *Science* **2008**, *322*, 1365–1368. [CrossRef]

14. Bah, A.; Forman-Kay, J.D. Modulation of Intrinsically Disordered Protein Function by Post-translational Modifications. *J. Boil. Chem.* **2016**, *291*, 6696–6705. [CrossRef]

15. Launay, H.; Receveur-Bréchot, V.; Carrière, F.; Gontero, B. Orchestration of algal metabolism by protein disorder. *Arch. Biochem. Biophys.* **2019**, *672*, 108070. [CrossRef]

16. Mallo, G.V.; Fiedler, F.; Calvo, E.L.; Ortiz, E.M.; Vasseur, S.; Keim, V.; Morisset, J.; Iovanna, J.L. Cloning and Expression of the Rat p8 cDNA, a New Gene Activated in Pancreas during the Acute Phase of Pancreatitis, Pancreatic Development, and Regeneration, and Which Promotes Cellular Growth. *J. Boil. Chem.* **1997**, *272*, 32360–32369. [CrossRef]

17. Chowdhury, U.R.; Samant, R.S.; Fodstad, O.; Shevde, L.A. Emerging role of nuclear protein 1 (NUPR1) in cancer biology. *Cancer Metastasis Rev.* **2009**, *28*, 225–232. [CrossRef] [PubMed]

18. Goruppi, S.; Iovanna, J.L. Stress-inducible Protein p8 Is Involved in Several Physiological and Pathological Processes. *J. Boil. Chem.* **2009**, *285*, 1577–1581. [CrossRef]

19. Cano, C.; Hamidi, T.; Sandi, M.J.; Iovanna, J.L. Nupr1: The Swiss-knife of cancer. *J. Cell. Physiol.* **2010**, *226*, 1439–1443. [CrossRef] [PubMed]

20. Malicet, C.; Giroux, V.; Vasseur, S.; Dagorn, J.C.; Neira, J.L.; Iovanna, J.L. Regulation of apoptosis by the p8/prothymosin alpha complex. *Proc. Natl. Acad. Sci. USA* **2006**, *103*, 2671–2676. [CrossRef]

21. Encinar, J.A.; Mallo, G.V.; Mizyrycki, C.; Giono, L.E.; González-Ros, J.M.; Rico, M.; Cánepa, E.T.; Moreno, S.; Neira, J.L.; Iovanna, J.L. Human p8 is a HMG-I/Y-like protein with DNA binding activity enhanced by phosphorylation. *J. Boil. Chem.* **2000**, *276*, 2742–2751. [CrossRef]

22. Aguado-Llera, D.; Hamidi, T.; Doménech, R.; Pantoja-Uceda, D.; Gironella, M.; Santoro, J.; Velázquez-Campoy, A.; Neira, J.L.; Iovanna, J.L. Deciphering the binding between Nupr1 and MSL1 and Their DNA-Repairing Activity. *PLoS ONE* **2013**, *8*, e78101. [CrossRef] [PubMed]

23. Neira, J.L.; Bintz, J.; Arruebo, M.; Rizzuti, B.; Bonacci, T.; Vega, S.; Lanas, A.; Velázquez-Campoy, A.; Iovanna, J.L.; Abián, O. Identification of a Drug Targeting an intrinsically disordered protein involved in pancreatic adenocarcinoma. *Sci. Rep.* **2017**, *7*, 39732. [CrossRef] [PubMed]

24. Santofimia-Castaño, P.; Rizzuti, B.; Pey, A.L.; Soubeyran, P.; Vidal, M.; Urrutia, R.; Iovanna, J.L.; Neira, J.L. Intrinsically disordered chromatin protein NUPR1 binds to the C-terminal region of Polycomb RING1B. *Proc. Natl. Acad. Sci. USA* **2017**, *114*, E6332–E6341. [CrossRef] [PubMed]

25. Valacco, M.P.; Varone, C.L.; Malicet, C.; Cánepa, E.T.; Iovanna, J.L.; Moreno, S. Cell growth-dependent subcellular localization of p8. *J. Cell. Biochem.* **2006**, *97*, 1066–1079. [CrossRef]

26. Jäkel, S.; Mingot, J.-M.; Schwarzmaier, P.; Hartmann, E.; Görlich, D. Importins fulfil a dual function as nuclear import receptors and cytoplasmic chaperones for exposed basic domains. *EMBO J.* **2002**, *21*, 377–386. [CrossRef]

27. Marvaldi, L.; Panayotis, N.; Alber, S.; Dagan, S.Y.; Okladnikov, N.; Koppel, I.; Di Pizio, A.; Song, D.-A.; Tzur, Y.; Terenzio, M.; et al. Importin α3 regulates chronic pain pathways in peripheral sensory neurons. *Science* **2020**, *369*, 842–846. [CrossRef]

28. Gill, S.C.; Von Hippel, P.H. Calculation of protein extinction coefficients from amino acid sequence data. *Anal. Biochem.* **1989**, *182*, 319–326. [CrossRef]

29. Danielsson, J.; Jarvet, J.; Damberg, P.; Gräslund, A. Translational diffusion measured by PFG-NMR on full length and fragments of the Alzheimer Aβ(1-40) peptide. Determination of hydrodynamic radii of random coil peptides of varying length. *Magn. Reson. Chem.* **2002**, *40*, S89–S97. [CrossRef]

30. Neira, J.L.; Hornos, F.; Bacarizo, J.; Camara-Artigas, A.; Gómez, J. The monomeric species of the regulatory domain of Tyrosine Hydroxylase has a low conformational stability. *Biochemistry* **2016**, *55*, 3418–3431. [CrossRef]

31. Benjwal, S.; Verma, S.; Röhm, K.; Gursky, O. Monitoring protein aggregation during thermal unfolding in circular dichroism experiments. *Protein Sci.* **2006**, *15*, 635–639. [CrossRef] [PubMed]

32. Santofimia-Castaño, P.; Xia, Y.; Lan, W.; Zhou, Z.; Huang, C.; Peng, L.; Soubeyran, P.; Velázquez-Campoy, A.; Abian, O.; Rizzuti, B.; et al. Ligand-based design identifies a potent NUPR1 inhibitor exerting anticancer activity via necroptosis. *J. Clin. Investig.* **2019**, *129*, 2500–2513. [CrossRef] [PubMed]

33. Cavanagh, J.; Fairbrother, W.J.; Palmer, A.G.; Skelton, N.J. *Protein NMR Spectroscopy: Principles and Practice*; Academic Press: New York, NY, USA, 1996.

34. Wilkins, D.K.; Grimshaw, S.B.; Receveur, V.; Dobson, C.M.; Jones, J.A.; Smith, L.J. Hydrodynamic radii of native and denatured proteins measured by pulse field gradient NMR techniques. *Biochemistry* **1999**, *38*, 16424–16431. [CrossRef] [PubMed]

35. Marion, D.; Wüthrich, K. Application of phase sensitive two-dimensional correlated spectroscopy (COSY) for measurements of 1H-1H spin-spin coupling constants in proteins. *Biochem. Biophys. Res. Commun.* **1983**, *113*, 967–974. [CrossRef]

36. Bax, A.; Davis, D.G. MLEV-17-based two-dimensional homonuclear magnetization transfer spectroscopy. *J. Magn. Reson.* **1985**, *65*, 355–360. [CrossRef]

37. Kumar, A.; Ernst, R.; Wüthrich, K. A two-dimensional nuclear Overhauser enhancement (2D NOE) experiment for the elucidation of complete proton-proton cross-relaxation networks in biological macromolecules. *Biochem. Biophys. Res. Commun.* **1980**, *95*, 1–6. [CrossRef]

38. Cavanagh, J.; Rance, M. Suppression of cross-relaxation effects in TOCSY spectra via a modified DIPSI-2 mixing sequence. *J. Magn. Reson.* **1992**, *96*, 670–678. [CrossRef]

39. Piotto, M.; Saudek, V.; Sklenář, V. Gradient-tailored excitation for single-quantum NMR spectroscopy of aqueous solutions. *J. Biomol. NMR* **1992**, *2*, 661–665. [CrossRef]

40. Wüthrich, K. *NMR of Proteins and Nucleic Acids*; John Wiley and Sons: New York, NY, USA, 1986.

41. Kjaergaard, M.; Brander, S.; Poulsen, F.M. Random coil chemical shift for intrinsically disordered proteins: Effects of temperature and pH. *J. Biomol. NMR* **2011**, *49*, 139–149. [CrossRef]

42. Kjaergaard, M.; Poulsen, F.M. Sequence correction of random coil chemical shifts: Correlation between neighbor correction factors and changes in the Ramachandran distribution. *J. Biomol. NMR* **2011**, *50*, 157–165. [CrossRef]

43. Bienkiewicz, E.A.; Lumb, K.J. Random-coil chemical shifts of phosphorylated amino acids. *J. Biomol. NMR* **1999**, *15*, 203–206. [CrossRef] [PubMed]

44. Hendus-Altenburger, R.; Fernandes, C.B.; Bugge, K.; Kunze, M.B.A.; Boomsma, W.; Kragelund, B.B. Random coil chemical shifts for serine, threonine and tyrosine phosphorylation over a broad pH range. *J. Biomol. NMR* **2019**, *73*, 713–725. [CrossRef] [PubMed]

45. Trott, O.; Olson, A.J. AutoDock Vina: Improving the speed and accuracy of docking with a new scoring function, efficient optimization, and multithreading. *J. Comput. Chem.* **2009**, *31*, 455–461. [CrossRef] [PubMed]

46. Nakada, R.; Matsuura, Y. Crystal structure of importin–α bound to the nuclear localization signal of Epstein–Barr virus EBNA–LP protein. *Protein Sci.* **2017**, *26*, 1231–1235. [CrossRef] [PubMed]

47. Forli, S.; Huey, R.; Pique, M.E.; Sanner, M.F.; Goodsell, D.S.; Olson, A.J. Computational protein–ligand docking and virtual drug screening with the AutoDock suite. *Nat. Protoc.* **2016**, *11*, 905–919. [CrossRef]

48. Grande, F.; Rizzuti, B.; Occhiuzzi, M.A.; Ioele, G.; Casacchia, T.; Gelmini, F.; Guzzi, R.; Garofalo, A.; Statti, G. Identification by molecular docking of homoisoflavones from *Leopoldia comosa* as ligands of estrogen receptors. *Molecules* **2018**, *23*, 894. [CrossRef]

49. Grimsley, G.R.; Huyghues-Despointes, B.M.; Pace, C.N.; Scholtz, J.M. Measuring the Conformational Stability of a Protein by NMR. *Cold Spring Harb. Protoc.* **2006**, *2006*, 253–259. [CrossRef]

50. Lan, W.; Santofimia-Castaño, P.; Swayden, M.; Xia, Y.; Zhou, Z.; Audebert, S.; Camoin, L.; Huang, C.; Peng, L.; Jiménez-Alesanco, A.; et al. ZZW-115-dependent inhibition of NUPR1 nuclear translocation sensitizes cancer cells to genotoxic agents. *JCI Insight* **2020**, 138117. [CrossRef]

51. Neira, J.L.; López, M.B.; Sevilla, P.; Rizzuti, B.; Camara-Artigas, A.; Vidal, M.; Iovanna, J.L. The chromatin nuclear protein NUPR1L is intrinsically disordered and binds to the same proteins as its paralogue. *Biochem. J.* **2018**, *475*, 2271–2291. [CrossRef]

52. Santofimia-Castaño, P.; Rizzuti, B.; Abian, O.; Velázquez-Campoy, A.; Iovanna, J.L.; Neira, J.L. Amphipathic helical peptides hamper protein-protein interactions of the intrinsically disordered chromatin nuclear protein 1 (NUPR1). *Biochim. Biophys. Acta Gen. Subj.* **2018**, *1862*, 1283–1295. [CrossRef]

53. Miyatake, H.; Sanjoh, A.; Unzai, S.; Matsuda, G.; Tatsumi, Y.; Miyamoto, Y.; Dohmae, N.; Aida, Y. Crystal structure of human Importin-α1 (Rch1), revealing a potential autoinhibition mode involving homodimerization. *PLoS ONE* **2015**, *10*, e0115995. [CrossRef] [PubMed]

54. Sankhala, R.S.; Lokareddy, R.K.; Begum, S.; Pumroy, R.A.; Gillilan, R.E.; Cingolani, G. Three-dimensional context rather than NLS amino acid sequence determines importin α subtype specificity for RCC1. *Nat. Commun.* **2017**, *8*, 979. [CrossRef] [PubMed]

55. Pumroy, R.A.; Ke, S.; Hart, D.J.; Zacharie, U.; Cingolani, G. Molecular determinants for nuclear import if influenza A PB2 by importin alpha isoforms 3 and 7. *Structure* **2015**, *23*, 374–384. [CrossRef] [PubMed]

56. Junod, S.L.; Kelich, J.M.; Ma, J.; Yang, W. Nucleocytoplasmic transport of intrinsically disordered proteins studied by high–speed super–resolution microscopy. *Protein Sci.* **2020**, *29*, 1459–1472. [CrossRef]

57. Neira, J.L.; Correa, J.; Rizzuti, B.; Santofimia-Castaño, P.; Abián, O.; Velázquez-Campoy, A.; Fernandez-Megia, E.; Iovanna, J.L. Dendrimers as competitors of protein–protein interactions of the intrinsically disordered nuclear chromatin protein NUPR1. *Biomacromolecules* **2019**, *20*, 2567–2576. [CrossRef]

58. Yadahalli, S.; Neira, J.L.; Johnson, C.M.; Tan, Y.S.; Rowling, P.J.E.; Chattopadhyay, A.; Verma, C.; Itzhaki, L.S. Kinetic and thermodynamic effects of phosphorylation on p53 binding to MDM2. *Sci. Rep.* **2019**, *9*, 693. [CrossRef]

59. Bah, A.; Vernon, R.M.; Siddiqui, Z.; Krzeminski, M.; Muhandiram, R.; Zhao, C.W.; Sonenberg, N.; Kay, L.E.; Forman-Kay, J.D. Folding of an intrinsically disordered protein by phosphorylation as a regulatory switch. *Nature* **2014**, *519*, 106–109. [CrossRef]

60. Gandhi, N.S.; Landrieu, I.; Byrne, C.; Kukić, P.; Amniai, L.; Cantrelle, F.-X.; Wieruszeski, J.-M.; Mancera, R.L.; Jacquot, Y.; Lippens, G. A Phosphorylation-induced turn defines the Alzheimer's disease AT8 antibody epitope on the Tau protein. *Angew. Chem. Int. Ed.* **2015**, *54*, 6819–6823. [CrossRef]

61. Beck, D.A.C.; Alonso, D.O.V.; Inoyama, D.; Daggett, V. The intrinsic conformational propensities of the 20 naturally occurring amino acids and reflection of these propensities in proteins. *Proc. Natl. Acad. Sci. USA* **2008**, *105*, 12259–12264. [CrossRef]

62. Muñoz, V.; Serrano, L. Intrinsic secondary structure propensities of the amino acids, using statistical φ-ψ matrices: Comparison with experimental scales. *Proteins: Struct. Funct. Bioinform.* **1994**, *20*, 301–311. [CrossRef]

63. Gibbs, E.B.; Lu, F.; Portz, B.; Fisher, M.J.; Medellin, B.P.; Laremore, T.N.; Zhang, Y.S.; Gimour, D.S.; Showalter, S.A. Phosphorylation induces sequence-specific conformational switches in the RNA polymerase II C-terminal domain. *Nat. Commun.* **2017**, *8*, 15233. [CrossRef] [PubMed]

64. Xiang, S.; Gapsys, V.; Kim, H.-Y.; Bessonov, S.; Hsiao, H.-H.; Möhlmann, S.; Klaukien, V.; Ficner, R.; Becker, S.; Urlaub, H.; et al. Phosphorylation drives a dynamic switch in Serine/Arginine-rich proteins. *Structure* **2013**, *21*, 2162–2174. [CrossRef] [PubMed]

65. Turner, A.L.; Watson, M.; Wilkins, O.G.; Cato, L.; Travers, A.; Thomas, J.O.; Stott, K. Highly disordered histone H1–DNA model complexes and their condensates. *Proc. Natl. Acad. Sci. USA* **2018**, *115*, 11964–11969. [CrossRef] [PubMed]

66. Banavali, N.K.; Roux, B. Anatomy of a structural pathway for activation of the catalytic domain of Src kinase Hck. *Proteins: Struct. Funct. Bioinform.* **2007**, *67*, 1096–1112. [CrossRef]
67. Espinoza-Fonseca, L.M.; Kast, D.; Thomas, D.D. Molecular dynamics simulations reveal a disorder-to-order transition on phosphorylation of smooth muscle myosin. *Biophys. J.* **2007**, *93*, 2083–2090. [CrossRef]
68. Hendus-Altenburger, R.; Lambrughi, M.; Terkelsen, T.; Pedersen, S.F.; Papaleo, E.; Lindorff-Larsen, K.; Kragelund, B.B. A phosphorylation-motif for tuneable helix stabilisation in intrinsically disordered proteins–Lessons from the sodium proton exchanger 1 (NHE1). *Cell. Signal.* **2017**, *37*, 40–51. [CrossRef]
69. Chu, I.M.; Hengst, L.; Slingerland, J.M. The Cdk inhibitor p27 in human cancer: Prognostic potential and relevance to anticancer therapy. *Nat. Rev. Cancer* **2008**, *8*, 253–267. [CrossRef]
70. He, Y.; Chen, Y.; Mooney, S.M.; Rajagopalan, K.; Bhargava, A.; Sacho, E.; Weninger, K.; Bryan, P.N.; Kulkarni, P.; Orban, J. Phosphorylation-induced conformational ensemble switching in an intrinsically disordered cancer/testis antigen*. *J. Boil. Chem.* **2015**, *290*, 25090–25102. [CrossRef]
71. Zimmermann, M.T.; Williams, M.M.; Klee, E.W.; Lomberk, G.L.; Urrutia, R.A. Modeling post–translational modifications and cancer–associated mutations that impact the heterochromatin protein 1α–importin α heterodimers. *Proteins: Struct. Funct. Bioinform.* **2019**, *87*, 904–916. [CrossRef]

Article

Molecular Context-Dependent Effects Induced by Rett Syndrome-Associated Mutations in MeCP2

David Ortega-Alarcon [1,†], Rafael Claveria-Gimeno [1,2,3,†,‡], Sonia Vega [1], Olga C. Jorge-Torres [4], Manel Esteller [4,5,6,7], Olga Abian [1,2,3,8,9,*] and Adrian Velazquez-Campoy [1,3,8,9,10,*]

[1] Institute of Biocomputation and Physics of Complex Systems (BIFI), Joint Units IQFR-CSIC-BIFI, and GBsC-CSIC-BIFI, Universidad de Zaragoza, 50018 Zaragoza, Spain; dortega@bifi.es (D.O.-A.); rafacg@bifi.es (R.C.-G.); svega@bifi.es (S.V.)

[2] Instituto Aragonés de Ciencias de la Salud (IACS), 50009 Zaragoza, Spain

[3] Instituto de Investigación Sanitaria Aragón (IIS Aragón), 50009 Zaragoza, Spain

[4] Josep Carreras Leukaemia Research Institute (IJC), 08916 Badalona, Spain; ojorge@carrerasresearch.org (O.C.J.-T.); mesteller@carrerasresearch.org (M.E.)

[5] Centro de Investigacion Biomedica en Red Cancer (CIBERONC), 28029 Madrid, Spain

[6] Institucio Catalana de Recerca i Estudis Avançats (ICREA), 08010 Barcelona, Spain

[7] Physiological Sciences Department, School of Medicine and Health Sciences, University of Barcelona (UB), l'Hospitalet de Llobregat, 08907 Barcelona, Spain

[8] Centro de Investigación Biomédica en Red en el Área Temática de Enfermedades Hepáticas y Digestivas (CIBERehd), 28029 Madrid, Spain

[9] Departamento de Bioquímica y Biología Molecular y Celular, Universidad de Zaragoza, 50009 Zaragoza, Spain

[10] Fundación ARAID, Gobierno de Aragón, 50009 Zaragoza, Spain

* Correspondence: oabifra@unizar.es (O.A.); adrianvc@unizar.es (A.V.-C.); Tel.: +34-876-555-417 (O.A.); +34-976-762-996 (A.V.-C.)

† Contributed equally.

‡ Current address: Certest Biotec S.L., 50840 Zaragoza, Spain; rclaveria@certest.es.

Received: 11 October 2020; Accepted: 8 November 2020; Published: 10 November 2020

Abstract: Methyl-CpG binding protein 2 (MeCP2) is a transcriptional regulator and a chromatin-binding protein involved in neuronal development and maturation. Loss-of-function mutations in MeCP2 result in Rett syndrome (RTT), a neurodevelopmental disorder that is the main cause of mental retardation in females. MeCP2 is an intrinsically disordered protein (IDP) constituted by six domains. Two domains are the main responsible elements for DNA binding (methyl-CpG binding domain, MBD) and recruitment of gene transcription/silencing machinery (transcription repressor domain, TRD). These two domains concentrate most of the RTT-associated mutations. R106W and R133C are associated with severe and mild RTT phenotype, respectively. We have performed a comprehensive characterization of the structural and functional impact of these substitutions at molecular level. Because we have previously shown that the MBD-flanking disordered domains (N-terminal domain, NTD, and intervening domain, ID) exert a considerable influence on the structural and functional features of the MBD (Claveria-Gimeno, R. et al. Sci Rep. **2017**, *7*, 41635), here we report the biophysical study of the influence of the protein scaffold on the structural and functional effect induced by these two RTT-associated mutations. These results represent an example of how a given mutation may show different effects (sometimes opposing effects) depending on the molecular context.

Keywords: Methyl-CpG-binding protein 2 (MeCP2); Rett syndrome; intrinsically disordered protein (IDP); protein stability; protein-DNA interaction; isothermal titration calorimetry (ITC)

---

1. Introduction

Methyl-CpG binding protein 2 (MeCP2) is an intrinsically disordered protein (IDP) involved in early stages of neuronal development, differentiation, maturation, and synaptic plasticity control [1]. Although it was identified as a methyl-dependent chromatin binding protein and an epigenetic methylation reader, and, therefore, associated to gene silencing, recent evidences suggest it could be considered as a transcriptional regulator whose primary role is recruiting co-repressor complexes to methylated sites and contributing to decreasing transcriptional noise [2].

MeCP2 exhibits a promoter-specific dsDNA interaction required for finely tuning gene transcription, but it also binds massively to heterochromatin when acting as a chromatin architecture remodeling factor. From the initial embryonic development stages, MeCP2 gradually replaces histone 1 as a sort of nucleosomal linker [3–5]. The possibility to establish different types of interaction with DNA together with its ability to interact with other many biological partners (RNA, structural and transcriptional proteins, nucleosomal elements) and its central role as an important network interaction hub within gene transcription regulation networks, as well as the additional regulatory level of MeCP2 activity through post-translational modifications are made possible thanks to its modular, dynamic and adaptive structure [6,7].

Abnormal MeCP2 activity leads to disease [2,8,9]. MeCP2 point mutations or deletions causing activity loss are associated with Rett syndrome (RTT). RTT is the main cause of mental retardation in females (1:10,000 births), exhibiting a clinically broad expression phenotype gradation. RTT shares features with other neurological diseases from the autistic spectrum. Importantly, duplication of mecp2 gene results in overexpression of MeCP2 and leads to MeCP2 duplication syndrome (MDS), another much rarer disorder affecting males and, strikingly, sharing phenotypic features with RTT, such as severe intellectual disability and impaired motor function.

Each of the six domains MeCP2 is either completely or partially disordered: N-terminal domain (NTD), methyl binding domain (MBD), intervening domain (ID), transcriptional repression domain (TRD), C-terminal domain α (CTDα), and C-terminal domain β (CTDβ) (Supplemental Figure S1) [3,10]. Because of the importance of the interaction of MeCP2 with the nuclear co-receptor co-repressor (NCoR), an additional NCoR/SMRT interaction domain (NID) is often considered between TRD and CTDα [11]. Most of MeCP2 polypeptide chain ($\geq$60%) lacks well-defined secondary/tertiary structure. Flexible, disordered regions facilitate structural rearrangements necessary for exposing different interaction motifs and adapting to the many interacting partners, as well as the giving rise to the allosteric regulation through which the protein conformational landscape is modulated by ligand binding.

The most important domains are MBD, initially associated with methylated CpG (mCpG) DNA binding, and TRD, associated with transcription repression activities [12,13]. Most RTT-associated mutations are concentrated within these two domains, including missense and nonsense mutations, insertions, duplications, and deletions [14]. Nevertheless, only eight missense and nonsense mutations (R106W, R133C, T158M, R168X, R255X, R270X, R294X and R306C) account for approximately 70% of all mutations in RTT [15]. In particular, R133C, T158M, and R106W (in increasing order for phenotype severity and disease burden) represent 5%, 12%, and 3% of RTT cases [16,17].

MBD is the best characterized domain in MeCP2. MBD structure basically consists of a wedge-shaped structured core containing a 3-stranded anti-parallel β-sheet with an α-helix on the C-terminal side, with two unstructured regions flanking this core [16,17]. MBD is considered to be directly involved in maintaining the global organization of the protein through interactions with other domains through inter-domain coupling [5,18,19]. Mutations in this domain would have an impact on the local and the global stability in MeCP2 [3,18].

In a previous biophysical study of three MeCP2 variants (MBD, and NTD-MBD, and NTD-MBD-ID), we established that the isolated MBD might not be the appropriate construct to study and assay its dsDNA binding features, because the presence of NTD and ID increased considerably the dsDNA binding affinity and the structural stability, besides adding a second, functionally independent dsDNA binding site [20]. Here we report a biophysical study of the structural stability and the dsDNA interaction of mutant variants containing the substitutions R106W and R133C, two main RTT-mutations. These mutations were selected because they consist in an arginine substitution by a bulkier or a smaller residue, they are located in different positions regarding the dsDNA binding interface, and they correspond to different disease severity and burden levels. According to the results presented here, the inclusion of those substitutions into different protein constructions (MBD and NTD-MBD-ID) results in different structural and functional effects, highlighting the importance of selecting an appropriate molecular context (i.e., protein construction) when evaluation mutational effects, and emphasizing, in particular for MeCP2, the potential interdomain interaction in intrinsically disordered proteins [18,19].

2. Materials and Methods

2.1. Plasmid Construction

MeCP2 variants from isoform were expressed in *E. coli* using a pET30b plasmid. The different protein variants were obtained by inserting appropriate substitutions: MBD, MBD R106W, MBD R133C, NTD-MBD-ID, NTD-MBD-ID R106W, and NTD-MBD-ID R133C (Supplemental Figure S1). An N-terminal polyhistidine-tag was inserted for quick purification, and it was removed through an inserted PreScission Protease cleavage site. Appropriate expression was assessed by sequencing analysis: Sanger sequencing using a BigDye Terminator v3.1 Cycle Sequencing Kit (Life Technologies, Carlsbad, CA, USA) in an Applied Biosystems 3730/DNA Analyzer (Thermo Fisher Scientific, Waltham, MA, USA).

2.2. Protein Expression and Purification

Protein variants (MBD, MBD R106W, MBD R133C, NTD-MBD-ID, NTD-MBD-ID R106W, NTD-MBD-ID R133C) were expressed and purified following identical procedures. Plasmids were transformed into BL21 (DE3) Star *E. coli* strain. Cultures were grown in 150 mL of LB/kanamycin (50 µg/mL) media at 37 °C overnight. Then, 4 L of LB/kanamycin (25 µg/mL) were inoculated (1:100 dilution) and incubated under the same conditions until reaching an OD ($\lambda = 600$ nm) of 0.6. Protein expression was induced with 1 mM isopropyl 1-thio-β-D-galactopyranoside (IPTG) at 18 °C overnight. Cells were sonicated in ice and benzonase (Merck-Millipore, Madrid, Spain) was added (20 U/mL) to remove nucleic acids. Proteins were purified using metal affinity chromatography employing a HiTrap TALON column (GE-Healthcare Life Sciences, Barcelona, Spain) with two washing steps: buffer sodium phosphate 50 mM, pH 7, NaCl 300 mM, and buffer sodium phosphate 50 mM, pH 7, NaCl 800 mM. Elution was performed applying an imidazole 10–150 mM elution gradient. Protein purity was evaluated by SDS-PAGE.

The polyhistidine-tag was removed by processing with GST-tagged PreScission Protease in protease buffer (50 mM Tris-HCl, 150 mM NaCl, pH 7.5) at 4 °C for 4 h. Progress of the proteolytic processing was monitored by SDS-PAGE. In the final step the protein was further purified with a combination of two affinity chromatographic steps to remove the polyhistidine-tag (HiTrap TALON column) and the GST-tagged PreScission Protease (GST TALON column, from GE-Healthcare Life Sciences, Barcelona, Spain). Purity and homogeneity were evaluated by SDS-PAGE and size-exclusion chromatography. Storage buffer consisted of Tris 50 mM pH 7.0 and pooled samples were kept at −80 °C. The identity of all proteins was checked by mass spectrometry (4800plus MALDI-TOF/MS, from Applied Biosystems-Thermo Fisher Scientific, Waltham, MA, USA). Potential DNA contamination was always estimated by UV absorption 260/280 ratio. Because a single tryptophan is located in MBD, an extinction coefficient of 11,460 M^{-1} cm^{-1} at 280 nm was employed for all variants, except for the R106W mutants for which a value of 16,960 M^{-1} cm^{-1} was applied.

Stability and binding assays were performed at different pH and buffer conditions (Tris 50 mM pH 7–9, NaCl 0–150 mM; Pipes 50 mM, pH 7; Phosphate 50 mM, pH 7). When needed, buffer exchange was done employing a 3 or 10 kDa-pore size ultrafiltration device (Amicon centrifugal filter, Merck-Millipore, Madrid, Spain) at 4000 rpm and 4 °C.

2.3. Double-Stranded DNA

HPLC-purified methylated and unmethylated 45-bp single-stranded DNA (ssDNA) oligomers corresponding to the promoter IV of the mouse brain-derived neurotrophic factor (BDNF) gene [18,19], were purchased from Integrated DNA Technologies. Two complementary pairs of DNA were used for DNA binding assays: forward unmethylated: 5′- GCCATGCCCTGGAACGGAACTCTCCTAATAAAAG-ATGTATCATTT-3′; reverse unmethylated: 5′- AAATGATACATCTTTTATTAGGAGAGTTCCGTTCC-AGGGCATGGC-3′; forward mCpG: 5′- GCCATGCCCTGGAA(5-Me)CGGAACTCTCCTAATAAA-AGATGTATCATTT-3′; reverse mCpG: 5′- AAATGATACATCTTTTATTAGGAGAGTTC(5-Me)CGTT-CCAGGGCATGGC-3′.

The ssDNA oligonucleotides were dissolved at a concentration of 0.5 mM, mixed at equimolar ratio, and annealed to obtain 45-bp double-stranded DNA (dsDNA) using a Stratagene Mx3005P qPCR real-time thermal cycler (Agilent Technologies, Santa Clara, CA, USA). The thermal annealing profile consisted of: (1) equilibration at 25 °C for 30 s; (2) heating ramp up to 99 °C; (3) equilibration at 99 °C for 1 min; and (4) 3-h cooling process down to 25 °C at a rate of 1 °C/3 min.

2.4. Circular Dichroism

Circular dichroism spectra were recorded in a thermostated Chirascan spectrometer (Applied Photophysics, Leatherhead, UK) using a 0.1 cm (far-UV) or 0.4 cm (near-UV) path-length quartz cuvette (Hellma Analytics, Müllheim, Germany) with a bandwidth of 1 nm, a spectral resolution of 0.5 nm, and a response time of 5 s. Temperature was controlled by a Peltier unit and monitored using a temperature probe. The assays were performed in the far-UV (200–260 nm) and the near-UV (250–310 nm) ranges. Protein concentration was set at 10–50 µM, depending on the signal-to-noise ratio.

2.5. Fluorescence Spectroscopy

Protein thermal unfolding studies were performed in a Cary Eclipse fluorescence spectrophotometer (Varian—Agilent, Santa Clara, CA, USA) using a protein concentration of 5 µM and a 1 cm path-length quartz cuvette (Hellma Analytics, Müllheim, Germany). The temperature was controlled by a Peltier unit and monitored using a temperature probe, at a heating rate of 1 °C/min. Fluorescence emission spectra were recorded from 300 to 400 nm using an excitation wavelength of 290 nm and a bandwidth of 5 nm. Assays were performed and at the emission wavelength of 330 nm (maximal protein spectral change along the unfolding). A simple two-state unfolding model was considered for analyzing the assays:

$$F(T) = \frac{(A_N + B_N T) + (A_U + B_U T)exp\left(-\frac{\Delta G(T)}{RT}\right)}{1 + exp\left(-\frac{\Delta G(T)}{RT}\right)}$$

$$\Delta G(T) = \Delta H(T_m)\left(1 - \frac{T}{T_m}\right) + \Delta C_P\left(T - T_m - T \ln \frac{T}{T_m}\right)$$

(1)

where $F(T)$ is the fluorescence signal at a given absolute temperature T, T_m is the unfolding temperature, $\Delta H(T_m)$ is the unfolding enthalpy (at the T_m), ΔC_P is the unfolding heat capacity, and $\Delta G(T)$ is the stabilization Gibbs energy (which is a temperature function). The adjustable parameters A_N, B_N, A_U, and B_U are instrumental parameters defining the pre- (native) and post-transition (unfolded) regions in the unfolding trace. The stabilizing effect upon dsDNA interaction was assessed performing thermal denaturations of the different proteins (at 5 µM) in the presence of methylated and unmethylated DNA (at 10 µM) under the same conditions.

2.6. Isothermal Titration Calorimetry (ITC)

The interaction between the different proteins and dsDNA was studied in an Auto-iTC200 (MicroCal, Malvern-Panalytical, Malvern, UK). dsDNA (50 μM) in the injecting syringe was titrated into protein in the calorimetric cell (3–5 μM). Series of 2 μL-injections of titrant with a time-spacing of 150 s were programmed, maintaining a stirring speed of 750 rpm, and a reference power of 10 μcal/s. The association constant, K_B, and the observed enthalpy of binding, $\Delta H_{B,obs}$, were estimated through non-linear regression of the experimental data employing a single ligand binding site model (1:1 protein:dsDNA stoichiometry) or a two ligand binding sites model (1:2 protein:dsDNA stoichiometry) implemented in Origin (OriginLab, Northampton, MA, USA) [21,22]. The dissociation constant K_d was calculated as the inverse of $K_{B,obs}$, and the binding Gibbs energy and entropy were calculated applying standard well-known relationships: $\Delta G = -RT \ln K_B$, $\Delta G = \Delta H - T\Delta S$.

The number of protons released from or uptaken by the protein-dsDNA complex upon dsDNA binding, Δn_H, was determined, according to [23–25]:

$$\Delta H_{B,obs} = \Delta H + \Delta n_H \Delta H_{buffer} \tag{2}$$

where ΔH is the buffer-independent binding enthalpy, and ΔH_{buffer} is the ionization enthalpy of the buffer. Titrations were performed in buffers with different ionization enthalpies (Tris, 11.35 kcal/mol; Pipes, 2.67 kcal/mol; and phosphate, 0.86 kcal/mol) [26] in order to estimate the buffer-independent thermodynamic parameters (ΔH and Δn_H) from linear regression using Equation (2). From ΔG and ΔH, the buffer-independent binding entropy can be readily calculated. The parameter Δn_H may be non-zero if ligand binding results in changes in the proton dissociation constant of certain ionizable residues (either in the protein or the ligand) as a consequence of changes in their microenvironment upon complex formation. The association binding constant K_B will be not affected by the buffer ionization as long as the pK_a of the buffer is close to the experimental pH. However, the observed binding enthalpy (and, therefore, the observed entropic contribution) will contain an additional contribution from buffer ionization as indicated above. The experimental strategy allows removing the extrinsic contribution from buffer ionization. Noticeably, Δn_H has practical utility since it reports the change in binding affinity as a result of a (moderate) change in pH, according to Wyman's linkage relationships [27]:

$$\Delta n_H = -\left(\frac{\partial \log K_B}{\partial pH}\right)_P \tag{3}$$

3. Results

3.1. Mutation R106W Induces Larger Structural Rearrangements Compared to R133C

RTT-associated MeCP2 variants were expressed in *E. coli* and successfully purified by affinity chromatography, allowing us to study the potential destabilizing impact of the selected mutations on the MBD conformation and its interaction with dsDNA.

Far-UV and near-UV CD spectra were recorded to assess whether R106W and R133C mutations might disrupt MBD secondary and tertiary structure. Being tryptophan bulkier than cysteine (which is even smaller than arginine), and being R106 located buried inside the MBD while R133 is solvent exposed, a larger rearrangement would be expected for R106W mutation. In addition, assays with MBD and NTD-MBD-ID variants were carried out to determine the effect of flanking domains might exert on the destabilizing impact of these RTT-associated mutations.

Far-UV CD spectra of MBD (Figure 1) exhibited two regions typical from β-sheet and random-coil (around 208–210 nm) and α-helix (around 222 nm). MBD mutants showed conserved secondary structure, exhibiting a similar proportion of α-helix and β-sheet to wild-type MBD, but a smaller intensity of the signal could be appreciated, which might caused by disrupting effects that affected equally to α-helix and β-sheet elements (Figure 1). Furthermore, the MBD R106W mutant showed a positive band around 230 nm typically produced by the presence of an additional tryptophan residue [28]. The far-UV CD spectra of the NTD-MBD-ID mutants exhibited a lower intensity (50% reduction in MRE, molar residue ellipticity) compared to the MBD spectra when normalized by the number of residues, indicating that the flanking domains, NTD and ID, are disordered and hardly contribute to the CD signal. Also, in the presence of NTD and ID the location of the minimum of the R106W CD signal was significantly shifted to larger wavelengths (minimum at 218 nm), while the minimum of R133C CD signal remained at lower wavelengths minimum at 213 nm), showing that these mutations had a considerable impact on NTD-MBD-ID protein secondary structure. Even the wild-type NTD-MBD-ID showed a shifted minimum at larger wavelength (around 215 nm). R106W also exhibited a CD band around 230 nm, compared to the wild type, although it was sensibly smaller than that of MBD R106W. Regarding the near-UV spectra, the R106W mutants exhibit larger alterations than those observed for R133C mutants, whose spectra are quite similar to those of wild-type variants, indicating that in the R106W the environment of the aromatic residues is altered. Similarly, normalization of the spectra with the number of residues results in smaller signal, in agreement with a negligible contribution to the spectra from NTD and ID.

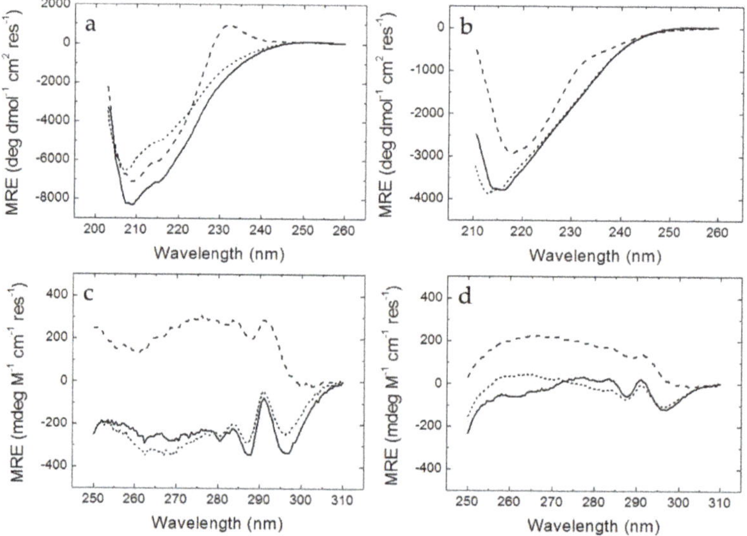

Figure 1. Far-UV (**a,b**) and near-UV (**c,d**) circular dichroism spectra of wild-type and mutant Methyl-CpG binding protein 2 (MeCP2) methyl-CpG binding domain (MBD) (**a,c**) and NTD-MBD-ID (**b,d**): wild-type (continuous line), R106W (dashed line), and R133C (dotted line). Spectra were recorded at pH 7.

3.2. RTT-Associated Mutations Alter Protein Stability and the dsDNA-Induced Stabilization Effect

3.2.1. Stability Changes on MBD Induced by RTT-Associated Mutations

Because of the low content in secondary structure and the small change in the CD signal accompanying the thermal denaturation process, fluorescence spectroscopy was employed for thermal unfolding assays. All protein variants exhibited a well-defined thermal unfolding transition (Figures 2 and 3), which, together with the CD data, indicate that all protein constructs present a well-folded conformation, likely corresponding to the structured region in MBD.

Except for NTD-MBD-ID R106W, the T_m decreased with pH and increased with ionic strength (Figure 2 and Table 1), as observed with the wild-type variants [20]. This indicates that the unfolding process is coupled to the preferential interaction of protons and salt ions with the folded conformation of the protein variants (i.e., unfolding is accompanied by the release of protons and salt ions).

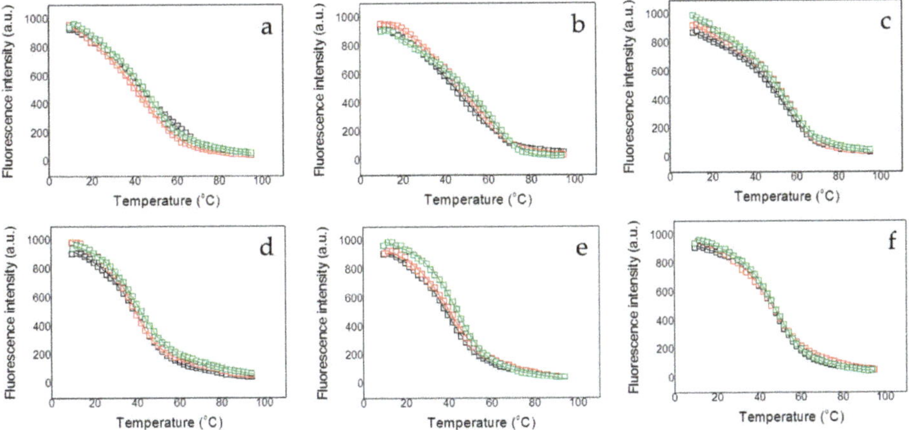

Figure 2. Fluorescence thermal denaturations for the MBD R106W (**a–c**), and MBD R133C (**d–f**) under different conditions. The influence of the pH was assessed (pH 7, black squares; pH 8, red squares; and pH 9, green squares) (**a,d**). The influence of the presence of dsDNA at pH 7 was assessed (absence of dsDNA, black squares; presence of unmethylated CpG-dsDNA, red squares; and presence of methylated mCpG-dsDNA, green squares) at pH 7 and low ionic strength (**b,e**) and high ionic strength (**c,f**). All unfolding traces could be fitted employing a two-state unfolding model (continuous lines) according to Equation (1).

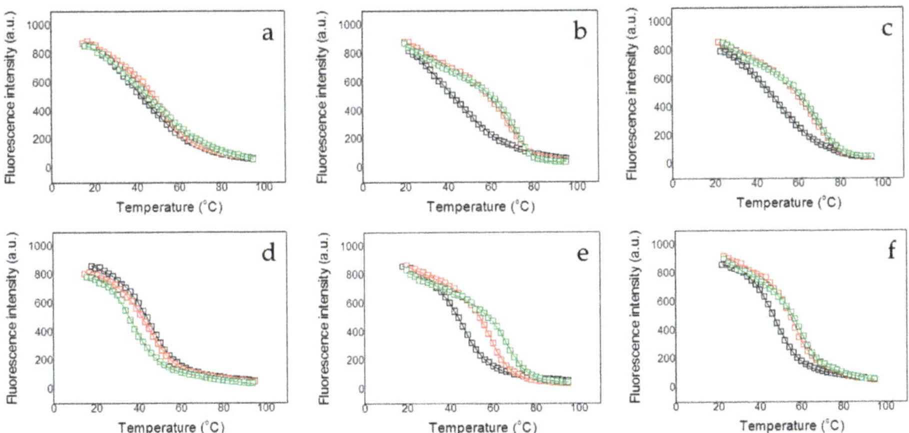

Figure 3. Fluorescence thermal denaturations for the NTD-MBD-ID R106W (**a–c**), and NTD-MBD-ID R133C (**d–f**) under different conditions. The influence of the pH was assessed (pH 7, black squares; pH 8, red squares; and pH 9, green squares) (**a,d**). The influence of the presence of dsDNA at pH 7 was assessed (absence of dsDNA, black squares; presence of unmethylated CpG-dsDNA, red squares; and presence of methylated mCpG-dsDNA, green squares) at pH 7 and low ionic strength (**b,e**) and high ionic strength (**c,f**). All unfolding traces could be fitted employing a two-state unfolding model (continuous lines) according to Equation (1).

Table 1. Thermal stability of the different MeCP2 variants under different conditions.

		T_m (°C)	$\Delta H(T_m)$ (kcal/mol)
[a] MBD	pH 7	38.4 ± 0.2	29 ± 1
	pH 8	36.9 ± 0.3	33 ± 2
	pH 9	30.8 ± 0.3	27 ± 1
	pH 7, NaCl 150 mM	46.4 ± 0.4	32 ± 1
[b] MBD R106W	pH 7	47.3 ± 0.2	21 ± 1
	pH 8	45.8 ± 0.2	20 ± 1
	pH 9	46.7 ± 0.2	23 ± 1
	pH 7, NaCl 150 mM	54.9 ± 0.2	28 ± 1
[b] MBD R133C	pH 7	39.2 ± 0.2	26 ± 1
	pH 8	38.4 ± 0.3	33 ± 2
	pH 9	38.2 ± 0.2	25 ± 1
	pH 7, NaCl 150 mM	47.0 ± 0.1	29 ± 2
[a] NTD-MBD-ID	pH 7	46.2 ± 0.2	37 ± 1
	pH 8	45.9 ± 0.3	48 ± 3
	pH 9	45.4 ± 0.2	53 ± 2
	pH 7, NaCl 150 mM	49.8 ± 0.1	38 ± 1
[b] NTD-MBD-ID R106W	pH 7	43.1 ± 0.3	23 ± 1
	pH 8	52.8 ± 0.3	30 ± 2
	pH 9	54.3 ± 0.3	34 ± 2
	pH 7, NaCl 150 mM	45.2 ± 0.2	37 ± 1
[b] NTD-MBD-ID-TRD R133C	pH 7	45.8 ± 0.2	33 ± 1
	pH 8	43.8 ± 0.2	31 ± 1
	pH 9	34.5 ± 0.2	30 ± 1
	pH 7, NaCl 150 mM	47.2 ± 0.1	37 ± 2

[a] Previous work [20]. [b] This work.

Each MBD mutation has a different impact on protein stability (Figure 2 and Table 1). Thus, R106W mutation exhibited a stabilizing effect increasing the T_m value in all assayed experimental conditions compared to wild-type MBD. In contrast, R133C mutation did exhibit little impact on protein stability with regard to wild-type MBD, just a slight stabilization in agreement with previous results [18,19]; it seems that this mutation does not affect the protein stability, and, very likely, its deleterious effect might be related to the protein functionality. The stabilizing effect caused by salt ions was similar in both R106W and R133C mutants. MBD R106W and R133C MBD mutants interact preferentially with salt ions in the native state, but very likely salt ions may have an additional charge-screening effect and contribute to the increase stability at high ionic strength by diminishing repulsive interactions between positively charged groups.

One of the most striking results for wild-type MBD was that, at any pH and ionic strength, the structural stability gradually increased with the addition of the disordered domains NTD and ID [20]. The stability of NTD-MBD-ID mutants was also assessed to determine the contribution of those disordered regions, through specific or unspecific effects, to the structural stability (Figure 3 and Table 1). Contrary to wild-type MBD, the addition of NTD and ID may increase or decrease the structural stability of the mutant variant depending on the pH and the ionic strength. For example, at pH 7 and low ionic strength, addition of MBD-flanking domains lowers the stability in the R106W mutant, but raises the stability in the R133C mutant; however, at high ionic strength, addition of MBD-flanking domains lowers the stability of the R106W, while the R133C undergoes no stability change. However, taking NTD-MBD-ID as a reference, both R106W and R133C lowered the stability, in reasonable agreement with previous reported results given the differences in the experimental conditions [18,19]. In general, R133C mutants unfolding parameters are closer to wild-type variants, compared to

R106W mutants. Nevertheless, a stabilizing effect caused by R106W mutation can still be observed at higher pH values, indicating a stronger pH dependency of this mutant in the unfolding process.

3.2.2. Stability Changes in RTT-Associated Mutant Proteins Induced by ds-DNA Binding

The interaction of MBD and NTD-MBD-ID mutant variants with dsDNA was indirectly determined by assessing the stabilizing effect induced by unmethylated CpG-dsDNA and methylated mCpG-dsDNA. Thermal denaturations were performed for all MeCP2 protein variants employing the same protocol used for the dsDNA-free variants, in order to determine the apparent thermodynamic parameters for the unfolding of the protein-dsDNA complex (Figures 2 and 3 and Tables 2 and 3).

Table 2. Thermal stability of the different MeCP2 variants in the presence of unmethylated (CpG-) and methylated (mCpG-) dsDNA at pH 7.

		T_m (°C)	$\Delta H(T_m)$ (kcal/mol)
[a] MBD		38.4 ± 0.3	29 ± 1
	CpG-dsDNA	48.9 ± 0.3	38 ± 2
	mCpG-dsDNA	56.5 ± 0.3	44 ± 2
[b] MBD R106W		47.3 ± 0.2	21 ± 1
	CpG-dsDNA	60.7 ± 0.2	17 ± 1
	mCpG-dsDNA	64.3 ± 0.2	28 ± 2
[b] MBD R133C		39.2 ± 0.2	26 ± 1
	CpG-dsDNA	39.4 ± 0.1	24 ± 1
	mCpG-dsDNA	43.5 ± 0.2	29 ± 2
[a] NTD-MBD-ID		46.2 ± 0.2	37 ± 2
	CpG-dsDNA	64.5 ± 0.1	60 ± 2
	mCpG-dsDNA	71.2 ± 0.2	86 ± 4
[b] NTD-MBD-ID R106W		43.1 ± 0.2	23 ± 1
	CpG-dsDNA	70.3± 0.2	47 ± 2
	mCpG-dsDNA	72.8 ± 0.2	62 ± 3
[b] NTD-MBD-ID R133C		45.8 ± 0.2	34 ± 2
	CpG-dsDNA	59.1 ± 0.1	45 ± 2
	mCpG-dsDNA	67.8 ± 0.2	51 ± 3

[a] Previous work [20]. [b] This work.

Table 3. Thermal stability of the different MeCP2 variants in the presence of unmethylated (CpG-) and methylated (mCpG-) dsDNA at pH 7 and high ionic strength (NaCl 150 mM).

		T_m (°C)	$\Delta H(T_m)$ (kcal/mol)
[a] MBD		46.4 ± 0.4	32 ± 1
	CpG-dsDNA	48.3 ± 0.3	34 ± 2
	mCpG-dsDNA	49.5 ± 0.3	35 ± 2
[b] MBD R106W		54.9 ± 0.2	27 ± 1
	CpG-dsDNA	57.4 ± 0.2	34 ± 2
	mCpG-dsDNA	58.6 ± 0.2	33 ± 2
[b] MBD R133C		47.0 ± 0.1	29 ± 1
	CpG-dsDNA	46.6 ± 0.1	28 ± 1
	mCpG-dsDNA	47.7 ± 0.2	29 ± 1
[a] NTD-MBD-ID		49.8 ± 0.1	38 ± 2
	CpG-dsDNA	65.6 ± 0.2	55 ± 2
	mCpG-dsDNA	66.4 ± 0.2	63 ± 3

Table 3. *Cont.*

		T_m (°C)	$\Delta H(T_m)$ (kcal/mol)
[b] NTD-MBD-ID R106W		45.2 ± 0.1	23 ± 1
	CpG-dsDNA	68.5 ± 0.2	33 ± 2
	mCpG-dsDNA	70.2 ± 0.2	41 ± 2
[b] NTD-MBD-ID-TRD R133C		47.2 ± 0.1	37 ± 2
	CpG-dsDNA	58.4 ± 0.2	46 ± 2
	mCpG-dsDNA	60.5 ± 0.2	50 ± 2

[a] Previous work [20]. [b] This work.

In general, dsDNA increased the stability in all variants, as can be observed in the values in T_m compared to those for the dsDNA-free variants, at each experimental condition (Table 2). High ionic strength diminished the extent of the dsDNA stabilization effect (Table 3) through a double mechanism: ionic strength increases protein variants stability and decreases dsDNA binding affinity [20]. MBD R106W is preferentially stabilized by dsDNA, while MBD R133C shows a limited ability to bind dsDNA, because the stabilizing effect induced by dsDNA is very small (Figure 2 and Tables 2 and 3). A small stabilization effect can be observed for MBD R133C in presence of methylated dsDNA; therefore, MBD R133C might conserve its ability to bind dsDNA, but with a much lower affinity. Wild-type MBD was stabilized by both types of dsDNA, but the stabilizing effect induced by methylated dsDNA was significantly larger. However, MBD R106W showed more similar stabilizing effects for methylated and unmethylated dsDNA.

In the constructs containing the two MBD-flanking domains, NTD-MBD-ID, methylated dsDNA always induced a larger stabilization effect than unmethylated dsDNA (Figure 3 and Tables 2 and 3), reflecting that the preferential interaction or specificity of MBD towards methylated dsDNA is maintained. Similarly to previous data [21], the MBD-flanking domains, NTD and ID, not only increase (in general) the thermal stability of dsDNA-free MBD mutants, but they also strengthen the stabilizing effect induced by dsDNA binding, and enhance the discriminating power regarding dsDNA methylation as observed by the extent of the stabilization effect. NTD-MBD-ID R106W showed a large stabilizing effect induced by dsDNA, exhibiting a slight difference between both methylated and unmethylated dsDNA effects. NTD-MBD-ID R133C showed an interesting behavior: whereas MBD R133C hardly exhibited a dsDNA-induced stabilizing effect, the addition of the MBD-flanking disordered domains rescued the ability to interact with dsDNA as indicated by the considerable dsDNA-induced stabilization effect. For both MBD and NTD-MBD-ID variants, the substitution R106W induces larger dsDNA stabilization effects compared to the wild-type variant, whereas the substitution R133C induces a smaller stabilization effects. This will be discussed later on.

Although the stabilization effect induced by dsDNA illustrates the fact that MeCP2 mutant constructs are able to interact with dsDNA, this observable cannot be easily employed to measure and quantify binding affinities and other thermodynamic parameters for the interaction in a straightforward manner. The extent of the stabilization effect caused by the presence of dsDNA (ΔT_m) on a given protein conformation depends on the binding affinity, the binding stoichiometry, and the concentration of (free) dsDNA. Furthermore, the binding enthalpy and the binding heat capacity strongly influence the extent of the dsDNA-induced stabilization effect. This makes not possible to estimate binding affinities properly and to establish an affinity ranking based on ΔT_m values. Therefore, further ITC experiments were required to accomplish a biophysical characterization of the interaction between MeCP2 mutants and methylated and unmethylated dsDNA, in order to determine binding affinities and the enthalpic/entropic contributions to the binding.

3.3. RTT-Associated Mutations Affect dsDNA Interaction Differently Depending on the MeCP2 Construction

3.3.1. Interaction of MBD Mutants with ds-DNA

Contrary to MBD R106W, which shows considerable binding affinity for dsDNA (Figure 4 and Table 4), MBD R133C did not show any interaction with dsDNA under any tested experimental condition (Supplemental Figure S2); this cannot be explained by a loss of structure due to the mutation, because the previous results provided evidence for MBD R133C being properly (partially) folded. The interaction of MBD R106W with dsDNA was characterized by moderate affinity (K_d in the submicromolar range), slightly higher than that of wild-type MBD, and exhibiting a much more favorable entropic contribution to the binding and a much more unfavorable binding enthalpy (Table 4). Opposite to what occurs with wild-type MBD, the formation of the MBD R106W dsDNA complex is accompanied by the uptake of about 2–3 protons from the bulk solution (i.e. at least three ionizable groups are involved in the proton exchange process) (Table 4). Hence, regarding the pH dependency of the binding affinity, increasing the pH in 1 unit would cause a 400-fold decrease in the binding affinity, whereas in the case of wild-type MBD it would cause a 160-fold increase in binding affinity. These results indicated that, although the affinity is comparable or slightly better for MBD R106W, the mode of interaction with dsDNA might be quite different compared to wild-type MBD. In addition, the ability to discriminate between unmethylated and methylated dsDNA was diminished.

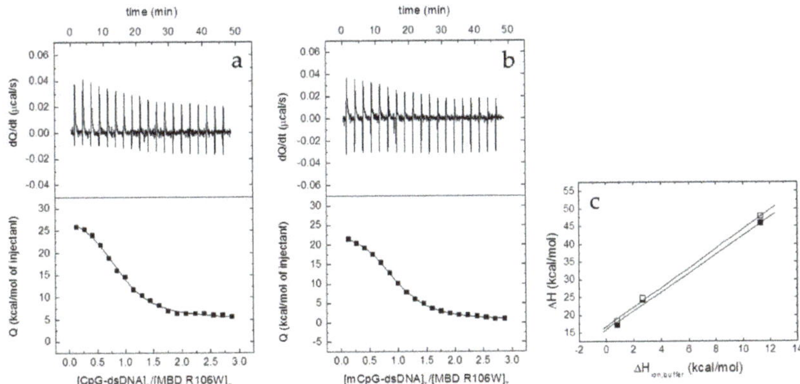

Figure 4. Interaction of the MBD R106W with unmethylated CpG-dsDNA (**a**) and methylated mCpG-dsDNA (**b**) by ITC in Pipes 50 mM, pH 7, 20 °C. Upper plots show the thermogram (raw thermal power as a function of time) and the lower plots show the binding isotherm (ligand-normalized heat effects as a function of the molar ratio). Non-linear least-squares analysis using a single binding site model allowed estimating the observed binding affinity and enthalpy (continuous lines). (**c**) Equation (2) was employed for estimating the buffer-independent binding enthalpy: CpG-dsNA (open squares) and mCpG-dsDNA (closed squares). From linear regression, the intercept with the *y*-axis (observed enthalpy extrapolated at zero buffer ionization enthalpy) provides the buffer-independent interaction enthalpy ΔH, and the slope provides Δn_H.

Table 4. Buffer-independent binding parameters for the interaction of the different MeCP2 variants with unmethylated (CpG-) and methylated (mCpG-) dsDNA at pH 7 and 20 °C.

		K_d (nM)	ΔG (kcal/mol)	ΔH (kcal/mol)	$-T\Delta S$ (kcal/mol)	Δn_H
[a] MBD	CpG-DNA	450	−8.5	0.8	−9.3	−2.4
	mCpG-DNA	240	−8.9	1.5	−10.4	−2.1

Table 4. *Cont.*

		K_d (nM)	ΔG (kcal/mol)	ΔH (kcal/mol)	$-T\Delta S$ (kcal/mol)	Δn_H
[b] MBD R106W	CpG-DNA	110	−9.3	16.7	−26.0	2.7
	mCpG-DNA	95	−9.4	16.0	−25.4	2.6
[b] NTD-MBD R133C	CpG-DNA	n.i.				
	mCpG-DNA	n.i.				
[a] NTD-MBD-ID	CpG-DNA	1.9	−11.7	−54.6	42.9	−0.1
		250	−8.9	−7.6	−1.3	−2.9
	mCpG-DNA	0.56	−12.4	−48.4	36.0	−0.1
		62	−9.7	−2.1	−7.6	−1.3
[b] NTD-MBD-ID R106W	CpG-DNA	27	−10.2	15.6	−25.8	1.4
		320	−8.7	−25.6	16.9	−1.6
	mCpG-DNA	23	−10.2	22.6	−32.8	1.4
		430	−8.5	−16.7	8.2	−1.7
[b] NTD-MBD-ID R133C	CpG-DNA	2.5	−11.5	−49.1	37.6	−2.9
		83	−9.5	−11.8	2.3	−0.7
	mCpG-DNA	2.1	−11.6	−45.0	33.4	−2.9
		62	−9.7	−10.0	0.3	−0.5

Two dissociation constants K_d (for high-affinity site in MBD and low-affinity site in intervening domain (ID)) are provided for NTD-MBD-ID variants. Relative error in K_d is 15%, absolute error in ΔG is 0.1 kcal/mol, absolute error in ΔH and $-T\Delta S$ is 0.3 kcal/mol, and absolute error in Δn_H is 0.2. [a] Previous work [20]. [b] This work. n.i.: no interaction was observed (at any experimental condition).

3.3.2. Interaction of NTD-MBD-ID Mutants with ds-DNA

It was previously shown that the presence of ID, a small unstructured domain located flanking MBD in its C-terminal position, modified completely the dsDNA interaction: a second interaction site appeared together with a marked increase in the binding affinity of MBD for dsDNA [19,20]. The interaction of the two mutants, NTD-MBD-ID R106W and R133C, could be observed by ITC (Figures 5 and 6, Table 4).

The presence of ID increased the binding affinity of the MBD site in NTD-MBD-ID R106W (taking MBD R106W as a reference), but that increase (3-fold) is small compared to that observed for wild-type (400-fold increase) (Table 4). However, taking NTD-MBD-ID as a reference, the R106W mutation decreased the dsDNA affinity. In NTD-MBD-ID R106W, the binding affinity of the MBD interaction site (high-affinity site) remained in the submicromolar range, comparable to that exhibited by the ID interaction site (low-affinity site), whose binding affinity is only 10-fold lower. Overall, the thermodynamic signature of the interaction was completely opposite to that of the wild type. The R106W mutation strongly disturbs the interaction with dsDNA at the MBD binding site, but also at the ID binding site. In NTD-MBD-ID R133C the addition of the ID domain not only provided an additional dsDNA binding site in ID with a binding affinity in the submicromolar range, but also recovered dsDNA-binding ability in the binding site in MBD with a dramatically increased affinity in the low nanomolar range (Table 4). NTD-MBD-ID R133C did not exhibit large variations in binding affinity regarding the wild type variant, showing similar binding affinities for both the MBD and the ID interaction sites. Strikingly, except the net number of exchanged protons, this mutant exhibited a thermodynamic binding profile similar to that of the wild-type variant. The presence of an additional dsDNA binding site in NTD-MBD-ID variants is responsible, together with the higher affinity, for the much larger dsDNA-induced stabilization effect observed on those variants, compared to the smaller stabilization effects observed for the MBD constructs.

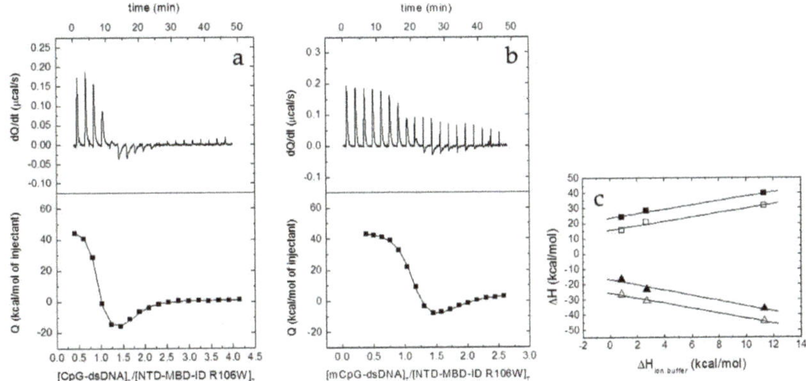

Figure 5. Interaction of the NTD-MBD-ID R106W with unmethylated CpG-dsDNA (**a**) and methylated mCpG-dsDNA (**b**) by ITC in Tris 50 mM, 20 °C, pH 7. Upper plots show the thermogram (raw thermal power as a function of time) and the lower plots show the binding isotherm (ligand-normalized heat effects as a function of the molar ratio). Non-linear least-squares analysis using a two binding sites model allowed estimating the observed binding affinity and enthalpy (continuous lines). (**c**) Equation (2) was employed for estimating the buffer-independent binding enthalpy for both sites (high affinity, squares; low affinity, triangles) for CpG-dsNA (open symbols) and mCpG-dsDNA (closed symbols). From linear regression, the intercept with the *y*-axis (observed enthalpy extrapolated at zero buffer ionization enthalpy) provides the buffer-independent interaction enthalpy ΔH, and the slope provides Δn_H.

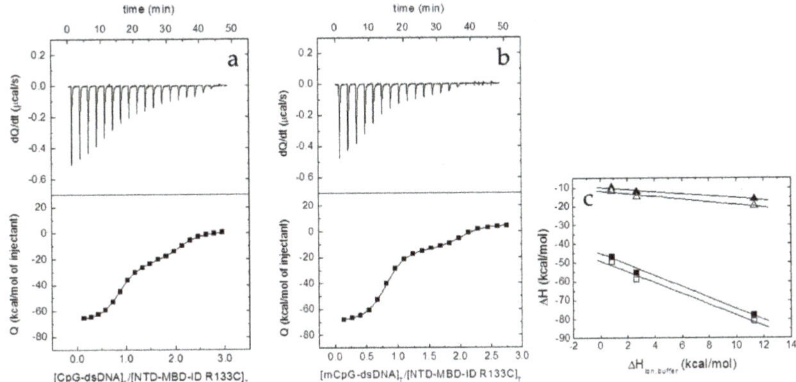

Figure 6. Interaction of the NTD-MBD-ID R133C with unmethylated CpG-dsDNA (**a**) and methylated mCpG-dsDNA (**b**) by ITC in Tris 50 mM, 20 °C, pH 7. Upper plots show the thermogram (raw thermal power as a function of time) and the lower plots show the binding isotherm (ligand-normalized heat effects as a function of the molar ratio). Non-linear least-squares analysis using a two binding sites model allowed estimating the observed binding affinity and enthalpy (continuous lines). (**c**) Equation (2) was employed for estimating the buffer-independent binding enthalpy for both sites (high affinity, squares; low affinity, triangles) for CpG-dsNA (open symbols) and mCpG-dsDNA (closed symbols). From linear regression, the intercept with the *y*-axis (observed enthalpy extrapolated at zero buffer ionization enthalpy) provides the buffer-independent interaction enthalpy ΔH, and the slope provides Δn_H.

4. Discussion

Disordered regions in proteins are characterized by a biased amino acid composition, where residues exhibiting considerable propensity to be exposed to the solvent (polar and charged amino acids) predominate [29]. They may influence protein conformation and function through steric effects, or exerting long-distance attractive or repulsive electrostatic interactions due to their highly polar/charged character, or making contacts with other structured regions affecting the global stability and the dynamics of the protein, as well as modulating the interaction with a binding partner. Thus, even lacking a well-defined structure, disordered regions may contribute to the overall stability of the protein, as it happens in MeCP2. Related to that, we have recently reported that: (1) NTD and ID, the two completely disordered MBD-flanking domains significantly increase the thermal stability of MBD [20]; and (2) the two differentially expressed MeCP2 isoforms (E1 and E2) as a result of differential splicing and differing in just a few amino acids at the N-terminal part of the completely disordered NTD, differ in their thermal stability and functional capabilities [30]. Thus, it may be possible that the conformation and/or the dynamics of MBD is altered by presence of the two disordered flanking domains, resulting in a different stability and different affinity toward binding partners.

While it is reasonable to expect that the structural and functional impact of point mutations located on structured regions may be predicted with certain reliability, the impact of those located on or close to disordered regions may be more difficult to assess. The two mutations studied in this work, R106W and R133C, are some of the most relevant clinically associated with RTT. They are not located in disordered regions, but in the structured region of the MeCP2 MBD. However, the MBD is very dynamic and susceptible to many environmental factors (pH, temperature, solutes, ligands …), being considered as a key element able to interact with or allosterically regulate the other functional domains [5,18,19]. Both mutations show some similarities and many dissimilarities: (1) both involve the substitution of an arginine residue, but R106W involves the substitution by a bulkier aromatic hydrophobic residue, while R133C involves the substitution by a smaller polar aliphatic residue; (2) R106 is located far from the DNA binding interface, while R133 is located in the DNA binding interface (Figure 7); (3) R106 establishes many interactions with many surrounding residues (in particular, four hydrogen bonding residues: M94, D156, T158, and V159), while R133 interacts with fewer residues (only one hydrogen bonding residue: E137) (Figure 7); and (4) R106 does not interact with DNA, while R133 interacts with DNA through hydrogen bonds and van der Waals contacts (Figure 7). Therefore, the impact of both substitutions is expected to be structurally and functionally different. In fact, if the main rotamers for tryptophan and cysteine are introduced in positions 106 and 133, respectively, all W106 rotamers clash with neighboring residues, while C133 shows no clashes at all, indicating that R106W substitution would result in considerable structural distortion in the vicinity of that position to accommodate such substitution.

As indicated above, the purpose of this work was to gain insight into the relationship between the phenotypic effect and the molecular effect of RTT-associated mutations by assessing the impact of two clinically relevant substitutions in MeCP2 MBD. MBD variants containing the two R106W and R133C mutations were studied regarding their structural stability and dsDNA interaction. In addition, because we have reported before an allosteric coupling between MBD and ID, in which the presence of ID dramatically increased the MBD dsDNA binding affinity and contributed an additional dsDNA binding site, we wanted to address whether the location of those mutations on different scaffolds (MBD or NTD-MBD-ID) would result in different structural and/or functional properties. The experimental strategy consisted of a combination of spectroscopic (CD and fluorescence) and calorimetric (ITC), taking advantage of their strengths and overcoming their limitations. Thus, CD and fluorescence are suitable for gathering coarse-grained structural information, and ITC is the gold-standard for determining binding affinity and providing a complete thermodynamic description of biomolecular interactions. In addition, contrary to other techniques, ITC is appropriate for studying biological interactions with more than one binding site, where the interplay between binding affinity and enthalpy makes easy to observe different binding processes occurring at different locations in a macromolecule.

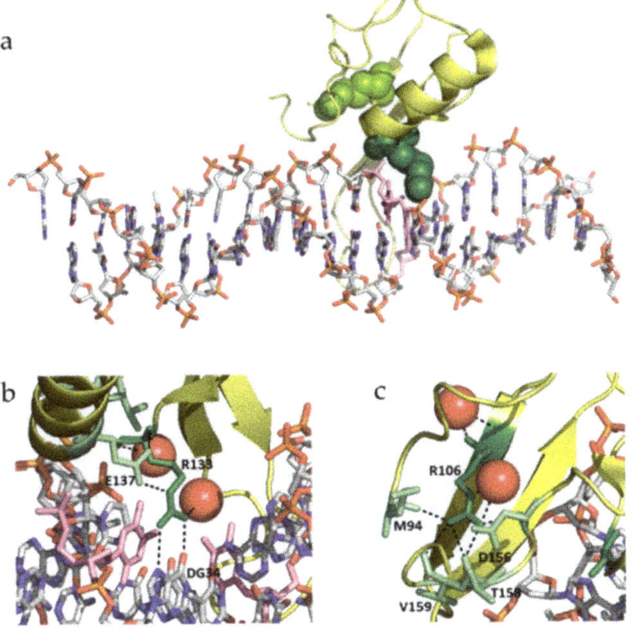

Figure 7. (**a**) Structure of MeCP2 MBD (yellow) in complex with methylated mCpG-dsDNA (CPK view) (pdb id: 3C2I). Residues R106 (light green spheres) and R133 (dark green spheres) are shown. Methylated cytosines are shown as pink sticks (**b**) Close view of R133 (dark green), showing the elements establishing hydrogen bonds with that residue: Glu137 (light green), deoxyguanosine-34 (CPK view), and two water molecules (red spheres). Methylated cytosines are shown in pink sticks. (**c**) Close view of R106 (dark green), showing the elements establishing hydrogen bonds with that residue: Met84, Asp156, T158, V159, and two water molecules (red spheres).

From the results presented here, it is apparent that the impact of R106W and R133C substitutions on the structural stability and the dsDNA binding capability depends on the molecular context, i.e., the scaffold (MBD or NTD-MBD-ID) in which the substitutions are introduced. Thus, to highlight some of the most important findings: (1) R106W and R133C substitutions increase the thermal stability of MBD, but decrease the thermal stability of NTD-MBD-ID (Figure 8); (2) high ionic strength induces a large stabilization in MBD wild-type and mutants R106W and R133C, maintaining the same stability ranking, but a minor stabilization could be observed for NTD-MBD-ID variants (Figure 8); (3) R106W induces an increase in dsDNA binding affinity in MBD, but a decrease in dsDNA binding affinity in NTD-MBD-ID, compared to their respective wild-type variants; and 4) R133C abolishes dsDNA binding in MBD, but behaves similar to the wild-type variant in NTD-MBD-ID in terms of dsDNA affinity and methyl-dependent discrimination.

The large stabilization observed in NTD-MBD-ID R106W when bound to dsDNA compared to the small stabilization observed for NTD-MBD-ID R133C, taking wild-type NTD-MBD-ID as a reference, may be considered a largely unexpected result (Tables 2 and 3). The higher dsDNA affinity for the R133C mutant should have induced a larger stabilization extent for that mutant when bound to dsDNA, because the stabilization energy provided by dsDNA binding is equal to $+RT\ln(1+[\mathrm{dsDNA}]/K_d)$. Thus, the extent of the stabilization effect caused by the presence of dsDNA on protein conformation (quantified as increase in stability energy or increase in T_m) depends on the dsDNA binding affinity, the binding stoichiometry, and the concentration of dsDNA. But the binding affinity is dependent on temperature, and, as a consequence, it will change along the thermal denaturation process. The temperature dependency of the binding affinity will be determined by the dsDNA binding enthalpy (and the binding heat capacity), which might not be the same for each interaction (as it occurs for R106W and R133C variants) and will further modulate the overall extent of the ligand-induced stabilization effect. The R106W variant exhibits a strongly endothermic dsDNA binding to the high-affinity site, indicating that, as the temperature starts increasing during the thermal denaturation process, initially the binding affinity and the strength of the complex would increase (according to the van't Hoff equation) until a temperature in which the binding enthalpy becomes zero (the binding heat capacity is expected to be negative, as found for MBD and NTD-MBD-ID), and then the binding affinity decreases from that temperature. On the contrary, the R133C variant exhibits a strongly exothermic dsDNA binding, and the binding affinity and the strength of the complex would continuously decrease from the beginning of the thermal denaturation process. Therefore, the temperature evolution of K_d would be different for the two mutants and the stabilization extent would also be different. This is a nice example of two molecules (R106W and R133C variants) binding to a common molecule with different binding affinities, but exerting different stabilization effects: the higher affinity interaction is associated with a smaller stabilization effect.

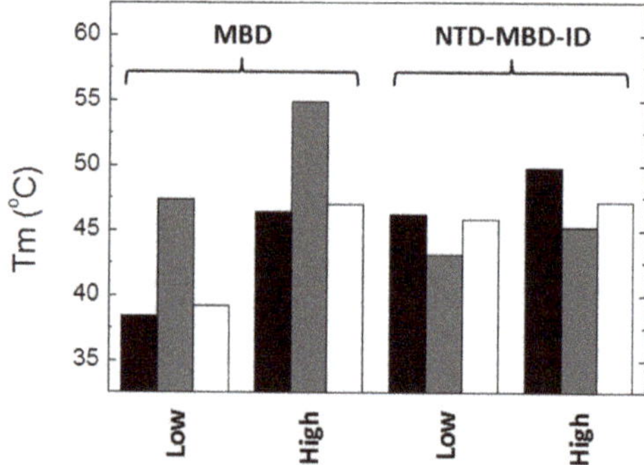

Figure 8. Unfolding temperature for MBD and NTD-MBD-ID variants (wild-type, black bars; R106W, gray bars; R133C, white bars) at low and high ionic strength (NaCl 150 mM).

There are several intriguing facts derived from the experimental results previously shown here. First, how R106W substitution, which is far from the dsDNA binding interface, could affect dsDNA binding and increase its affinity? Of course, even located far from the dsDNA binding interface, the large structural rearrangements resulting from R106W substitution could very likely propagated to distal regions in MBD thanks to its intrinsic structural plasticity. Second, how R133C substitution of a main player in the dsDNA interaction would result in abolishment of interaction for isolated MBD, but almost no effect in NTD-MBD-ID. And third, how these molecular findings can be related to the phenotypic outcomes associated with those mutations: What is the final consequence of R106W substitution at molecular level? Is R106W substitution interfering with the interaction of MeCP2 with other biological partners through the surface residues around R106? What is the final consequence of R133C substitution at molecular level? Is R133C substitution causing an overlooked rearrangement that interferes with other interactions? According to the current classification of RTT mutations, R106W is associated to a severe phenotype, whereas R133C is associated to a mild phenotype [31]. Interestingly, from the evidence gathered in this work, we expect larger functional alterations due to R106W substitution.

5. Conclusions

MeCP2 is a potential pharmacological protein target associated with RTT (caused by defective MeCP2 activity) and MDS (caused by excess of MeCP2 activity), two neurological disorders with similar phenotypic features. MeCP2 is mainly involved in neuronal development and maturation, and synaptic plasticity. While the in vivo effect of MeCP2 duplication is still difficult to explain and correlate with cellular events, the in vivo effect of RTT-associated mutations and their connection with molecular and cellular events may be even more challenging. A valuable strategy consists of gathering experimental evidence on the structural and functional impact of those RTT-associated mutations. Isolated MBD, full-length MeCP2, or other constructs have been previously employed as the protein scaffolds for studying those mutations. Because different MeCP2 constructs may behave differently [20,30], a different impact from RTT-associated substitutions might also be expected depending on the molecular context or protein scaffold in which they are introduced. We have provided here evidence for such phenomenon involving R106W and R133C substitutions. This finding underscores the importance of selecting an appropriate protein construction when assessing the effect of a given mutation, being even more important for intrinsically disordered proteins.

Each MeCP2 mutation associated to RTT may cause different perturbations on protein structure, stability and functionality, depending on the disrupted intra- e intermolecular interactions. The environment of the mutation is crucial and strongly influences the potential deleterious impact caused on MeCP2 functionality, thus modulating MeCP2 ability to interact with dsDNA (binding affinity and methyl-dependent discrimination) or other biological partners (RNA and proteins), and further conditioning the ability to undergo functionally-related posttranslational modifications. Mutations can also produce detrimental effects in regions located far apart from them through allosteric coupling. In fact, the ID binding site was fairly compromised in terms of affinity by MBD mutations (Table 4), revealing that the MBD protein structure might be indispensable for ID-dsDNA interaction. Thus, the R106W substitution in NTD-MBD-ID not only affected the MBD interaction site, but its influence was extended to the ID interaction site, causing a non-negligible reduction in binding affinity.

Supplementary Materials: The following are available online at http://www.mdpi.com/2218-273X/10/11/1533/s1, Figure S1: Domain structure of MeCP2, Figure S2: Calorimetric titration of MBD R133C with methylated mCpG-dsDNA.

Author Contributions: Conceptualization, M.E., O.A. and A.V.-C.; methodology, O.A. and A.V.-C.; software, O.A. and A.V.-C.; validation, D.O.-A., R.C.-G., S.V., O.C.J.-T., M.E., O.A. and A.V.-C.; formal analysis, D.O.-A., R.C.-G., S.V., O.C.J.-T., M.E., O.A. and A.V.-C.; investigation, D.O.-A., R.C.-G., S.V. and O.C.J.-T.; resources, M.E., O.A. and A.V.-C.; data curation, D.O.-A. and R.C.-G.; writing—original draft preparation, O.A. and A.V.-C.; writing—review and editing, D.O.-A., R.C.-G., S.V., O.C.J.-T., M.E., O.A. and A.V.-C.; visualization, D.O.-A., R.C.-G., S.V., O.C.J.-T., M.E., O.A. and A.V.-C.; supervision, M.E., O.A. and A.V.-C.; project administration, O.A. and A.V.-C.; funding acquisition, O.A: and A.V.-C. All authors have read and agreed to the published version of the manuscript.

Funding: This research was funded by the Spanish Ministry of Economy and Competitiveness and European ERDF Funds (MCIU/AEI/FEDER, EU) (BFU2016-78232-P to A.V.C.; BES-2017-080739 to D.O.A.); Miguel Servet Program from Instituto de Salud Carlos III (CPII13/00017 to O.A.); Fondo de Investigaciones Sanitarias from Instituto de Salud Carlos III and European Union (ERDF/ESF, "Investing in your future") (PI15/00663 and PI18/00349 to O.A.); Diputación General de Aragón (Protein Targets and Bioactive Compounds Group E45_17R to A.V.C. and Digestive Pathology Group B25_17R to O.A.); and the Centro de Investigación Biomédica en Red en Enfermedades Hepáticas y Digestivas (CIBERehd).

Conflicts of Interest: The authors declare no conflict of interest. The funders had no role in the design of the study; in the collection, analyses, or interpretation of data; in the writing of the manuscript; or in the decision to publish the results.

References

1. Ausio, J.; de Martínez Paz, A.; Esteller, M. MeCP2: The long trip from a chromatin protein to neurological disorders. *Trends Mol. Med.* **2014**, *20*, 487–498. [PubMed]

2. Tillotson, R.; Bird, A. The Molecular Basis of MeCP2 Function in the Brain. *J. Mol. Biol.* **2020**, *432*, 1602–1623. [CrossRef]

3. Adkins, N.L.; Georgel, P.T. MeCP2: Structure and function. *Biochem. Cell Biol.* **2011**, *89*, 1–11. [CrossRef] [PubMed]

4. Nan, X.; Campoy, F.; Bird, A. MeCP2 Is a Transcriptional Repressor with Abundant Binding Sites in Genomic Chromatin. *Cell* **1997**, *88*, 471–481. [CrossRef] [PubMed]

5. Hansen, J.C.; Ghosh, R.P.; Woodcock, C.L. Binding of the Rett syndrome protein, MeCP2, to methylated and unmethylated DNA and chromatin. *IUBMB Life* **2010**, *62*, 732–738. [CrossRef]

6. Guy, J.; Cheval, H.; Selfridge, J.; Bird, A. The Role of MeCP2 in the Brain. *Annu. Rev. Cell Dev. Biol.* **2011**, *27*, 631–652. [CrossRef]

7. Claveria-Gimeno, R.; Abian, O.; Velazquez-Campoy, A.; Ausio, J. MeCP2 ... Nature's wonder protein or medicine's most feared one? *Curr. Genet. Med. Rep.* **2016**, *4*, 180–194.

8. Amir, R.E.; Veyver, I.B.V.D.; Wan, M.; Tran, C.Q.; Francke, U.; Zoghbi, H.Y. Rett syndrome is caused by mutations in X-linked MECP2, encoding methyl-CpG-binding protein 2. *Nat. Genet.* **1999**, *23*, 185–188. [CrossRef]

9. Neul, J.L. The relationship of Rett syndrome and MECP2 disorders to autism. *Dialog. Clin. Neurosci.* **2012**, *14*, 253–262.

10. Hite, K.C.; Adams, V.H.; Hansen, J.C. Recent advances in MeCP2 structure and function. *Biochem. Cell Biol.* **2009**, *87*, 219–227. [CrossRef]

11. Kokura, K.; Kaul, S.C.; Wadhwa, R.; Nomura, T.; Khan, M.; Shinagawa, T.; Yasukawa, T.; Colmenares, C.; Ishii, S. The Ski Protein Family Is Required for MeCP2-mediated Transcriptional Repression. *J. Biol. Chem.* **2001**, *276*, 34115–34121. [CrossRef] [PubMed]

12. Nan, X.; Meehan, R.R.; Bird, A. Dissection of the methyl-CpG binding domain from the chromosomal protein MeCP2. *Nucleic Acids Res.* **1993**, *21*, 4886–4892. [CrossRef] [PubMed]

13. Hall, J.A.; Georgel, P.T. CHD proteins: A diverse family with strong ties. *Biochem. Cell Biol.* **2007**, *85*, 463–476. [CrossRef] [PubMed]

14. Shah, R.R.; Bird, A.P. MeCP2 mutations: Progress towards understanding and treating Rett syndrome. *Genome Med.* **2017**, *9*, 1–4. [CrossRef]

15. Kyle, S.M.; Vashi, N.; Justice, M.J. Rett syndrome: A neurological disorder with metabolic components. *Open Biol.* **2018**, *8*, 170216. [CrossRef]

16. Brown, K.; Selfridge, J.; Lagger, S.; Connelly, J.; De Sousa, D.; Kerr, A.; Webb, S.; Guy, J.; Merusi, C.; Koerner, M.V.; et al. The molecular basis of variable phenotypic severity among common missense mutations causing Rett syndrome. *Hum. Mol. Genet.* **2016**, *25*, 558–570. [CrossRef]

17. Neul, J.L.; Fang, P.; Barrish, J.; Lane, J.; Caeg, E.B.; Smith, E.O.; Zoghbi, H.Y.; Percy, A.; Glaze, D.G. Specific mutations in methyl-CpG-binding protein 2 confer different severity in Rett syndrome. *Neurology* **2008**, *70*, 1313–1321. [CrossRef]

18. Ghosh, R.P.; Horowitz-Scherer, R.A.; Nikitina, T.; Gierasch, L.M.; Woodcock, C.L. Rett Syndrome-causing Mutations in Human MeCP2 Result in Diverse Structural Changes That Impact Folding and DNA Interactions. *J. Biol. Chem.* **2008**, *283*, 20523–20534. [CrossRef]

19. Ghosh, R.P.; Nikitina, T.; Horowitz-Scherer, R.A.; Gierasch, L.M.; Uversky, V.N.; Hite, K.; Hansen, J.C.; Woodcock, C.L. Unique Physical Properties and Interactions of the Domains of Methylated DNA Binding Protein 2. *Biochemistry* **2010**, *49*, 4395–4410. [CrossRef]

20. Clavería-Gimeno, R.; Lanuza, P.M.; Morales-Chueca, I.; Torres, O.D.L.C.J.; Vega, S.; Abian, O.; Esteller, M.; Velazquez-Campoy, A. The intervening domain from MeCP2 enhances the DNA affinity of the methyl binding domain and provides an independent DNA interaction site. *Sci. Rep.* **2017**, *7*, 41635. [CrossRef]

21. Vega, S.; Abian, O.; Velazquez-Campoy, A. A unified framework based on the binding polynomial for characterizing biological systems by isothermal titration calorimetry. *Methods* **2015**, *76*, 99–115. [CrossRef] [PubMed]

22. Freire, E.; Schön, A.; Velazquez-Campoy, A. Chapter 5 Isothermal Titration Calorimetry. *Methods Enzymol.* **2009**, *455*, 127–155. [CrossRef] [PubMed]

23. Eftink, M.; Biltonen, R.L. Thermodynamics of interacting biological systems. In *Biological Microcalorimetry*; Beezer, A.E., Ed.; Academic Press: London, UK, 1980; pp. 343–412.

24. Hinz, H.J.; Shiao, D.D.F.; Sturtevant, J.M. Calorimetric investigation of inhibitor binding to rabbit muscle aldolase. *Biochemistry* **1971**, *10*, 1347–1352. [CrossRef] [PubMed]

25. Gómez, J.; Freire, E. Thermodynamic Mapping of the Inhibitor Site of the Aspartic Protease Endothiapepsin. *J. Mol. Biol.* **1995**, *252*, 337–350. [CrossRef]

26. Goldberg, R.N.; Kishore, N.; Lennen, R.M. Thermodynamic Quantities for the Ionization Reactions of Buffers. *J. Phys. Chem. Ref. Data* **2002**, *31*, 231–370. [CrossRef]

27. Wyman, J. Linked Functions and Reciprocal Effects in Hemoglobin: A Second Look. *Adv. Protein Chem.* **1964**, *19*, 223–286. [CrossRef]

28. Vuilleumier, S.; Sancho, J.; Loewenthal, R.; Fersht, A.R. Circular dichroism studies of barnase and its mutants: Characterization of the contribution of aromatic side chains. *Biochemistry* **1993**, *32*, 10303–10313. [CrossRef]

29. Uversky, V.N. Unusual biophysics of intrinsically disordered proteins. *Biochim. Biophys. Acta Proteins Proteom.* **2013**, *1834*, 932–951. [CrossRef]

30. De Paz, A.M.; Khajavi, L.; Martin, H.; Claveria-Gimeno, R.; Dieck, S.T.; Cheema, M.S.; Sanchez-Mut, J.V.; Moksa, M.M.; Carles, A.; Brodie, N.I.; et al. MeCP2-E1 isoform is a dynamically expressed, weakly DNA-bound protein with different protein and DNA interactions compared to MeCP2-E2. *Epigenetics Chromatin* **2019**, *12*, 1–16. [CrossRef]

31. Cuddapah, V.A.; Pillai, R.B.; Shekar, K.V.; Lane, J.B.; Motil, K.J.; Skinner, S.A.; Tarquinio, D.C.; Glaze, D.G.; McGwin, G.; Kaufmann, W.E.; et al. Methyl-CpG-binding protein 2 (MECP2) mutation type is associated with disease severity in Rett syndrome. *J. Med. Genet.* **2014**, *51*, 152–158. [CrossRef]

Publisher's Note: MDPI stays neutral with regard to jurisdictional claims in published maps and institutional affiliations.

Article

Degradation of Intrinsically Disordered Proteins by the NADH 26S Proteasome

Peter Tsvetkov *, Nadav Myers, Julia Adler and Yosef Shaul *

Department of Molecular Genetics, Weizmann Institute of Science, Rehovot 76100, Israel; nadav.myers@weizmann.ac.il (N.M.); julia.adler@weizmann.ac.il (J.A.)
* Correspondence: ptsvetko@broadinstitute.org (P.T.); yosef.shaul@weizmann.ac.il (Y.S.); Tel.: +972-8934-2320 (Y.S.)

Received: 22 October 2020; Accepted: 2 December 2020; Published: 7 December 2020

Abstract: The 26S proteasome is the endpoint of the ubiquitin- and ATP-dependent degradation pathway. Over the years, ATP was regarded as completely essential for 26S proteasome function due to its role in ubiquitin-signaling, substrate unfolding and ensuring its structural integrity. We have previously reported that physiological concentrations of NADH are efficient in replacing ATP to maintain the integrity of an enzymatically functional 26S PC. However, the substrate specificity of the NADH-stabilized 26S proteasome complex (26S PC) was never assessed. Here, we show that the binding of NADH to the 26S PC inhibits the ATP-dependent and ubiquitin-independent degradation of the structured ODC enzyme. Moreover, the NADH-stabilized 26S PC is efficient in degrading intrinsically disordered protein (IDP) substrates that might not require ATP-dependent unfolding, such as p27, Tau, c-Fos and more. In some cases, NADH-26S proteasomes were more efficient in processing IDPs than the ATP-26S PC. These results indicate that in vitro, physiological concentrations of NADH can alter the processivity of ATP-dependent 26S PC substrates such as ODC and, more importantly, the NADH-stabilized 26S PCs promote the efficient degradation of many IDPs. Thus, ATP-independent, NADH-dependent 26S proteasome activity exemplifies a new principle of how mitochondria might directly regulate 26S proteasome substrate specificity.

Keywords: proteostasis; ubiquitin independent degradation; intrinsically disordered proteins; NADH-26S proteasome

---

1. Introduction

The function of the 26S proteasome complex (26S PC) is considered to be completely dependent on ATP availability and hydrolysis [1–3]. This is largely due to the multiple roles of ATP in the process of ubiquitin-dependent degradation by the 26S PC. The 26S PC is composed of the 20S catalytic complex and the 19S regulatory particle that contains six ATPases, Psmc1-6, residing at the interface between the 19S and 20S PCs [4,5]. ATP binding and hydrolysis in the catalytic cycle of the 26S proteasome were shown to regulate the ubiquitin processing of the substrate, protein unfolding, and also to maintain the integrity of the 26S proteasomal complex [6,7]. In the absence of ATP, the 26S proteasome dissociates quite rapidly into the 20S and 19S particles [1,3,8]. However, 26S PC integrity is achieved also by CTP, UTP and ADP [1,9], by the unnatural nucleotides ATPγS and AMPPNP [10,11], and also by proteasome inhibitors [9]. In addition, proteins such as Ecm29 [12] act to stabilize 26S PC. As such, although the functions of the proteasome that require the hydrolysis of ATP cannot be replaced by other metabolites, the ATP function in stabilizing the 26S PC can be substituted by other natural and artificial metabolic molecules.

There is increasing evidence suggesting that altered cellular metabolism is associated with both proteasome function and integrity [13–15]. We have previously shown that NADH maintains 26S PC integrity in the absence of ATP. NADH is a key metabolic molecule that couples redox regulation to cellular metabolism, serving as both a shuttle of electrons from glycolysis and TCA cycle metabolism to the electron transfer chain, and also serving as a substrate for many antioxidant enzymes [16,17]. We have shown that NADH specifically interacts with a distinct NADH binding box in the N-terminus of Psmc1, a 19S AAA-ATPase subunit [18]. Furthermore, using differential sensitivity of NADH-26S PC to high levels of $MgCl_2$, we showed that NADH-26S PC is detected in a number of mouse tissues. There are three distinct functions of ATP in 26S proteasome activity: maintaining stability, promoting ubiquitin processing and the unfolding of the substrates. This raises the possibility that proteins that are not ubiquitinated and do not need unfolding can be degraded by a stabilized and functional proteasome, even in the absence of ATP. Indeed, this was shown with ATPγS (an artificial non-hydrolysable form of ATP)-stabilized 26S PC, that could degrade disordered proteins such as casein, p21 and oxidized proteins [10,11,19].

About a third of eukaryotic proteins are intrinsically disordered proteins (IDPs) or consist of large disordered regions (IDRs) [20–22]. IDPs/IDRs, by large, are more labile proteins with short half-lives [23], and as many of them were shown to be degraded in an ATP-independent manner by the 20S proteasome [24–27], suggesting that at least a subset of IDPs do not require unfolding for proteasome degradation. In this work, we set out to determine how NADH affects 26S proteasome degradation and if NADH-stabilized 26S PCs can promote the ubiquitin-independent degradation of IDPs/IDRs.

2. Materials and Methods

2.1. Proteasomal Complex Stability Assay

We used the protocol previously reported by us [18], with minor modifications. In short, proteasomes from NIH3T3 cells or purified 26S proteasomes from rabbit muscles in Deg Buffer (50 mM Tris 7.5, 150 mM NaCl, 5 mM $MgCl_2$) were supplemented with either ATP or NADH in the presence or absence of 8 mU/µL apyrase (Sigma, St. Louis, MO, USA). After incubation at 37 °C for the indicated time points, the reaction was loaded on nondenaturing 4% polyacrylamide gel for monitoring proteasomal complex stability.

2.2. Nondenaturing PAGE

Proteasomal samples were loaded on a nondenaturing 4% polyacrylamide gel as previously reported [18]. After blotting to nitrocellulose membranes, immunoblotting was conducted using the indicated antibodies.

2.3. Proteasomal Activity

To measure proteasomal activity, the hydrolysis of Suc-LLVY-AMC was quantified as described in the manufacturer's protocol (Biomol, USA).

2.4. ^{35}S In Vitro Translated Proteins and Purification

In vitro translation ^{35}S Methionine labeled flag-Yap1, flag-Taz, flag-c-Fos and ODC were subjected to immunoprecipitation with flag-beads (Sigma) in Deg buffer, as previously reported [28]. After 1 h incubation at 4 °C, the flag beads were washed three times with Deg buffer. The proteins retained on the beads were eluted with the addition of Deg buffer containing 100 µg/mL of flag peptide (Sigma) and incubation at 37 °C for 20 min.

2.5. In Vitro Degradation Assay

In Vitro translated ^{35}S Methionine-labeled either crude or purified proteins, and recombinant proteins were incubated in the presence or absence of 1 µg of purified 26S proteasomes, as previously described [24,28]. The reactions were conducted for the indicated times at 37 °C with the supplementation of either 2 mM ATP, ATPγS or NADH. Reactions were stopped with the addition of Laemmli sample buffer and heated at 95 °C for 5 min. The products were separated by polyacrylamide-SDS gel. [35]S Methionine-labeled proteins were detected by autoradiography. The purified proteins were transferred to nitrocellulose membranes and detected with the indicated antibodies.

3. Results

3.1. NADH Inhibits ATP-Dependent, 26S Proteasome Degradation of ODC

Previously, we have reported that the 26S proteasome can be stabilized by NADH [18] and (Figure 1a). However, the functionality of such NADH-stabilized 26S proteasome was not determined. As such, we decided to systematically address the role of NADH in the regulation of 26S proteasome substrate degradation (Figure 1b). To uncouple the degradation process by the 26S PC from the ubiquitin regulation process, we utilized the antizyme (Az)-mediated ODC degradation system that has been extensively characterized in the context of ubiquitin-independent degradation by the 26S proteasome [29]. Az binds ODC monomerizes and targets ODC monomer to the 26S PC in the process of initiating ATP-dependent 26S proteasomal degradation (Figure 1c) [30,31]. ^{35}S-Methionine labeled ODC was incubated with Az in rabbit reticulocyte extracts reported to support ODC degradation. As expected, ODC was completely degraded with the supplementation of ATP to the mix (Figure 1d). In the absence of ATP or in the presence of a non-hydrolysable form of ATP, ATPγS, no Az-mediated degradation of ODC was observed (Figure 1d). The supplementation of NADH in the absence of ATP was not sufficient to induce ODC degradation. However, when NADH was added in the presence of ATP, it resulted in the inhibition of Az-mediated degradation of ODC in a dose-dependent manner (Figure 1e,f). The inhibitory effect of NADH was alleviated by increasing amounts of ATP (Figure 1g). Thus, NADH binding to the 26S proteasome has an inhibitory effect on ubiquitin-independent degradation of ODC, possibly by generating NADH-26S PC.

3.2. ATPγS-26S PC Promotes the Degradation of IDPs

The unfolding of globular proteins for their degradation is mediated by hydrolysis of ATP by the 26S PC ATPase subunits [7]. ATPγS is a form of ATP that is almost non-hydrolysable, and as such inhibits the proteasome functions that rely on the ATP hydrolysis of ATP such as protein unfolding. Some unfolded and denatured proteins can be degraded by the ATPγS-stabilized 26S PC (in the absence of ATP) [11]. Here, we experimentally addressed the question whether IDPs, which inherently lack a defined structure and thus are independent of the unfolding step during degradation, can be degraded by the ATPγS 26S PC.

The key transcription regulators Taz, Yap1 and c-Fos are all predicted to be highly disordered (Figure 2a–c). To test their susceptibility to ATPγS-26S PC, we in-vitro translated and ^{35}S-Metionine-labeled flag-tagged versions of these IDPs. To prevent any undesired effects that can arise from the reticulocyte mix, we further immune-purified these proteins by flag affinity purification followed by flag peptide elution, resulting in purified, ^{35}S-Methionine-labeled IDPs. Each of these IDPs was separately incubated with either ATP-26S or ATPγS-26S PCs. ATPγS-stabilized 26S PC was efficient in degrading all three tested IDPs (Taz, Yap, c-Fos) (Figure 2d). This was not due to residual ATP in the ATPγS 26S PC, as degradation of Yap1 and Taz (Figure 2e) remained efficient even in the presence of apyrase, an enzyme that converts ATP and ADP to AMP [9,10,18]. The ability of the ATPγS-26S PC to mediate IDP degradation is consistent with the reports that ATPγS mediates degradation of proteins lacking a defined structure [10,11,32], and further suggest that hydrolysable ATP is not required for IDP degradation by the 26S PC.

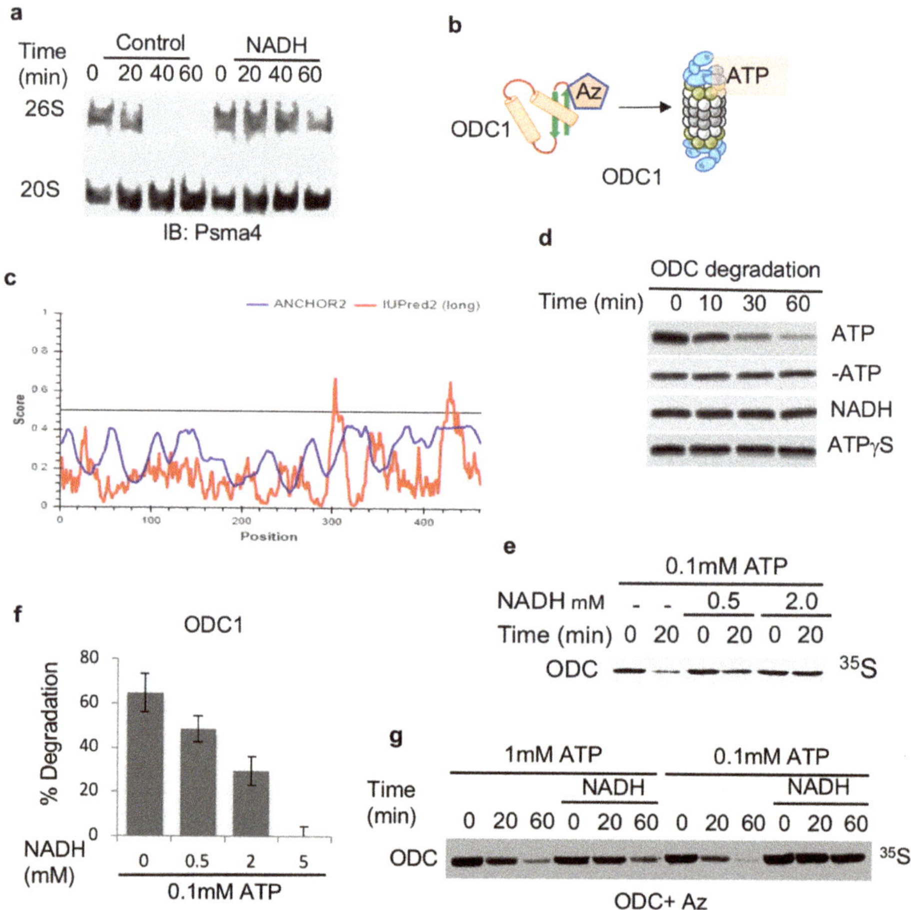

Figure 1. ODC is degraded by ATP-26S but not by NADH-26S proteasome. (**a**) NADH stabilizes the 26S PC. Proteasomes fractionated from NIH3T3 cells were incubated at 37 °C for indicated times in the presence or absence of 2 mM NADH. (**b**) Scheme/model (**c**) ODC (UniProt P11926) is, by large, an ordered protein based on the scores of prediction output using IUPred2A (the red curve) and ANCHOR2 (the blue curve) prediction programs. (**d**) NADH-26S proteasomes cannot induce ODC degradation by Az. ODC degradation in rabbit reticulocyte extract by Az was examined in the presence or absence of 2 mM ATP and an ATP-regenerating system (ATP). This reaction was also conducted under removal of the ATP by Apyrase (-ATP). Similar reactions were conducted in the presence of 2 mM NADH with Apyrase (NADH) and 2 mM ATPγS with Apyrase (ATPγS). (**e**) NADH represses ODC degradation. In vitro translated ^{35}S Methionine-labeled ODC was subjected to degradation in reticulocyte lysate in the presence of Antizyme (Az). ODC degradation was examined in the presence of 0.1 mM ATP and increasing concentrations of NADH at 37 °C for 20 min. (**f**) The data from at least three independent experiments were averaged and shown with their standard deviation. (**g**) The NADH-mediated inhibition of ODC degradation is alleviated by higher ATP concentration. ODC degradation was examined in the presence of 1 mM or 0.1 mM ATP in the presence or absence of 5 mM NADH for the indicated time points at 37 °C. The level of ^{35}S-labeled ODC degradation was visualized by autoradiography following SDS-PAGE and quantified.

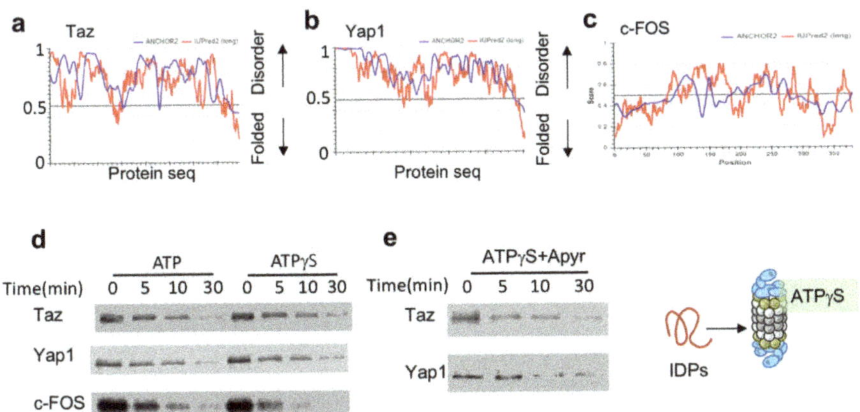

Figure 2. IDPs are degraded by ATPγS 26S PC (a to c). (**a**) Taz (UniProt Q4V7E6). (**b**) Yap (UniProt P146937) and (**c**) c-Fos (UniProt P01100), are intrinsically disordered based on the scores of prediction output using IUPred2A (the red curve) and ANCHOR2 (the blue curve) prediction programs. (**d**) Taz, Yap1 and c-Fos are degraded by ATPγS 26S PC. In vitro translated [35]S Methionine labeled and purified Taz, Yap1 and c-Fos were subjected to degradation by the 26S proteasome in the presence or absence of either 2 mM ATP or ATPγS. (**e**) In the presence of ATPγS the elimination of residual ATP by preincubation with 5 mu/µL apyrase did not inhibit the 26S ability to degrade Taz and Yap1.

3.3. NADH-26S Proteasomes Can Degrade IDPs

Next, we set out to determine if NADH-stabilized 26S PC are capable of facilitating the degradation of IDPs. Initially, we used bacterially expressed and purified p27, a highly disordered protein [33,34] (Figure 3a). NADH 26S-PC were very efficient at promoting the degradation of p27, kinetically even faster than ATP-26S PC (Figure 3b). To validate that the destabilization of p27 is due to proteasome-mediated degradation, we incubated p27 either alone or in the presence of the proteasome and a proteasome inhibitor MG132. As expected, MG132 completely blocked the degradation of p27 by the NADH-26S proteasome (Figure 3c). In our degradation reactions, NADH was not oxidized (data not shown) validating that NADH is a 26S PC-stabilizing cofactor and not the substrate of an unknown enzyme that facilitates proteasome activity. NADH (and NAD[+]) also did not have an effect on the catalytic activity of purified 20S proteasomes (Figure 3d). However, NADH induced an increase in the catalytic activity of the 26S proteasome (Figure 3e). These results suggest that NADH does not directly affect the catalytic activity of the proteasome but possibly affects the gating of the 26S proteasome complex. Together, these data suggest that NADH-26S PC is active in degradation of p27 IDP.

To generalize the proposed model of IDP being highly efficiently degraded by the NADH-26S PC, we examined the in vitro translated and purified ^{35}S-Methionine-labeled Yap and c-Fos proteins. Yap protein was efficiently degraded by the NADH-26S PC as compared to ATP-26S PC (Figure 3f). Purified in vitro translated ^{35}S-Methionine c-Fos was also efficiently degraded by NADH-26S PC with similar time kinetics of ATPγS-26S, but more efficiently than ATP-26S activity (Figure 3g). The observation that ATPγS is more active than ATP in inducing degradation of IDPs by the 26S PC is consistent with the published reports [11]. Unlike ODC, c-Fos degradation was not inhibited by increasing concentration of NADH (Figure 3h,i). We further analyzed the bacterially expressed and purified tau protein as another highly disordered protein substrate (Figure 3j). Tau protein was efficiently degraded by both NADH- and ATPγS- 26S PCs but was not efficiently processed by ATP-26S PC (Figure 3k). Taken together, our data indicate that IDPs are highly susceptible to degradation by the NADH-26S PC in a similar fashion as observed for ATPγS-26S PC and, in some cases, the degradation of IDPs by the NADH-PC is more efficient than that observed for ATP-26S PC.

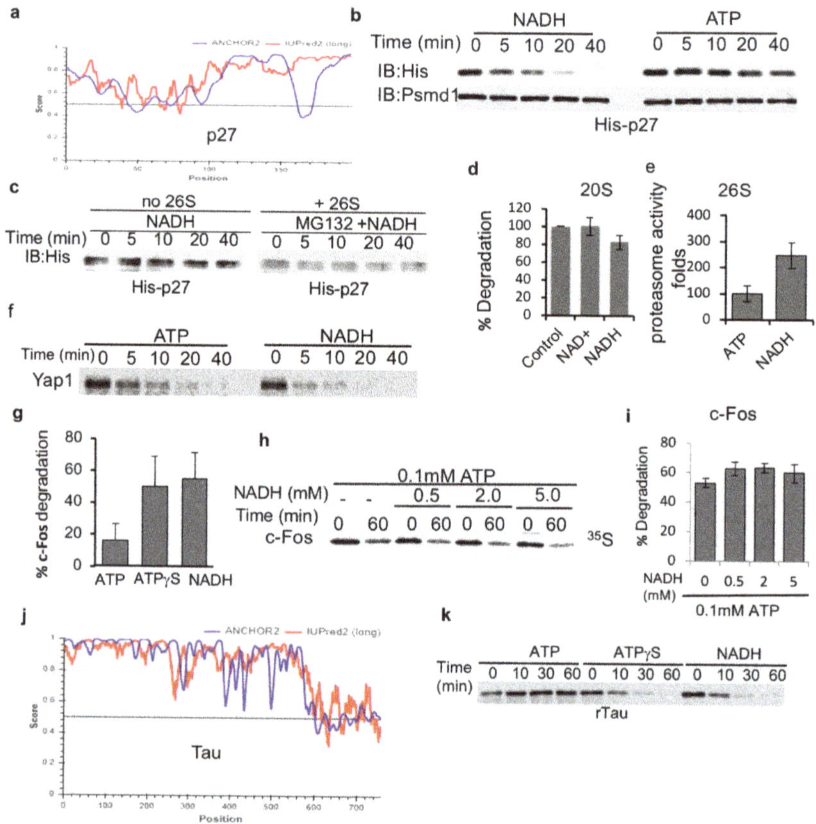

Figure 3. IDPs are susceptible to NADH-26S PC degradation. (**a**) p27 (UniProt P46527) is a disordered protein based on the prediction output using IUPred2A (red curve) and ANCHOR2 (blue curve) prediction programs. (**b**) Recombinant p27 is susceptible to degradation by the NADH-26S proteasome. Purified recombinant p27 protein was subjected to 26S proteasomal degradation. Time kinetics were analyzed for p27 degradation in the presence of either 2 mM ATP or NADH. His-p27 was detected by immunoblotting with anti-His antibody. (**c**) p27 is stable in the presence of NADH alone or in the presence of 26S proteasomes, 2 mM NADH and 25 µM of the proteasomal inhibitor MG132. (**d**) NADH has minor inhibitory effect on the purified 20S proteasome mediated degradation of synthetic substrate of the Chymotrypsin-like activity as measured based on the hydrolysis of Suc-LLVY-AMC substrate. (**e**) NADH induces the 26S proteasomal activity as measured by the cleavage of the Suc-LLVY-AMC peptide by purified 26S proteasomes in the presence of either 1 mM ATP or NADH. (**f**) In vitro translated purified proteins are degraded by the NADH-26S proteasome. In vitro translated ^{35}S Methionine labeled flag-Yap1 was purified by flag IP followed by flag elution in the degradation buffer (see Section 2). Yap1 degradation kinetics by the 26S proteasome were analyzed in the presence of either 2 mM NADH or ATP. (**g**) Degradation of purified c-Fos was analyzed following 30 min incubation with 26S proteasomes in the presence of 2 mM ATP, ATPγS or NADH. (**h**) In vitro translated ^{35}S Methionine-Fos degradation was examined in the presence of 0.1 mM ATP and increasing concentrations of NADH at 37 °C for 60 min. (**i**) The data from at least three independent experiments, as shown in (**h**) were averaged and shown with their standard deviation. (**j**) Tau (UniProt P10636) is a highly disordered protein as demonstrated by the scores of prediction output using IUPred2A (red curve) and ANCHOR2 (blue curve) prediction programs. (**k**) NADH and ATPγS induce 26S proteasomal degradation of recombinant tau protein (rTau) degradation by the 26S proteasome. Degradation was examined in the presence of either 2 mM ATP, ATPγS or NADH.

4. Discussion

We show here that NADH can directly regulate substrate specificity of the 26S proteasome in vitro. Our findings here elaborate on our previous observation, showing that functional 26S proteasome can be stabilized by NADH [18]. As there is no straightforward way to distinguish the functionality of NADH-26S from the ATP-26S PC in the context of the cell, we chose a reductive in vitro approach to test our hypothesis that NADH-26S PC is capable of facilitating the degradation of proteins that do not require the ATP-dependent functions of the proteasome such as ubiquitin processing and unfolding. As such, intrinsically disordered proteins (IDPs) are the perfect candidates, as many of them are readily degraded by the 20S CP in vitro [27,28]. We show here that IDPs are also substrates of this new form of 26S PC, namely the NADH-26S PC.

Our findings illuminate that NADH is a regulator of proteasome function, as it can promote the degradation of IDPs but also inhibit the natural function of ATP-dependent degradation of the 26S proteasome, a function that can affect substrate specificity in the context of the cell. The possibility that NADH competes for ATP-binding sites, resulting in competitive inhibition of the ATP-dependent functions of the 26S PC, was ruled out [18]. The other possibility is that NADH binds the N-terminus of PSMC1 (as previously shown) and possibly other 19S subunits, inducing an allosteric effect that results in inhibition of the directional cycling of the ATP in the hexametric ATPase ring [18]. Recent structural analysis of the ATPγS-26S reveals a strong structural rearrangement of many of the 19S subunits in the ATPγS-bound state, resulting in higher alignment of the ATPase ring with the gate of the 20S core particle, suggesting a state with facilitated translocation of the substrate [35]. This is in agreement with what we observed here with ATPγS-26S degradation of IDPS. The possibility that NADH exerts a similar allosteric structural shift can be addressed with Cryo-EM analysis in the future. However, many of the predicted NADH binding motifs in the 19S subunits are conserved in vertebrates, but not in yeast [18].

The emerging picture is that a non-hydrolysable ATP is sufficient for 26S complex formation/stabilization, whereas ATP hydrolysis is essential for the process of unfolding of the structured substrates [7,10,11,32,36,37]. The finding that IDPs are degraded by the 26S PC in the absence of ATP hydrolysis lends further support to this model. Interestingly however, we also observed that ATP-26S PC was actually inefficient in degrading certain IDPs. This might mean that ATP has a role in allosterically gating the 20S catalytic complex or that the binding of ATP masks certain IDP to target the 26S proteasome.

Whether NADH-26S PC has a physiological role is an important question. Analyzing the effect of various concentrations of ATP and NADH on IDP and ODC degradation led us to conclude that the formation of NADH-26S PC is reversible and depends on the NADH/ATP ratio. Under normal ATP concentrations at 1–10 mM [38], the NADH concentration of 10–100 µM is required to efficiently form the NADH-26S PC. These physiological concentrations suggest that, in the cells, a certain fraction of the 26S PC is of the NADH type. Cellular NADH level is determined by the electron transfer chain (ETC) functionality [14,39]. Inefficient ETC activity results in higher NADH levels and lower levels of ATP-26S PC [14], and under this condition, the NADH-26S PC level is expected to increase. Under this condition, mitochondrial biogenesis is compromised by NADH-dependent degradation of PGC-1α, a transcription co-activator regulating mitochondrial biogenesis [40]. NADH also has an indirect role in inhibiting 20S PC-mediated IDP degradation via NQO1 [41]. Overall, the emerging picture is that defective ETC remodels the IDP degradation process to be more prone to NADH 26S PC degradation (Figure 4).

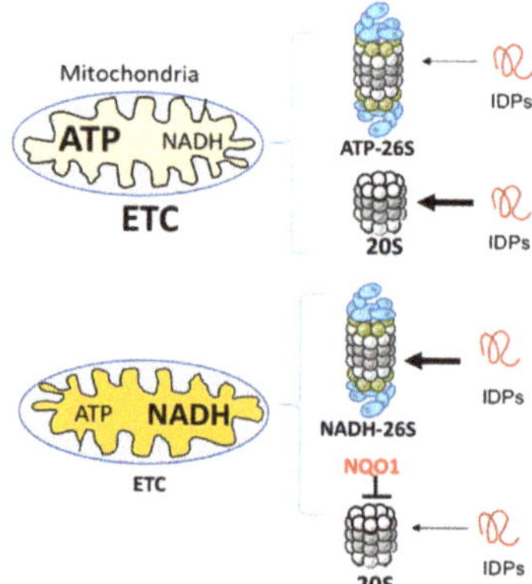

Figure 4. Mitochondria physiology and IDP degradation. Mitochondria with efficient electron transfer chain (ETC) activity (**upper**) generate a high level of ATP and low level of NADH. An opposite picture is obtained with mitochondria with low, or malfunctioning ETC (**lower**). The ATP and NADH levels regulate the formation of different types of 26S PC. Under high levels of NADH, NADH-26S PC is formed that is active in IDP degradation (the thick arrow). However, NADH through NQO1 has a repressive function in inhibiting IDP degradation by the 20S proteasome [42].

5. Conclusions

Proteasomal degradation of intrinsically disordered proteins (IDPs) or proteins consist of large disordered regions (IDRs) is not exclusively mediated via the classical ubiquitin-26S proteasome pathway but also subjected to ubiquitin-independent degradation. Here we investigated the process of IDP/IDR degradation via unique class of 26S proteasome that is free of ATP. Two different such 26S proteasomes were investigated; the non-hydrolysable ATPγS-26S and the recently reported NADH-26S proteasomes. We show here that both are in vitro active in degradation of IDP/IDR but not of ODC structured protein. The finding that physiological metabolite like NADH uniquely regulates IDP/IDR degradation exemplifies a new principle of how mitochondria, the key organelle in NADH production, regulate IDP/IDR homeostasis.

Author Contributions: Conceptualization, P.T. and Y.S.; methodology, J.A. and N.M.; validation, J.A. and N.M.; formal analysis, P.T.; writing—original draft preparation, P.T. and Y.S.; writing—review and editing, Y.S.; supervision, Y.S. All authors have read and agreed to the published version of the manuscript.

Funding: Supported by a research grant from the Israel Science Foundation (Grant no. 1591/15).

Conflicts of Interest: The authors declare no conflict of interest. The funders had no role in the design of the study; in the collection, analyses, or interpretation of data; in the writing of the manuscript, or in the decision to publish the results.

References

1. Hough, R.; Pratt, G.; Rechsteiner, M. Purification of two high molecular weight proteases from rabbit reticulocyte lysate. *J. Biol. Chem.* **1987**, *262*, 8303–8313. [PubMed]
2. Tanaka, K.; Ichihara, A. Involvement of proteasomes (multicatalytic proteinase) in ATP-dependent proteolysis in rat reticulocyte extracts. *FEBS Lett.* **1988**, *236*, 159–162. [CrossRef]
3. Waxman, L.; Fagan, J.M.; Goldberg, A.L. Demonstration of two distinct high molecular weight proteases in rabbit reticulocytes, one of which degrades ubiquitin conjugates. *J. Biol. Chem.* **1987**, *262*, 2451–2456. [PubMed]
4. Glickman, M.H.; Rubin, D.M.; Coux, O.; Wefes, I.; Pfeifer, G.; Cjeka, Z.; Baumeister, W.; A Fried, V.; Finley, D. A subcomplex of the proteasome regulatory particle required for ubiquitin-conjugate degradation and related to the COP9-Signalosome and eIF. *Cell* **1998**, *94*, 615–623. [CrossRef]

5. Schmidt, M.; Finley, D. Regulation of proteasome activity in health and disease. *Biochim. Biophys. Acta (BBA)-Bioenerg.* **2014**, *1843*, 13–25. [CrossRef]

6. Bard, J.A.; Goodall, E.A.; Greene, E.R.; Jonsson, E.; Dong, K.C.; Martin, A. Structure and function of the 26S Proteasome. *Annu. Rev. Biochem.* **2018**, *87*, 697–724. [CrossRef]

7. Collins, G.A.; Goldberg, A.L. The logic of the 26S proteasome. *Cell* **2017**, *169*, 792–806. [CrossRef]

8. Shyu, Y.J.; Liu, H.; Deng, X.; Hu, C.-D. Identification of new fluorescent protein fragments for bimolecular fluorescence complementation analysis under physiological conditions. *Biotechniques* **2006**, *40*, 61–66. [CrossRef]

9. Kleijnen, M.F.; Roelofs, J.; Park, S.; Hathaway, N.A.; Glickman, M.H.; King, R.W.; Finley, D. Stability of the proteasome can be regulated allosterically through engagement of its proteolytic active sites. *Nat. Struct. Mol. Biol.* **2007**, *14*, 1180–1188. [CrossRef]

10. Liu, C.-W.; Li, X.; Thompson, D.; Wooding, K.; Chang, T.-L.; Tang, Z.; Yu, H.; Thomas, P.J.; DeMartino, G.N. ATP binding and ATP hydrolysis play distinct roles in the function of 26s proteasome. *Mol. Cell* **2006**, *24*, 39–50. [CrossRef]

11. Smith, D.M.; Kafri, G.; Cheng, Y.; Ng, D.; Walz, T.; Goldberg, A.L. ATP binding to PAN or the 26S ATPases causes association with the 20S proteasome, gate opening, and translocation of unfolded proteins. *Mol. Cell* **2005**, *20*, 687–698. [CrossRef] [PubMed]

12. Leggett, D.S.; Hanna, J.; Borodovsky, A.; Crosas, B.; Schmidt, M.; Baker, R.T.; Walz, T.; Ploegh, H.; Finley, D. Multiple associated proteins regulate proteasome structure and function. *Mol. Cell* **2002**, *10*, 495–507. [CrossRef]

13. Livnat-Levanon, N.; Kevei, É.; Kleifeld, O.; Krutauz, D.; Segref, A.; Rinaldi, T.; Erpapazoglou, Z.; Cohen, M.; Reis, N.; Hoppe, T.; et al. Reversible 26S proteasome disassembly upon mitochondrial stress. *Cell Rep.* **2014**, *7*, 1371–1380. [CrossRef] [PubMed]

14. Meul, T.; Berschneider, K.; Schmitt, S.; Mayr, C.H.; Mattner, L.F.; Schiller, H.B.; Yazgili, A.S.; Wang, X.; Lukas, C.; Schlesser, C.; et al. Mitochondrial regulation of the 26S proteasome. *Cell Rep.* **2020**, *32*, 108059. [CrossRef] [PubMed]

15. Tsvetkov, P.; Detappe, A.; Cai, K.; Keys, H.R.; Brune, Z.; Ying, W.; Thiru, P.; Reidy, M.; Kugener, G.; Rossen, J.; et al. Author Correction: Mitochondrial metabolism promotes adaptation to proteotoxic stress. *Nat. Chem. Biol.* **2019**, *15*, 757. [CrossRef]

16. Ying, W. NAD+/NADH and NADP+/NADPH in cellular functions and cell death: Regulation and biological consequences. *Antioxid. Redox Signal.* **2008**, *10*, 179–206. [CrossRef]

17. Spinelli, J.B.; Haigis, M.C. The multifaceted contributions of mitochondria to cellular metabolism. *Nat. Cell Biol.* **2018**, *20*, 745–754. [CrossRef]

18. Tsvetkov, P.; Myers, N.; Eliav, R.; Adamovich, Y.; Hagai, T.; Adler, J.; Navon, A.; Shaul, Y. NADH binds and stabilizes the 26S proteasomes independent of ATP. *J. Biol. Chem.* **2014**, *289*, 11272–11281. [CrossRef]

19. Smith, D.M.; Fraga, H.; Reis, C.; Kafri, G.; Goldberg, A.L. ATP binds to proteasomal ATPases in pairs with distinct functional effects, implying an ordered reaction cycle. *Cell* **2011**, *144*, 526–538. [CrossRef]

20. Babu, M.M. The contribution of intrinsically disordered regions to protein function, cellular complexity, and human disease. *Biochem. Soc. Trans.* **2016**, *44*, 1185–1200. [CrossRef]

21. Dunker, A.K.; Obradovic, Z.; Romero, P.; Garner, E.C.; Brown, C.J. Intrinsic protein disorder in complete genomes. *Genome Inf. Ser. Workshop Genome Inf.* **2000**, *11*, 161–171.

22. Uversky, V.N.; Dunker, A.K. Biochemistry: Controlled chaos. *Science* **2008**, *322*, 1340–1341. [CrossRef] [PubMed]

23. Gsponer, J.; Futschik, M.E.; Teichmann, S.A.; Babu, M.M. Tight regulation of unstructured proteins: From transcript synthesis to protein degradation. *Science* **2008**, *322*, 1365–1368. [CrossRef] [PubMed]

24. Tsvetkov, P.; Asher, G.; Paz, A.; Reuven, N.; Sussman, J.L.; Silman, I.; Shaul, Y. Operational definition of intrinsically unstructured protein sequences based on susceptibility to the 20S proteasome. *Proteins Struct. Funct. Bioinform.* **2008**, *70*, 1357–1366. [CrossRef] [PubMed]

25. Asher, G.; Reuven, N.; Shaul, Y. 20S proteasomes and protein degradation "by default". *BioEssays* **2006**, *28*, 844–849. [CrossRef] [PubMed]

26. Liu, C.-W.; Corboy, M.J.; DeMartino, G.N.; Thomas, P.J. Endoproteolytic activity of the proteasome. *Science* **2003**, *299*, 408–411. [CrossRef]

27. Myers, N.; Olender, T.; Savidor, A.; Levin, Y.; Reuven, N.; Shaul, Y. The disordered landscape of the 20S proteasome substrates reveals tight association with phase separated granules. *Proteomics* **2018**, *18*, e1800076. [CrossRef]

28. Tsvetkov, P.; Myers, N.; Moscovitz, O.; Sharon, M.; Prilusky, J.; Shaul, Y. Thermo-resistant intrinsically disordered proteins are efficient 20S proteasome substrates. *Mol. Biosyst.* **2012**, *8*, 368–373. [CrossRef]

29. Murakami, Y.; Matsufuji, S.; Kameji, T.; Hayashi, S.-I.; Igarashi, K.; Tamura, T.; Tanaka, K.; Ichihara, A. Ornithine decarboxylase is degraded by the 26S proteasome without ubiquitination. *Nature* **1992**, *360*, 597–599. [CrossRef]

30. Kahana, C. Protein degradation, the main hub in the regulation of cellular polyamines. *Biochem. J.* **2016**, *473*, 4551–4558. [CrossRef]

31. Coffino, P. Antizyme, a mediator of ubiquitin-independent proteasomal degradation. *Biochimie* **2001**, *83*, 319–323. [CrossRef]

32. Li, X.; DeMartino, G.N. Variably modulated gating of the 26S proteasome by ATP and polyubiquitin. *Biochem. J.* **2009**, *421*, 397–404. [CrossRef] [PubMed]

33. Ou, L.; Waddell, M.B.; Kriwacki, R.W. Mechanism of cell cycle entry mediated by the intrinsically disordered protein P27Kip1. *ACS Chem. Biol.* **2012**, *7*, 678–682. [CrossRef]

34. Tsytlonok, M.; Hemmen, K.; Hamilton, G.; Kolimi, N.; Felekyan, S.; Seidel, C.A.; Tompa, P.; Sanabria, H. Specific conformational dynamics and expansion underpin a multi-step mechanism for specific binding of p27 with Cdk2/Cyclin A. *J. Mol. Biol.* **2020**, *432*, 2998–3017. [CrossRef] [PubMed]

35. Asher, G.; Tsvetkov, P.; Kahana, C.; Shaul, Y. A mechanism of ubiquitin-independent proteasomal degradation of the tumor suppressors p53 and p73. *Genes Dev.* **2005**, *19*, 316–321. [CrossRef]

36. Henderson, A.; Erales, J.; Hoyt, M.A.; Coffino, P. Dependence of proteasome processing rate on substrate unfolding. *J. Biol. Chem.* **2011**, *286*, 17495–17502. [CrossRef]

37. Sauer, R.T.; Baker, T.A. AAA+ proteases: ATP-Fueled machines of protein destruction. *Annu. Rev. Biochem.* **2011**, *80*, 587–612. [CrossRef]

38. Ataullakhanov, F.I.; Vitvitsky, V.M. What determines the intracellular ATP concentration. *Biosci. Rep.* **2002**, *22*, 501–511. [CrossRef]

39. Chandel, N.S. Evolution of mitochondria as signaling organelles. *Cell Metab.* **2015**, *22*, 204–206. [CrossRef]

40. Adamovich, Y.; Shlomai, A.; Tsvetkov, P.; Umansky, K.B.; Reuven, N.; Estall, J.L.; Spiegelman, B.M.; Shaul, Y. The protein level of PGC-1, a key metabolic regulator, is controlled by NADH-Nqomol. *Cell. Biol.* **2013**, *33*, 2603–2613. [CrossRef]

41. Shaul, Y.; Tsvetkov, P.; Reuven, N. IDPs and protein degradation in the cell. In *Instrumental Analysis of Intrinsically Disordered Proteins*; Vladimir, N.U., Sonia, L., Eds.; John Wiley & Sons: Hoboken, NJ, USA, 2010; pp. 1–36.

42. Śledź, P.; Unverdorben, P.; Beck, F.; Pfeifer, G.; Schweitzer, A.; Förster, F.; Baumeister, W. Structure of the 26S proteasome with ATP-S bound provides insights into the mechanism of nucleotide-dependent substrate translocation. *Proc. Natl. Acad. Sci. USA* **2013**, *110*, 7264–7269. [CrossRef] [PubMed]

Publisher's Note: MDPI stays neutral with regard to jurisdictional claims in published maps and institutional affiliations.

SSP is based on the calculation of secondary chemical shifts, i.e., the difference between the measured and random coil chemical shifts (RCCS). We have used the corrected shifts for IDPs, POTENCI RCCS, for SSP analysis as POTENCI takes into account the effect of neighboring residues as well as the experimental conditions such as temperature, pH, and buffer conditions [72]. Our criterion to use SSP values is based on the lack of reporting of all measurable NMR parameters mentioned above, whereas for every IDP investigation, the backbone resonances are assigned (although only some are deposited in the biological magnetic resonance bank (BMRB) database) (Table 1). As such, we have summarized six reports in Figure 1 even though there have been more NMR studies on αS. As $^1H^N$ and ^{15}N chemical shifts are sensitive to experimental conditions and ^{13}C chemical shifts can be easily re-referenced and are good indicators of α-helices and β-sheets, we only used Cα, Cβ, and C′ chemical shifts for the SSP analysis (Figure 1).

Table 1. A list of nuclear magnetic resonance (NMR) studies on human αS. Seven reports deposited the assigned chemical shifts to the BMRB database, as shown in the third column.

Year	Sample Condition	BMRB	ref
2001	~100 µM αS, 100 mM NaCl, 10 mM Na$_2$HPO$_4$, pH 7.4, 283 K		[28]
2003	0.3 mM αS, 20 mM sodium phosphate, 50 mM SDS, pH 7.4, 298 K		[73]
	1 mM αS, 20 mM phosphate, 0.5 mM EDTA, 200 mM NaCl, 10% D$_2$O, pH 6.5, 285.5 K	5744 d	
	0.2 mM αS, PBS buffer, pH 7.4, 263 K	6968	[74]
	...phosphate, 140 mM NaCl, pH 2.5, 10% D$_2$O, 288 K		[35]
	...% D$_2$O, pH 3.0 & pH 7.4, 288 K	16342	[75]
			[76]
			[27]

Review

Salient Features of Monomeric Alpha-Synuclein Revealed by NMR Spectroscopy

Do-Hyoung Kim [1], Jongchan Lee [2], K. H. Mok [3,4], Jung Ho Lee [2,*] and Kyou-Hoon Han [5,*]

1   Core Facility Management Center, Division of KRIBB Strategic Projects, Korea Research Institute of Bioscience and Biotechnology (KRIBB), Daejeon 34141, Korea; organic2@kribb.re.kr
2   Department of Chemistry, Seoul National University, Seoul 08826, Korea; lejon0605@snu.ac.kr
3   Trinity Biomedical Sciences Institute (TBSI), School of Biochemistry & Immunology, Trinity College Dublin, The University of Dublin, Dublin 2, Ireland; mok1@tcd.ie
4   Centre for Research on Adaptive Nanostructures and Nanodevices (CRANN), Dublin 2, Ireland
5   Genome Editing Research Center, Division of Biomedical Sciences, Korea Research Institute of Bioscience and Biotechnology (KRIBB), Daejeon 34141, Korea
*   Correspondence: jungho.lee@snu.ac.kr (J.H.L.); khhan600@kribb.re.kr (K.-H.H.); Tel.: +82-2-880-4363 (J.H.L.); +82-42-860-4250 (K.-H.H.)

Received: 15 January 2020; Accepted: 4 March 2020; Published: 10 March 2020

Abstract: Elucidating the structural details of proteins is highly valuable and important for the proper understanding of protein function. In the case of intrinsically disordered proteins (IDPs), however, obtaining the structural details is quite challenging, as the traditional structural biology tools have only limited use. Nuclear magnetic resonance (NMR) is a unique experimental tool that provides ensemble conformations of IDPs at atomic resolution, and when studying IDPs, a slightly different experimental strategy needs to be employed than the one used for globular proteins. We address this point by reviewing many NMR investigations carried out on the α-synuclein protein, the aggregation of which is strongly correlated with Parkinson's disease.

Keywords: alpha-synuclein; NMR; secondary structure propensity; pre-structured motifs (PreSMos); intrinsically disordered protein

1. Introduction

Alpha-synuclein (αS) is a presynaptic terminal protein that is localized at the nuclear envelope and presynaptic nerve terminals [1,2]. This small 14 kDa protein is important for the normal function and maintenance of synapses [3]. Clinically, it is strongly correlated with the pathogenesis of Parkinson's disease (PD), a neurodegenerative movement disorder associated with the degeneration of dopaminergic neurons in substantia nigra [4], and familial early onset PD is often associated with the overexpression and mutations of αS [5–7]. Age-dependent motor dysfunction can be caused by neuronal fibrillar αS deposits known as Lewy bodies [8,9], the diagnostic hallmark of PD being spherical protein inclusions found in the cytoplasm of nigral neurons in the brains of PD patients.

The fibrillary aggregates of αS have a characteristic cross-β structure consisting of β-sheets, where the individual β-strands are perpendicular to the axis of the fibril [10–12]. These fibrillary aggregates are morphologically similar to the amyloid fibrils found in Alzheimer's disease neuritic plaques and in deposits associated with other amyloidogenic diseases [13,14]. In addition to fibrils, advances in the structural elucidation of αS oligomers have been made recently [15,16]. Theories of (i) pore formation followed by membrane leakage [17,18], (ii) receptor-mediated mechanisms [19,20], and (iii) cellular protection by binding with extracellular chaperones [21] have been discussed in terms of the underlying molecular pathology of PD.

Although human αS is composed of 140-amino acid residues, it does not form a stable globular structure [22]. In fact, it is a well-known member of the so-called intrinsically disordered proteins (IDPs), unorthodox proteins that do not form well-defined three-dimensional structures under non-denaturing physiologcal (or near-physiological) conditions [23–25]. The primary structure of αS can be separated into three parts: (i) the N-terminal region (residues 1–60) has a series of 11-amino acid repeats with a conserved KTKEGV motif that, upon binding to synthetic lipid vesicles or detergent micelles in vitro, adopts a highly helical conformation [26–28]; (ii) the residues 61–95 that contain two additional KTKEGV repeats and the hydrophobic amyloidogenic NAC (non-amyloid-β component) region, are known to be involved in the formation of amyloid fibrils both in vitro and in vivo [14,29]; and (iii) the highly acidic C-terminal region (residues 96–140) is responsible for the overall net negative charge.

The critical early step in the fibrillation of αS is believed to involve conformational transition from the native monomeric αS into an aggregation-prone partially folded intermediate [30–32]. The observation that the truncation of the acidic C-terminus accelerates fibril formation in vitro [33], and the aggregated αS found in Lewy bodies has a truncated C-terminus [34], suggests that αS aggregation is slowed down by intermolecular electrostatic repulsions among the negatively charged C-terminal regions. As the partially folded oligomeric intermediates that are formed along the αS fibril formation pathway are known to be cytotoxic, we need detailed information on the conformational characteristics of the αS monomer, i.e., whether the monomer may have any peculiar conformational features that would enhance formation of oligomeric intermediates. Such knowledge should also shed light on how this protein performs its normal function.

The structural features of an IDP are described in two levels, one at a global level and the other at a local level. The global conformation of an IDP is best described as an ensemble populated with rapidly interconverting conformers [35–38]. Ensemble description is useful for understanding the overall topology of an IDP, as it provides a radius of gyration and also information on transient long-range contacts. Such ensemble representation is generally applicable to any protein and has been used not only for IDPs but for the unfolded or partially-unfolded state of globular proteins [39–43].

Unlike globular proteins, the ensemble structures of IDPs are not superimposable and do not converge into a single tertiary structure. Although molecular dynamics (MD) simulation alone can produce a conformational ensemble of IDPs, a more accurate ensemble is obtained when the experimental restraints from nuclear magnetic resonance (NMR) measurements, such as residual dipolar coupling constants (RDCs) [44,45] and long-range distance restraints derived from paramagnetic relaxation enhancement (PRE) experiments [40,46,47] or small-angle x-ray scattering (SAXS) experiments [48], are incorporated. An interesting point is that the ensemble conformation of an IDP (e.g., αS and tau) can be more compact than a simple random coil [46,49]. In contrast to the global conformation, the local-level conformation of an IDP is described by transient local secondary structures, termed pre-structured motifs (PreSMos) (see below) [24,25,50] that highly resemble the residual secondary structures found as a folding initiation core in the partially unfolded state of a globular protein.

2. Pre-structured Motifs (PreSMos) in IDPs

One prominent feature that is observed in ~70% of all IDPs or intrinsically disordered regions (IDRs) that have been thoroughly characterized by NMR is that these proteins, although intrinsically unstructured, contain transient secondary structures known as pre-structured motifs (PreSMos) [24,25]. The term PreSMos was proposed, as many different descriptions have been coined that address fundamentally the same phenomenon—that certain regions of IDPs are pre-populated with secondary structures [24,25]. PreSMos are the target-binding fragments that are primed before actual target binding. Most PreSMos are alpha-helices, but in addition there can also exist left-handed polyproline II helices (PPII), β-turns, and β-strands, and these transient structures (pre-populated only ~30% on average) become stable secondary structures in their target bound state [23,51–53].

Interestingly, a few IDPs such as 4EBPs (eIF4E binding proteins) and VP16 transactivation domains (TAD) have been subjected to more than one NMR investigation and the results from different investigators agree well in terms of the presence and location of PreSMos. For example, in one NMR study on 4EBP1, the residues 56–63, which form the key binding interface to eIF4E, were found to form a helix PreSMo [52]. In another NMR study on 4EBP2, a homolog of 4EBP1 with a sequence homology of ~70%, the same residues were found to form a pre-structured helix [53]. Similarly, two independent NMR groups found the same helix PreSMo in VP16 TAD that encompass the residues 442–447 and 465–483 [54,55].

Common NMR parameters obtained from the NMR studies of IDPs—chemical shifts, interproton nuclear Overhauser effects (NOEs), R_1 and R_2 relaxation rates (occasionally incorporated into spectral density functions), ^{15}N-^{1}H heteronuclear NOEs, J coupling constants (mostly $^{3}J_{HNH\alpha}$ associated with a backbone torsion angle ϕ), temperature coefficients of backbone amide protons, and backbone amide-water proton exchange rates—can be used to determine if an IDP possesses a PreSMo [23–25]. A deviation of chemical shifts from random coil values indicates the presence of a secondary structure.

Short-range interproton NOEs, such as intraresidue $d_{\alpha N}(i, i)$ and sequential $d_{\alpha N}(i, i+1)$-type NOEs, are commonly observed in IDPs, whereas sequential $d_{NN}(i, i+1)$, medium-range $d_{\alpha N}(i, i+2)$, $d_{\alpha N}(i, i+3)$ and $d_{NN}(i, i+2)$ are observed when a PreSMo is present. The ratio of sequential $d_{\alpha N}(i, i+1)$ to sequential $d_{NN}(i, i+1)$ NOEs [56] and that of sequential $d_{\alpha N}(i, i+1)$ to intraresidue $d_{\alpha N}(i+1, i+1)$ NOEs [57,58] are excellent measures of the backbone torsion angles. Thus, the combined analysis of different types of interproton NOEs can show if an IDP contains a locally ordered secondary structure. However, one should be aware that relatively weak interproton NOEs are observe[...] secondary structures in IDPs are of a transient nature. In addition, long-r[...] absent in IDPs, as IDPs lack the sta[...] Whereas inte[...]

 *biomolecules*

Review

Salient Features of Monomeric Alpha-Synuclein Revealed by NMR Spectroscopy

Do-Hyoung Kim [1], Jongchan Lee [2], K. H. Mok [3,4], Jung Ho Lee [2,*] and Kyou-Hoon Han [5,*]

[1] Core Facility Management Center, Division of KRIBB Strategic Projects, Korea Research Institute of Bioscience and Biotechnology (KRIBB), Daejeon 34141, Korea; organic2@kribb.re.kr
[2] Department of Chemistry, Seoul National University, Seoul 08826, Korea; lejon0605@snu.ac.kr
[3] Trinity Biomedical Sciences Institute (TBSI), School of Biochemistry & Immunology, Trinity College Dublin, The University of Dublin, Dublin 2, Ireland; mok1@tcd.ie
[4] Centre for Research on Adaptive Nanostructures and Nanodevices (CRANN), Dublin 2, Ireland
[5] Genome Editing Research Center, Division of Biomedical Sciences, Korea Research Institute of Bioscience and Biotechnology (KRIBB), Daejeon 34141, Korea
* Correspondence: jungho.lee@snu.ac.kr (J.H.L.); khhan600@kribb.re.kr (K.-H.H.); Tel.: +82-2-880-4363 (J.H.L.); +82-42-860-4250 (K.-H.H.)

Received: 15 January 2020; Accepted: 4 March 2020; Published: 10 March 2020

Abstract: Elucidating the structural details of proteins is highly valuable and important for the proper understanding of protein function. In the case of intrinsically disordered proteins (IDPs), however, obtaining the structural details is quite challenging, as the traditional structural biology tools have only limited use. Nuclear magnetic resonance (NMR) is a unique experimental tool that provides ensemble conformations of IDPs at atomic resolution, and when studying IDPs, a slightly different experimental strategy needs to be employed than the one used for globular proteins. We address this point by reviewing many NMR investigations carried out on the α-synuclein protein, the aggregation of which is strongly correlated with Parkinson's disease.

Keywords: alpha-synuclein; NMR; secondary structure propensity; pre-structured motifs (PreSMos); intrinsically disordered protein

---

1. Introduction

Alpha-synuclein (αS) is a presynaptic terminal protein that is localized at the nuclear envelope and presynaptic nerve terminals [1,2]. This small 14 kDa protein is important for the normal function and maintenance of synapses [3]. Clinically, it is strongly correlated with the pathogenesis of Parkinson's disease (PD), a neurodegenerative movement disorder associated with the degeneration of dopaminergic neurons in substantia nigra [4], and familial early onset PD is often associated with the overexpression and mutations of αS [5–7]. Age-dependent motor dysfunction can be caused by neuronal fibrillar αS deposits known as Lewy bodies [8,9], the diagnostic hallmark of PD being spherical protein inclusions found in the cytoplasm of nigral neurons in the brains of PD patients.

The fibrillary aggregates of αS have a characteristic cross-β structure consisting of β-sheets, where the individual β-strands are perpendicular to the axis of the fibril [10–12]. These fibrillary aggregates are morphologically similar to the amyloid fibrils found in Alzheimer's disease neuritic plaques and in deposits associated with other amyloidogenic diseases [13,14]. In addition to fibrils, advances in the structural elucidation of αS oligomers have been made recently [15,16]. Theories of (i) pore formation followed by membrane leakage [17,18], (ii) receptor-mediated mechanisms [19,20], and (iii) cellular protection by binding with extracellular chaperones [21] have been discussed in terms of the underlying molecular pathology of PD.

Although human αS is composed of 140-amino acid residues, it does not form a stable globular structure [22]. In fact, it is a well-known member of the so-called intrinsically disordered proteins (IDPs), unorthodox proteins that do not form well-defined three-dimensional structures under non-denaturing physiologcal (or near-physiological) conditions [23–25]. The primary structure of αS can be separated into three parts: (i) the N-terminal region (residues 1–60) has a series of 11-amino acid repeats with a conserved KTKEGV motif that, upon binding to synthetic lipid vesicles or detergent micelles in vitro, adopts a highly helical conformation [26–28]; (ii) the residues 61–95 that contain two additional KTKEGV repeats and the hydrophobic amyloidogenic NAC (non-amyloid-β component) region, are known to be involved in the formation of amyloid fibrils both in vitro and in vivo [14,29]; and (iii) the highly acidic C-terminal region (residues 96–140) is responsible for the overall net negative charge.

The critical early step in the fibrillation of αS is believed to involve conformational transition from the native monomeric αS into an aggregation-prone partially folded intermediate [30–32]. The observation that the truncation of the acidic C-terminus accelerates fibril formation in vitro [33], and the aggregated αS found in Lewy bodies has a truncated C-terminus [34], suggests that αS aggregation is slowed down by intermolecular electrostatic repulsions among the negatively charged C-terminal regions. As the partially folded oligomeric intermediates that are formed along the αS fibril formation pathway are known to be cytotoxic, we need detailed information on the conformational characteristics of the αS monomer, i.e., whether the monomer may have any peculiar conformational features that would enhance formation of oligomeric intermediates. Such knowledge should also shed light on how this protein performs its normal function.

The structural features of an IDP are described in two levels, one at a global level and the other at a local level. The global conformation of an IDP is best described as an ensemble populated with rapidly interconverting conformers [35–38]. Ensemble description is useful for understanding the overall topology of an IDP, as it provides a radius of gyration and also information on transient long-range contacts. Such ensemble representation is generally applicable to any protein and has been used not only for IDPs but for the unfolded or partially-unfolded state of globular proteins [39–43].

Unlike globular proteins, the ensemble structures of IDPs are not superimposable and do not converge into a single tertiary structure. Although molecular dynamics (MD) simulation alone can produce a conformational ensemble of IDPs, a more accurate ensemble is obtained when the experimental restraints from nuclear magnetic resonance (NMR) measurements, such as residual dipolar coupling constants (RDCs) [44,45] and long-range distance restraints derived from paramagnetic relaxation enhancement (PRE) experiments [40,46,47] or small-angle x-ray scattering (SAXS) experiments [48], are incorporated. An interesting point is that the ensemble conformation of an IDP (e.g., αS and tau) can be more compact than a simple random coil [46,49]. In contrast to the global conformation, the local-level conformation of an IDP is described by transient local secondary structures, termed pre-structured motifs (PreSMos) (see below) [24,25,50] that highly resemble the residual secondary structures found as a folding initiation core in the partially unfolded state of a globular protein.

2. Pre-structured Motifs (PreSMos) in IDPs

One prominent feature that is observed in ~70% of all IDPs or intrinsically disordered regions (IDRs) that have been thoroughly characterized by NMR is that these proteins, although intrinsically unstructured, contain transient secondary structures known as pre-structured motifs (PreSMos) [24,25]. The term PreSMos was proposed, as many different descriptions have been coined that address fundamentally the same phenomenon—that certain regions of IDPs are pre-populated with secondary structures [24,25]. PreSMos are the target-binding fragments that are primed before actual target binding. Most PreSMos are alpha-helices, but in addition there can also exist left-handed polyproline II helices (PPII), β-turns, and β-strands, and these transient structures (pre-populated only ~30% on average) become stable secondary structures in their target bound state [23,51–53].

Interestingly, a few IDPs such as 4EBPs (eIF4E binding proteins) and VP16 transactivation domains (TAD) have been subjected to more than one NMR investigation and the results from different investigators agree well in terms of the presence and location of PreSMos. For example, in one NMR study on 4EBP1, the residues 56–63, which form the key binding interface to eIF4E, were found to form a helix PreSMo [52]. In another NMR study on 4EBP2, a homolog of 4EBP1 with a sequence homology of ~70%, the same residues were found to form a pre-structured helix [53]. Similarly, two independent NMR groups found the same helix PreSMo in VP16 TAD that encompass the residues 442–447 and 465–483 [54,55].

Common NMR parameters obtained from the NMR studies of IDPs—chemical shifts, interproton nuclear Overhauser effects (NOEs), R_1 and R_2 relaxation rates (occasionally incorporated into spectral density functions), ^{15}N-^1H heteronuclear NOEs, J coupling constants (mostly $^3J_{HNH\alpha}$ associated with a backbone torsion angle ϕ), temperature coefficients of backbone amide protons, and backbone amide-water proton exchange rates—can be used to determine if an IDP possesses a PreSMo [23–25]. A deviation of chemical shifts from random coil values indicates the presence of a secondary structure.

Short-range interproton NOEs, such as intraresidue $d_{\alpha N}(i, i)$ and sequential $d_{\alpha N}(i, i+1)$-type NOEs, are commonly observed in IDPs, whereas sequential $d_{NN}(i, i+1)$, medium-range $d_{\alpha N}(i, i+2)$, $d_{\alpha N}(i, i+3)$ and $d_{NN}(i, i+2)$ are observed when a PreSMo is present. The ratio of sequential $d_{\alpha N}(i, i+1)$ to sequential $d_{NN}(i, i+1)$ NOEs [56] and that of sequential $d_{\alpha N}(i, i+1)$ to intraresidue $d_{\alpha N}(i+1, i+1)$ NOEs [57,58] are excellent measures of the backbone torsion angles. Thus, the combined analysis of different types of interproton NOEs can show if an IDP contains a locally ordered secondary structure. However, one should be aware that relatively weak interproton NOEs are observed in IDPs as the secondary structures in IDPs are of a transient nature. In addition, long-range interproton NOEs are absent in IDPs, as IDPs lack the stable topology that leads to such NOEs.

Whereas interproton NOEs provide short-range (< 5Å) information, paramagnetic relaxation enhancement (PRE) provides long-range (up to ~25 Å) information that may be present in IDPs [46,59,60]. Furthermore, larger spin-spin relaxation rates (shorter T_2 values) are observed for the residues forming a locally-ordered segment, i.e., a PreSMo [23,25,52,61,62]. Similarly, a locally ordered protein backbone ^{15}N-^1H amide bond generates a positive ^{15}N-^1H heteronuclear NOE value. In addition, $^3J_{HNH\alpha}$ coupling constants of 6 Hz or lower will be observed for helix-forming residues, while $^3J_{HNH\alpha}$ values larger than 8 Hz will be observed for β-type conformations [63]. A small (< 5 ppb/deg) temperature coefficient of a backbone amide proton suggests that the proton is involved in hydrogen bonding, indicating a transient helix or β-type structure [23,54,62]. Residual dipolar couplings (RDCs) provide information on the structure and dynamics of bond orientations and are measured to assess the conformational details of IDPs [44,59].

Chemical shifts possess information on secondary structures and many analysis tools aim at assessing the conformation of IDPs from chemical shifts. First, the secondary structure propensity (SSP) algorithm uses the protein backbone chemical shifts ($^{13}C^\alpha$, $^{13}C^\beta$, $^{13}C'$, $^1H^\alpha$, $^1H^N$, and ^{15}N) to generate a residue-specific score ranging from 1 to −1, which corresponds to the fully α-helical and β-sheet structure of a well-ordered protein, respectively [64]. As the polyproline II helix is an important structural motif in IDPs, the δ2D algorithm estimates the relative population distribution of α-helices, polyproline II helices, β-sheets, and random coils from the backbone chemical shifts of IDPs [65]. If additional NMR parameters such as multiple J coupling values and local interproton NOEs are available for IDPs, maximum entropy Ramachandran map analysis (MERA) can be used to estimate the relative distribution of their dihedral angles on a Ramachandran map [66].

As mentioned in the introduction, another class of methods provides representative ensembles of IDPs. A statistical random coil generator (e.g., TraDES [67], Flexible-Meccano [68]) creates a pool of random conformers and an ensemble selection algorithm (e.g., ENSEMBLE [69], ASTEROIDS [70], BEGR [71]) chooses a subset of ensembles that best matches the experimental results. We used the SSP algorithm to estimate secondary structures as it is simple but powerful and only the backbone chemical shifts of αS were available for comparative analysis (Table 1).

SSP is based on the calculation of secondary chemical shifts, i.e., the difference between the measured and random coil chemical shifts (RCCS). We have used the corrected shifts for IDPs, POTENCI RCCS, for SSP analysis as POTENCI takes into account the effect of neighboring residues as well as the experimental conditions such as temperature, pH, and buffer conditions [72]. Our criterion to use SSP values is based on the lack of reporting of all measurable NMR parameters mentioned above, whereas for every IDP investigation, the backbone resonances are assigned (although only some are deposited in the biological magnetic resonance bank (BMRB) database) (Table 1). As such, we have summarized six reports in Figure 1 even though there have been more NMR studies on αS. As $^1H^N$ and ^{15}N chemical shifts are sensitive to experimental conditions and ^{13}C chemical shifts can be easily re-referenced and are good indicators of α-helices and β-sheets, we only used Cα, Cβ, and C' chemical shifts for the SSP analysis (Figure 1).

Table 1. A list of nuclear magnetic resonance (NMR) studies on human αS. Seven reports deposited the assigned chemical shifts to the BMRB database, as shown in the third column.

Year	Sample Condition	BMRB	ref
2001	~100 μM αS, 100 mM NaCl, 10 mM Na$_2$HPO$_4$, pH 7.4, 283 K		[28]
2003	0.3 mM αS, 20 mM sodium phosphate, 50 mM SDS, pH 7.4, 298 K	5744 [d]	[73]
2006	1 mM αS, 20 mM phosphate, 0.5 mM EDTA, 200 mM NaCl, 10% D$_2$O, pH 6.5, 285.5 K	6968	[74]
2008	0.2 mM αS, PBS buffer, pH 7.4, 263 K		[35]
2009	0.65 mM αS, 10 mM phosphate, 140 mM NaCl, pH 2.5, 10% D$_2$O, 288 K		[75]
2009	0.3 mM αS, 20 mM NaOAc, 100 mM NaCl, 10% D$_2$O, pH 3.0 & pH 7.4, 288 K	16342	[76]
2009	0.6 mM αS, 20mM Na$_2$HPO$_4$ (pH 6.0), 6% D$_2$O, 0,02% NaN$_3$, in phospholipids, 293 K		[27]
2010	0.6 mM wild-type αS, mutants (A30P, E46K, A53T), 20 mM Na$_2$HPO$_4$, pH 6.0, 6% D$_2$O, 0.02% NaN$_3$, in phospholipids, 293 K		[77]
2012	0.1 mM αS, 5 mM dioxane, 20 mM sodium phosphate buffer, pH 6, in phospholipids, 288 K		[78]
2013 [a]	1.7 mM αS, 10% D$_2$O, 90% H$_2$O, pH 6.2, 277 K	19257	[79]
2013 [b]	- mM αS, 20 mM Tris-HCl, pH 7, 100 mM NaCl, 10% D$_2$O, 288 K		[80]
2014	0.35 mM αS, 20 mM sodium phosphate, pH 6, 288 K		[81]
2014	50 μM αS, NaCl/sodium phosphate buffer, 5% glycerol, 288 K		[82]
2014	0.3 mM αS, 20 mM NaOAc, 100 mM NaCl, 10% D$_2$O, pH 3.0 & pH 7, 288 K		[83]
2015	0.5 mM/0.7 mM wild-type/H50Q αS, 10 mM sodium phosphate, pH 7.5, 100 mM NaCl, 5% D$_2$O, 0.01% NaN$_3$, 0.001% DSS, 283 K	25227	[84]
2015	~0.43 mM (6.6mg/mL) αS, 20 mM HEPES, 10% D$_2$O, pH 7.0, 277 K	26557	[85]
2016 [a]	0.4 mM αS, 20 mM sodium phosphate, 150 mM NaCl, pH 7.0, 283 K		[60]
2017 [c]	75 μM αS, PBS buffer, 0.02% NaN$_3$, pH 7.4, 310 K		[86]
2018	~1 mM αS, pH 5.0, 10% D$_2$O, 298 K		[87]
2018	1 mM αS, 20 mM phosphate, 200 mM NaCl, 0.5 mM EDTA, pH 6.5, 285.5 K, 295 K, 305 K, 315 K	27348	[88]

[a] In-cell NMR. [b] No information is available on the αS concentration. [c] Aggregation inhibition experiment. [d] BMRB 5744 is for a folded αS, and hence is not used for the calculation of secondary structure propensity (SSP) values.

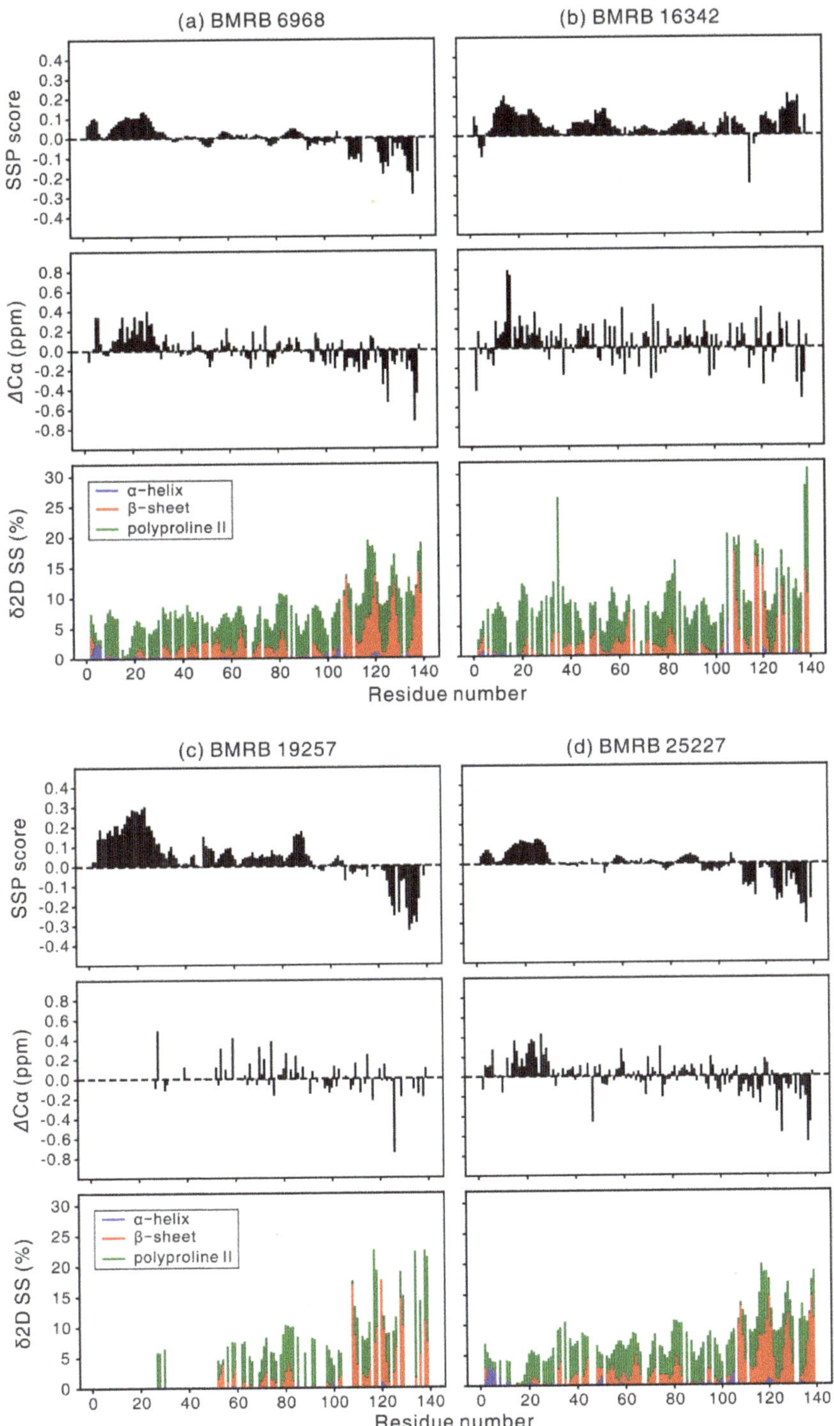

Figure 1. *Cont.*

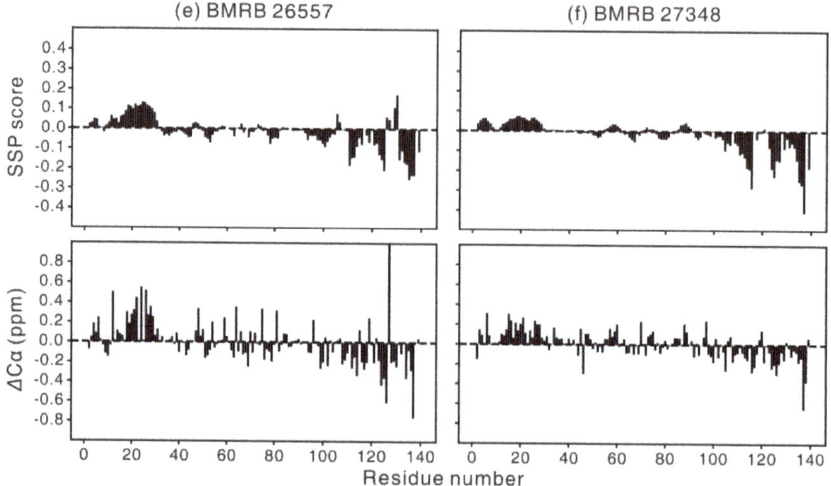

Figure 1. Comparison of the secondary structure propensity (SSP) scores, Cα secondary chemical shifts, and the secondary structure (SS) population from the δ2D analysis. The same ^{13}C chemical shifts were used as an input for the analysis. As at least three chemical shifts are required for the δ2D analysis, the δ2D percentage is not displayed for a residue, if even one of the Cα, Cβ, and C′ chemical shifts was unavailable from the BMRB database (e.g., at least one ^{13}C chemical shift information is missing for most residues of 1–50 in BMRB 19257, there was no C′ and Cβ chemical shift information for all residues in BMRB 26557 and 27348, respectively). Sample conditions for the six different NMR studies on α-synuclein (see Table 1) are (**a**) 1 mM αS, 20 mM phosphate, 0.5 mM EDTA, 200 mM NaCl, 10% D_2O, pH 6.5, 285.5 K (BMRB 6968), (**b**) 0.3 mM αS, 20 mM NaOAc, 100 mM NaCl, 10% D_2O, pH 3.0, 288 K (BMRB 16342), (**c**) 1.7 mM αS, 10% D_2O, 90% H_2O, 277 K, in-cell condition (BMRB 19257), (**d**) 0.5 mM αS, 10 mM sodium phosphate, pH 7.5, 100 mM NaCl, 5% D_2O, 283 K (BMRB 25227), (**e**) 0.43 mM αS, 20 mM HEPES, 10% D_2O, pH 7.0, 277 K (BMRB 26557), and (**f**) 1 mM αS, 200 mM NaCl, 0.5 mM EDTA, 20 mM phosphate, pH 6.5, 315 K (BMRB 27348).

3. Inconsistency

An early study by Eliezer et al. [28] described that the Cα chemical shifts for residues 6–37 in αS deviated more than ~0.3 ppm from RCCS, which indicates the existence of a helix PreSMo. In addition to this transient helix, four more transient helices, centered around residue numbers 44, 60, 84, and 100, were noted. The SSP values in Figure 1 seem to agree with this regarding the most prominent N-terminal helix formed by residues 10–30, even though the degree of pre-population of this helix does not match quantitatively among one another. Eliezer et al. [28] also showed that there are four more transient helices covering the NAC region. However, it was difficult to find such additional transient helices in our SSP analysis.

In Figure 1, two cases (Figure 1b and c) show two weak helices around residues 50 and 90. The SSP data obtained at low pH (Figure 1b) peculiarly show that there is an additional helix around residue 130 at the C-terminus, in contrast to the other five results, which have a β-type transient structure at the C-terminus. However, caution is advised when interpreting these results because, for example, the N-terminal helical propensities that are observed in all SSP plots (Figure 1) are not evident in the R_2 relaxation data [58,81]. Not every PreSMo shows a faster R_2 relaxation rate [23,25,52,61,62]. On the other hand, the concerted measurement of ^3J couplings implies that the ϕ angles have a very small helical tendency in the N-terminal 10–30 residues when the residue-specific average ϕ angles were compared in αS for each amino acid type [89].

An interesting observation is that the in-cell NMR data (Figure 1c) shows a higher population of the N-terminal helix than the *in vitro* results. This is probably due to differences in the sample conditions between in-cell NMR measurements and *in vitro* experiments, e.g., crowding effect, presence of lipid membranes. When αS was purified after it had been deliberately (by co-expressing an enzyme) N-terminal acetylated in *Escherichia coli* (*E. coli*) cells, the N-terminal residues of αS possessed enhanced helicity [78]. As it is known that αS can be N-terminal acetylated during and after translation in human cells, internally expressed or externally introduced αS in these cells is likely in an acetylated state [60]. However, because in-cell NMR data used to generate Figure 1c were obtained inside *E. coli* cells without enzymatic acetylation, we can safely rule out the effect of acetylation on αS conformation.

Another possibility is that the higher N-terminal helicity observed in this in-cell report is simply due to the fact that only carbonyl chemical shifts were used in computing the SSP scores in the N-terminal region as Cα and Cβ chemical shifts were not available. In addition to the discrepancy in the SSP values from different NMR studies (Figure 1), there is an intriguing point regarding the presence of the N-terminal transient helix. When αS was investigated by the δ2D algorithm, no pre-structured helix around residue 25 was found at the N-terminus [78]. Therefore, we have applied different computation tools to interpret chemical shifts in terms of the αS conformation. There were small-but-significant differences when different computational tools were employed (Figure 1), even when the same ^{13}C chemical shifts were used as an input.

As the Cα secondary chemical shift is by itself a good indicator of secondary structure, whose value is positive for α-helices and negative for β-type structures, we first compared the SSP scores to Cα secondary chemical shifts (Figure 1). They showed similar trends, although the SSP scores showed less fluctuations among adjacent residues. This observation can be ascribed to the algorithm of SSP, which averages the secondary chemical shifts from *i*-2 to *i*+2, and to the combined analysis of different nuclei that would reduce the observed error [64]. Next, the SSP scores were compared with the δ2D results using the same ^{13}C chemical shifts as an input (Figure 1). Some secondary structures were commonly observed in the two cases, as in the α-helix at residues 3–6 and β-sheet at the C-terminal region of BMRB 6968 and BMRB 25227.

However, the secondary structure patterns do not generally match, particularly for the α-helix (residues 10–30) that is clearly observed in the SSP analysis. This can be ascribed to the small populations of secondary structure in αS, redistribution of the α-helix population in SSP into the α-helix and PPII in δ2D [65], and the different RCCS employed in the two methods, namely, POTENCI for SSP and CamCoil [90] for δ2D. Taken together, caution is advised when interpreting chemical shift data in terms of conformation. This is true even when using NMR data other than chemical shifts, although collective analysis of independent data from NMR and other experiments would aid in accurate description of IDP conformation.

In NMR studies of globular proteins, slight differences in the NMR sample conditions (protein concentration, buffer, temperature, pH, etc.) do not significantly influence the overall 3D structure. The same is true even for IDPs as was seen in VP16 TAD and 4EBP1/2; slightly different sample conditions did not influence the results in terms of the presence and/or location of PreSMos. Then why do the results of different NMR studies on αS conformation not completely agree regarding the location and the degree of pre-population of transient structures?

4. The Effect of Environmental Conditions on αS Conformation

We believe that several factors, as described below, should be scrutinized prior to an NMR investigation of aggregation-prone proteins, such as αS.

4.1. Protein Concentration

In the early days of protein NMR experiments, often a very high (>10 mM) sample concentration was used in order to compensate for poor signal-to-noise ratios in low-field (<5 Tesla) NMR spectrometers [56]. Thus, a high concentration sample was inevitably used by NMR pulse sequence developers [91–93] to test their pulse schemes. The preparation of highly-concentrated protein samples was possible only because the investigators wisely chose a highly-soluble protein, e.g., bovine pancreatic trypsin inhibitor (BPTI). The measured NMR parameters with highly soluble proteins conformed well to, and complemented existing knowledge, e.g., the 3D structures known by x-ray crystallography or the properties that can be deduced from such structures.

In subsequent protein NMR studies, sample concentrations have decreased over the few decades to such an extent that they are now comparable to, or lower than ~1 mM (Table 1). In most investigations, the importance on whether a particular protein is truly a monomeric state cannot be overemphasized. Measuring the concentration dependence of mean residue ellipticity θ in circular dichroism experiments, and showing that this relationship is linear around the NMR sample concentrations, is one way of demonstrating the monomeric nature. Alternatively, analytical ultracentrifugation can be used to detect oligomerization in protein samples. A change of the NMR line widths and diffusion rates can also indicate the presence of oligomeric species.

Ensuring the monomeric state of a protein during data acquisition becomes critical when we deal with any protein with aggregation tendencies, such as αS. All SSP scores in Figure 1 show that the C-terminus of αS has a β-type structure except for the low pH case in Figure 1b. Is it possible that protein concentrations of 300 μM–1.7 mM impose a fibrillar β-type conformation in αS? Eliezer et al. [28] used a ~100 μM protein sample and observed only the α-helical propensities for αS. Does this suggest that NMR experiments at a very low protein concentration (<50 μM) are required to assess the truly monomeric state?

4.2. pH

The pH of a sample is also an important factor that can influence the conformation of proteins. In the case of globular proteins, provided that the charged states of surface-exposed amino acids have no impact on the overall globular topology, the 3D structures determined at different pH values are known to be quite similar. In the case of IDPs, however, most residues are fully exposed to the solvent. The charged states of surface hydrophilic residues are very likely to influence the topology, not to mention the local structures. An exception would be the residues that form PreSMos as they can transiently form hydrogen bonds and hence experience slow backbone amide-water proton exchange. In an extreme case of the spring-loaded mechanism adopted by influenza hemagglutinin, a flexible loop undergoes a drastic conformational change to become a helix that induces viral fusion with the cellular membrane [94]. Cho et al. examined the conformations of αS at pH 3 (Figure 1b) and pH 7.4 and the two were quite different with respect to the helical content of both C- and N-terminal parts [76]. This calls for an argument that the pH of an αS sample needs to be controlled with caution.

4.3. Temperature

For globular proteins which have a relatively rigid backbone topology, 3D structures determined at different temperatures are not grossly different. For IDPs, like αS, with a tendency to aggregate, performing NMR experiments at 277 K (Figure 1c,e) may lower the hydrophobic effect and thereby lead to conformational ensembles that are different from those found in the cellular environment. For globular proteins, performing an NMR experiment at temperatures that are within a non-denaturing range is acceptable. Yet, for an IDP whose oligomerization is directly related to the pathology of PD, temperature can have a significant effect on its conformation.

4.4. Buffer and Ionic Strength

Since the thermodynamic stability of globular proteins allows them to maintain their structural homeostasis, changes in the solvent buffers are considered unlikely to pose a problem in 3D structure determination. This may not be true for IDPs, as the counterion shielding of polar residues on the surface will influence the conformations of IDPs to a greater extent than of globular proteins. In one investigation, measurements done in rather high ionic strength (200 mM) and high temperature (42 °C) conditions (Figure 1f) resulted in the degree of pre-population of the N-terminal helix to be lower than for others. Note that the effect of ionic strengths on the protonation states of side chains is compensated by using POTENCI RCCS.

4.5. Lipid Membranes

The functional and pathogenic role of αS is closely related to its interaction with lipid membranes. For example, αS is involved in membrane remodeling [95], clustering [96], and maintaining the pool of synaptic vesicles [97], and chaperoning SNARE-complex assembly [98]. In addition, fibril formation of αS is influenced by lipid interaction, where the promotion and inhibition of αS aggregation depends on lipid conditions [99,100]. Therefore, it is important to investigate the detailed mode of interaction between αS and lipid membranes. When doing so, we wish to note the importance of being aware of the incomplete removal of detergents and lipids during the purification of the αS protein, as they can cause small line broadening in NMR resonances. Thus, lipids should be completely removed for studies on free αS, or the lipid conditions should be precisely controlled for studies on the interaction between αS and lipid membranes.

The αS-lipid interaction has been monitored by many biophysical tools. As there is a strong correlation between the extent of αS-lipid binding and lipid-induced αS helicity, circular dichroism has been the most popular method to measure αS-lipid binding [28,101]. In addition, fluorescence correlation spectroscopy [102] and fluorescence anisotropy [27] methods have been used to measure the change of translational and rotational diffusion of αS, respectively, in the presence of lipid vesicles. However, residue-level analysis can only be performed by NMR spectroscopy.

In the initial work by Eliezer and coworkers [28], interactions between αS and lipid vesicles were monitored by NMR spectroscopy. They observed a complete disappearance of residues 1–100 in the protein backbone ^{1}H-^{15}N correlation spectra with a molar excess of SDS micelles or small unilamellar vesicles (SUV) over αS, which indicates that the N-terminal region of αS binds tightly to the vesicle, while the C-terminal tail is not bound to the vesicle and preserves its disordered conformation. Furthermore, the structural details of αS bound to SDS micelles were monitored by NMR spectroscopy [103]. Residual dipolar couplings were measured to determine the two (V3-V37 and K45-T92) anti-parallel curved helices connected by a linker when αS was bound to a micelle.

In the presence of a small amount of SUV composed of phospholipids (lipid molecule/αS < 5, lipid vesicle/αS < 0.001), the condition that mimics the synaptic environment, due to the high concentration of αS at the synapse, only the change in intensity and line width of αS backbone NMR resonances was observed without any perturbation in chemical shift values [27]. Although a chemical shift analysis cannot be performed to assess the bound-state conformation, the residue-specific intensity profiles showed that the residues 1–100 were bound to the vesicle with multiple tight binding modes, whereas the 40 C-terminal residues remained flexible, as in the high lipid concentration condition. In addition, transferred NOE experiments showed that the 100 N-terminal residues form a helical structure upon binding to SUV [27]. The lysine sidechains of αS are protected from acetylation by N-succinimidyl acetate in a residue-position and in a lipid-concentration dependent manner, supporting the presence of multiple binding modes between αS and lipid membranes [104].

We can also think of a mechanism on how αS strongly binds to lipid membranes. Whether or not αS contains a transient helix at its N-terminus would change its binding mode to lipid membranes, as well as the associated thermodynamic and kinetic parameters. In the early days of IDP research, when IDPs were viewed to be completely unstructured without any trace of transient secondary structures, only the induced fit mechanism involving for example, a coil-to-helix transition [105], was used to explain IDP-target binding. With knowledge that dozens of IDPs (~70% of the IDPs characterized) contain transient secondary structures termed now as PreSMos (Pre-Structured Motifs) [24,25,50], an alternative target binding mechanism of IDPs, conformational selection, is now widely considered [106,107] (Figure 2).

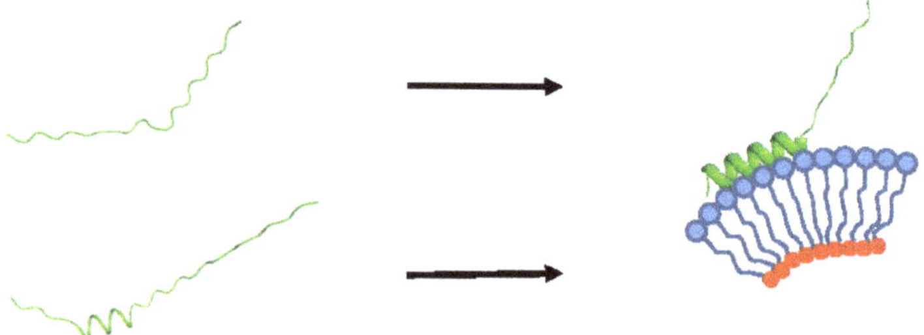

Figure 2. An illustration of the lipid-binding mode of αS. (top) Induced fit, (bottom) conformational selection. In the latter, an N-terminal helix is inserted merely to indicate the fact that a transient helix is present at the N-terminus of αS, i.e., the location of the helix is not exact. The thickness of the helix ribbon is adjusted to reflect the population of the helix.

5. Summary

In this review, we have presented secondary structure propensities calculated from the chemical shifts of αS deposited in the BMRB database [108]. Although the calculated propensities appear similar (Figure 1), there are subtle-yet-significant differences among the secondary structures and the degree of pre-population of the secondary structures. For example, there is a stronger helical propensity near the N-terminus of αS under cellular conditions (Figure 1c), whereas the pH 3 condition introduces a transient helical structure at the C-terminus (Figure 1b). SSP is a common tool; however, it is one of several methods to assess secondary structures in IDPs. Development of better experimental methods and more accurate analysis tools seem to be needed to improve the accuracy for describing IDP conformations.

Certain IDPs, including αS, are relatively more prone to aggregation than globular proteins. It is possible that αS represents one of those IDPs for which carrying out an NMR structural investigation should be performed with much more precaution than one might think, e.g., under a well-controlled condition resembling the native cellular settings as much as possible [109]. Certain sample conditions could modulate IDP conformations as exemplified here for the αS protein. Notwithstanding the plethora of previous S NMR investigations, it remains a tempting challenge to perform further investigations on αS under "authentic" conditions in order to obtain the genuine conformation of monomeric αS.

Author Contributions: K.-H.H., K.H.M., and J.H.L. wrote the paper; D.-H.K. made Table 1; J.L. and J.H.L made Figure 1. K.-H.H. and D.-H.K. made Figure 2. All authors have read and agreed to the published version of the manuscript.

Acknowledgments: This work was supported by a grant from National Research Council of Science and Technology (NST) by the Korea government (MSIT) (No. CCL-19-17-KRIBB) (K.-H.H.), a Marie Sklodowska Curie Initial Training Network (MSC ITN) "TRACT" (TRAining in Cancer mechanisms and Treatment) grant (K.H.M.), creative-pioneering researchers program through the Seoul National University (SNU) and the National Research Foundation of Korea (2019R1C1C1009685) (J.H.L.).

Conflicts of Interest: The authors declare no conflict of interest.

References

1. Maroteaux, L.; Campanelli, J.; Scheller, R. Synuclein: A neuron-specific protein localized to the nucleus and presynaptic nerve terminal. *J. Neurosci.* **1988**, *8*, 2804–2815. [PubMed]
2. Iwai, A.; Masliah, E.; Yoshimoto, M.; Ge, N.; Flanagan, L.; De Silva, H.R.; Kittel, A.; Saitoh, T. The precursor protein of non-Aβ component of Alzheimer's disease amyloid is a presynaptic protein of the central nervous system. *Neuron* **1995**, *14*, 467–475. [PubMed]
3. Burré, J. The synaptic function of α-synuclein. *J. Parkinsons Dis.* **2015**, *5*, 699–713. [PubMed]

4.   Poewe, W.; Seppi, K.; Tanner, C.M.; Halliday, G.M.; Brundin, P.; Volkmann, J.; Schrag, A.-E.; Lang, A.E. Parkinson disease. *Nat. Rev. Dis. Primers* **2017**, *3*, 17013.

5.   Singleton, A.; Farrer, M.; Johnson, J.; Singleton, A.; Hague, S.; Kachergus, J.; Hulihan, M.; Peuralinna, T.; Dutra, A.; Nussbaum, R. α-Synuclein locus triplication causes Parkinson's disease. *Science* **2003**, *302*, 841.

6.   Chartier-Harlin, M.-C.; Kachergus, J.; Roumier, C.; Mouroux, V.; Douay, X.; Lincoln, S.; Levecque, C.; Larvor, L.; Andrieux, J.; Hulihan, M. α-synuclein locus duplication as a cause of familial Parkinson's disease. *Lancet* **2004**, *364*, 1167–1169.

7.   Flagmeier, P.; Meisl, G.; Vendruscolo, M.; Knowles, T.P.; Dobson, C.M.; Buell, A.K.; Galvagnion, C. Mutations associated with familial Parkinson's disease alter the initiation and amplification steps of α-synuclein aggregation. *Proc. Natl. Acad. Sci. USA* **2016**, *113*, 10328–10333.

8.   Spillantini, M.G.; Schmidt, M.L.; Lee, V.M.Y.; Trojanowski, J.Q.; Jakes, R.; Goedert, M. α-Synuclein in Lewy bodies. *Nature.* **1997**, *388*, 839–840.

9.   Braak, H.; Del Tredici, K.; Rüb, U.; De Vos, R.A.; Steur, E.N.J.; Braak, E. Staging of brain pathology related to sporadic Parkinson's disease. *Neurobiol. Aging* **2003**, *24*, 197–211.

10. Kirschner, D.A.; Abraham, C.; Selkoe, D.J. X-ray diffraction from intraneuronal paired helical filaments and extraneuronal amyloid fibers in Alzheimer disease indicates cross-beta conformation. *Proc. Natl. Acad. Sci. USA* **1986**, *83*, 503–507.

11. Serpell, L.C.; Berriman, J.; Jakes, R.; Goedert, M.; Crowther, R.A. Fiber diffraction of synthetic α-synuclein filaments shows amyloid-like cross-β conformation. *Proc. Natl. Acad. Sci. USA* **2000**, *97*, 4897–4902. [PubMed]

12. Tuttle, M.D.; Comellas, G.; Nieuwkoop, A.J.; Covell, D.J.; Berthold, D.A.; Kloepper, K.D.; Courtney, J.M.; Kim, J.K.; Barclay, A.M.; Kendall, A. Solid-state NMR structure of a pathogenic fibril of full-length human α-synuclein. *Nat. Struct. Mol. Biol.* **2016**, *23*, 409. [PubMed]

13. Conway, K.A.; Harper, J.D.; Lansbury, P.T. Fibrils formed in vitro from α-synuclein and two mutant forms linked to Parkinson's disease are typical amyloid. *Biochemistry.* **2000**, *39*, 2552–2563. [PubMed]

14. El-Agnaf, O.M.A.; Jakes, R.; Curran, M.D.; Wallace, A. Effects of the mutations Ala30 to Pro and Ala53 to Thr on the physical and morphological properties of α-synuclein protein implicated in Parkinson's disease. *FEBS Lett.* **1998**, *440*, 67–70. [PubMed]

15. Chen, S.W.; Drakulic, S.; Deas, E.; Ouberai, M.; Aprile, F.A.; Arranz, R.; Ness, S.; Roodveldt, C.; Guilliams, T.; De-Genst, E.J. Structural characterization of toxic oligomers that are kinetically trapped during α-synuclein fibril formation. *Proc. Natl. Acad. Sci. USA* **2015**, *112*, E1994–E2003. [PubMed]

16. Varela, J.A.; Rodrigues, M.; De, S.; Flagmeier, P.; Gandhi, S.; Dobson, C.M.; Klenerman, D.; Lee, S.F. Optical structural analysis of individual α-synuclein oligomers. *Angew. Chem.* **2018**, *57*, 4886–4890.

17. Fusco, G.; Pape, T.; Stephens, A.D.; Mahou, P.; Costa, A.R.; Kaminski, C.F.; Schierle, G.S.K.; Vendruscolo, M.; Veglia, G.; Dobson, C.M. Structural basis of synaptic vesicle assembly promoted by α-synuclein. *Nat. Commun.* **2016**, *7*, 12563.

18. Fusco, G.; Chen, S.W.; Williamson, P.T.; Cascella, R.; Perni, M.; Jarvis, J.A.; Cecchi, C.; Vendruscolo, M.; Chiti, F.; Cremades, N. Structural basis of membrane disruption and cellular toxicity by α-synuclein oligomers. *Science* **2017**, *358*, 1440–1443.

19. Mao, X.; Ou, M.T.; Karuppagounder, S.S.; Kam, T.-I.; Yin, X.; Xiong, Y.; Ge, P.; Umanah, G.E.; Brahmachari, S.; Shin, J.-H. Pathological α-synuclein transmission initiated by binding lymphocyte-activation gene 3. *Science* **2016**, *353*, aah3374.

20. Ferreira, D.G.; Temido-Ferreira, M.; Miranda, H.V.; Batalha, V.L.; Coelho, J.E.; Szegö, É.M.; Marques-Morgado, I.; Vaz, S.H.; Rhee, J.S.; Schmitz, M. α-synuclein interacts with PrP C to induce cognitive impairment through mGluR5 and NMDAR2B. *Nat. Neurosci.* **2017**, *20*, 1569.

21. Whiten, D.R.; Cox, D.; Horrocks, M.H.; Taylor, C.G.; De, S.; Flagmeier, P.; Tosatto, L.; Kumita, J.R.; Ecroyd, H.; Dobson, C.M. Single-molecule characterization of the interactions between extracellular chaperones and toxic α-synuclein oligomers. *Cell Rep.* **2018**, *23*, 3492–3500. [PubMed]

22. Weinreb, P.H.; Zhen, W.; Poon, A.W.; Conway, K.A.; Lansbury, P.T. NACP, a protein implicated in Alzheimer's disease and learning, is natively unfolded. *Biochemistry.* **1996**, *35*, 13709–13715. [PubMed]

23. Lee, H.; Mok, K.H.; Muhandiram, R.; Park, K.-H.; Suk, J.-E.; Kim, D.-H.; Chang, J.; Sung, Y.C.; Choi, K.Y.; Han, K.-H. Local structural elements in the mostly unstructured transcriptional activation domain of human p53. *J. Biol. Chem.* **2000**, *275*, 29426–29432. [PubMed]

24. Lee, S.-H.; Kim, D.-H.; Han, J.; Cha, E.-J.; Lim, J.-E.; Cho, Y.-J.; Lee, C.; Han, K.-H. Understanding pre-structured motifs (PreSMos) in intrinsically unfolded proteins. *Curr. Protein Pept. Sci.* **2012**, *13*, 34–54.

25. Kim, D.-H.; Han, K.-H. PreSMo target-binding signatures in intrinsically disordered proteins. *Mol. Cells* **2018**, *41*, 889–899.

26. Uéda, K.; Fukushima, H.; Masliah, E.; Xia, Y.; Iwai, A.; Yoshimoto, M.; Otero, D.A.; Kondo, J.; Ihara, Y.; Saitoh, T. Molecular cloning of cDNA encoding an unrecognized component of amyloid in Alzheimer disease. *Proc. Natl. Acad. Sci. USA* **1993**, *90*, 11282–11286.

27. Bodner, C.R.; Dobson, C.M.; Bax, A. Multiple tight phospholipid-binding modes of α-synuclein revealed by solution NMR spectroscopy. *J. Mol. Biol.* **2009**, *390*, 775–790.

28. Eliezer, D.; Kutluay, E.; Bussell, R.; Browne, G. Conformational properties of α-synuclein in its free and lipid-associated states. *J. Mol. Biol.* **2001**, *307*, 1061–1073.

29. El-Agnaf, O.M.A.; Irvine, G.B. Aggregation and neurotoxicity of α-synuclein and related peptides. *Biochem. Soc. Trans.* **2002**, *30*, 559–565.

30. Uversky, V.N.; Li, J.; Fink, A.L. Evidence for a partially folded intermediate in α-synuclein fibril formation. *J. Biol. Chem.* **2001**, *276*, 10737–10744.

31. Uversky, V.N.; Fink, A.L. Conformational constraints for amyloid fibrillation: The importance of being unfolded. *Biochim. Biophys. Acta* **2004**, *1698*, 131–153.

32. Wetzel, R. For protein misassembly, it's the "I" decade. *Cell.* **1996**, *86*, 699–702. [PubMed]

33. Li, J.; Uversky, V.N.; Fink, A.L. Conformational behavior of human α-synuclein is modulated by familial Parkinson's disease point mutations A30P and A53T. *Neurotoxicology* **2002**, *23*, 553–567. [PubMed]

34. Conway, K.A.; Harper, J.D.; Lansbury, P.T. Accelerated in vitro fibril formation by a mutant α-synuclein linked to early-onset Parkinson disease. *Nat. Med.* **1998**, *4*, 1318–1320. [PubMed]

35. Wu, K.-P.; Kim, S.; Fela, D.A.; Baum, J. Characterization of conformational and dynamic properties of natively unfolded human and mouse alpha-synuclein ensembles by NMR: implication for aggregation. *J. Mol. Biol.* **2008**, *378*, 1104–1115.

36. Esteban-Martín, S.; Fenwick, R.B.; Salvatella, X. Refinement of ensembles describing unstructured proteins using NMR residual dipolar couplings. *J. Am. Chem. Soc.* **2010**, *132*, 4626–4632.

37. Jensen, M.R.; Salmon, L.c.; Nodet, G.; Blackledge, M. Defining conformational ensembles of intrinsically disordered and partially folded proteins directly from chemical shifts. *J. Am. Chem. Soc.* **2010**, *132*, 1270–1272.

38. Fisher, C.K.; Stultz, C.M. Constructing ensembles for intrinsically disordered proteins. *Curr. Opin. Struct. Biol.* **2011**, *21*, 426–431.

39. Barbar, E. NMR characterization of partially folded and unfolded conformational ensembles of proteins. *Biopolymers* **1999**, *51*, 191–207.

40. Lindorff-Larsen, K.; Kristjansdottir, S.; Teilum, K.; Fieber, W.; Dobson, C.M.; Poulsen, F.M.; Vendruscolo, M. Determination of an ensemble of structures representing the denatured state of the bovine acyl-coenzyme a binding protein. *J. Am. Chem. Soc.* **2004**, *126*, 3291–3299.

41. Kristjansdottir, S.; Lindorff-Larsen, K.; Fieber, W.; Dobson, C.M.; Vendruscolo, M.; Poulsen, F.M. Formation of native and non-native interactions in ensembles of denatured ACBP molecules from paramagnetic relaxation enhancement studies. *J. Mol. Biol.* **2005**, *347*, 1053–1062. [PubMed]

42. Marsh, J.A.; Forman-Kay, J.D. Structure and disorder in an unfolded state under nondenaturing conditions from ensemble models consistent with a large number of experimental restraints. *J. Mol. Biol.* **2009**, *391*, 359–374. [PubMed]

43. Huang, J.-r.; Grzesiek, S. Ensemble calculations of unstructured proteins constrained by RDC and PRE data: a case study of urea-denatured ubiquitin. *J. Am. Chem. Soc.* **2010**, *132*, 694–705. [PubMed]

44. Bernado, P.; Blanchard, L.; Timmins, P.; Marion, D.; Ruigrok, R.W.; Blackledge, M. A structural model for unfolded proteins from residual dipolar couplings and small-angle x-ray scattering. *Proc. Natl. Acad. Sci. USA* **2005**, *102*, 17002–17007. [PubMed]

45. Marsh, J.A.; Baker, J.M.; Tollinger, M.; Forman-Kay, J.D. Calculation of residual dipolar couplings from disordered state ensembles using local alignment. *J. Am. Chem. Soc.* **2008**, *130*, 7804–7805.

46. Dedmon, M.M.; Lindorff-Larsen, K.; Christodoulou, J.; Vendruscolo, M.; Dobson, C.M. Mapping long-range interactions in α-synuclein using spin-label NMR and ensemble molecular dynamics simulations. *J. Am. Chem. Soc.* **2005**, *127*, 476–477.

47. Ganguly, D.; Chen, J. Structural interpretation of paramagnetic relaxation enhancement-derived distances for disordered protein states. *J. Mol. Biol.* **2009**, *390*, 467–477.

48. Bernadó, P.; Mylonas, E.; Petoukhov, M.V.; Blackledge, M.; Svergun, D.I. Structural characterization of flexible proteins using small-angle X-ray scattering. *J. Am. Chem. Soc.* **2007**, *129*, 5656–5664.

49. Mukrasch, M.D.; Bibow, S.; Korukottu, J.; Jeganathan, S.; Biernat, J.; Griesinger, C.; Mandelkow, E.; Zweckstetter, M. Structural polymorphism of 441-residue tau at single residue resolution. *PLoS Biol.* **2009**, *7*, e1000034.

50. Kim, D.-H.; Han, K.-H. Transient secondary structures as general target-binding motifs in intrinsically disordered proteins. *Int. J. Mol. Sci.* **2018**, *19*, 3614.

51. Kussie, P.H.; Gorina, S.; Marechal, V.; Elenbaas, B.; Moreau, J.; Levine, A.J.; Pavletich, N.P. Structure of the MDM2 oncoprotein bound to the p53 tumor suppressor transactivation domain. *Science* **1996**, *274*, 948–953. [PubMed]

52. Kim, D.-H.; Lee, C.; Cho, Y.-J.; Lee, S.-H.; Cha, E.-J.; Lim, J.-E.; Sabo, T.M.; Griesinger, C.; Lee, D.; Han, K.-H. A pre-structured helix in the intrinsically disordered 4EBP1. *Mol. BioSyst.* **2015**, *11*, 366–369. [PubMed]

53. Lukhele, S.; Bah, A.; Lin, H.; Sonenberg, N.; Forman-Kay, J.D. Interaction of the eukaryotic initiation factor 4E with 4E-BP2 at a dynamic bipartite interface. *Structure.* **2013**, *21*, 2186–2196. [PubMed]

54. Kim, D.-H.; Lee, S.-H.; Nam, K.H.; Chi, S.-W.; Chang, I.; Han, K.-H. Multiple hTAF(II)31-binding motifs in the intrinsically unfolded transcriptional activation domain of VP16. *BMB Rep.* **2009**, *42*, 411–417. [PubMed]

55. Jonker, H.R.; Wechselberger, R.W.; Boelens, R.; Folkers, G.E.; Kaptein, R. Structural properties of the promiscuous VP16 activation domain. *Biochemistry* **2005**, *44*, 827–839. [PubMed]

56. Wüthrich, K. *NMR of Proteins and Nucleic Acids*; Wiley: New York, NY, USA, 1986.

57. Gagné, S.M.; Tsuda, S.; Li, M.X.; Chandra, M.; Smillie, L.B.; Sykes, B.D. Quantification of the calcium-induced secondary structural changes in the regulatory domain of troponin-C. *Protein Sci.* **1994**, *3*, 1961–1974. [PubMed]

58. Maltsev, A.S.; Grishaev, A.; Bax, A. Monomeric α-synuclein binds congo red micelles in a disordered manner. *Biochemistry* **2012**, *51*, 631–642.

59. Bertoncini, C.W.; Jung, Y.-S.; Fernandez, C.O.; Hoyer, W.; Griesinger, C.; Jovin, T.M.; Zweckstetter, M. Release of long-range tertiary interactions potentiates aggregation of natively unstructured α-synuclein. *Proc. Natl. Acad. Sci. USA* **2005**, *102*, 1430–1435.

60. Theillet, F.-X.; Binolfi, A.; Bekei, B.; Martorana, A.; Rose, H.M.; Stuiver, M.; Verzini, S.; Lorenz, D.; van Rossum, M.; Goldfarb, D.; et al. Structural disorder of monomeric α-synuclein persists in mammalian cells. *Nature* **2016**, *530*, 45–50.

61. Lee, C.; Kim, D.-H.; Lee, S.-H.; Su, J.; Han, K.-H. Structural investigation on the intrinsically disordered N-terminal region of HPV16 E7 protein. *BMB Rep.* **2016**, *49*, 431–436.

62. Kim, D.-H.; Wright, A.; Han, K.-H. An NMR study on the intrinsically disordered core transactivation domain of human glucocorticoid receptor. *BMB Rep.* **2017**, *50*, 522–527. [PubMed]

63. Markley, J.L.; Bax, A.; Arata, Y.; Hilbers, C.W.; Kaptein, R.; Sykes, B.D.; Wright, P.E.; Wüthrich, K. Recommendations for the presentation of NMR structures of proteins and nucleic acids. *Eur. J. Biochem.* **1998**, *256*, 1–15. [PubMed]

64. Marsh, J.A.; Singh, V.K.; Jia, Z.; Forman-Kay, J.D. Sensitivity of secondary structure propensities to sequence differences between α- and γ-synuclein: Implications for fibrillation. *Protein Sci.* **2006**, *15*, 2795–2804.

65. Camilloni, C.; De Simone, A.; Vranken, W.F.; Vendruscolo, M. Determination of secondary structure populations in disordered states of proteins using nuclear magnetic resonance chemical shifts. *Biochemistry* **2012**, *51*, 2224–2231. [PubMed]

66. Mantsyzov, A.B.; Shen, Y.; Lee, J.H.; Hummer, G.; Bax, A. MERA: a webserver for evaluating backbone torsion angle distributions in dynamic and disordered proteins from NMR data. *J. Biomol. NMR* **2015**, *63*, 85–95.

67. Feldman, H.J.; Hogue, C.W.V. A fast method to sample real protein conformational space. *Proteins* **2000**, *39*, 112–131.

68. Ozenne, V.; Bauer, F.; Salmon, L.; Huang, J.-r.; Jensen, M.R.; Segard, S.; Bernadó, P.; Charavay, C.; Blackledge, M. Flexible-meccano: a tool for the generation of explicit ensemble descriptions of intrinsically disordered proteins and their associated experimental observables. *Bioinformatics* **2012**, *28*, 1463–1470.

69. Krzeminski, M.; Marsh, J.A.; Neale, C.; Choy, W.-Y.; Forman-Kay, J.D. Characterization of disordered proteins with ENSEMBLE. *Bioinformatics* **2012**, *29*, 398–399.

70. Nodet, G.; Salmon, L.c.; Ozenne, V.; Meier, S.; Jensen, M.R.; Blackledge, M. Quantitative description of backbone conformational sampling of unfolded proteins at amino acid resolution from NMR residual dipolar couplings. *J. Am. Chem. Soc.* **2009**, *131*, 17908–17918.

71. Ytreberg, F.M.; Borcherds, W.; Wu, H.; Daughdrill, G.W. Using chemical shifts to generate structural ensembles for intrinsically disordered proteins with converged distributions of secondary structure. *Intrinsically Disord. Proteins* **2015**, *3*, e984565.

72. Nielsen, J.T.; Mulder, F.A. POTENCI: prediction of temperature, neighbor and pH-corrected chemical shifts for intrinsically disordered proteins. *J. Biomol. NMR* **2018**, *70*, 141–165. [PubMed]

73. Chandra, S.; Chen, X.; Rizo, J.; Jahn, R.; Südhof, T.C. A broken α-helix in folded α-synuclein. *J. Biol. Chem.* **2003**, *278*, 15313–15318. [PubMed]

74. Bermel, W.; Bertini, I.; Felli, I.C.; Lee, Y.-M.; Luchinat, C.; Pierattelli, R. Protonless NMR experiments for sequence-specific assignment of backbone nuclei in unfolded proteins. *J. Am. Chem. Soc.* **2006**, *128*, 3918–3919. [PubMed]

75. Wu, K.-P.; Weinstock, D.S.; Narayanan, C.; Levy, R.M.; Baum, J. Structural reorganization of α-synuclein at low pH observed by NMR and REMD simulations. *J. Mol. Biol.* **2009**, *391*, 784–796.

76. Cho, M.-K.; Nodet, G.; Kim, H.-Y.; Jensen, M.R.; Bernado, P.; Fernandez, C.O.; Becker, S.; Blackledge, M.; Zweckstetter, M. Structural characterization of α-synuclein in an aggregation prone state. *Protein Sci.* **2009**, *18*, 1840–1846.

77. Bodner, C.R.; Maltsev, A.S.; Dobson, C.M.; Bax, A. Differential phospholipid binding of α-synuclein variants implicated in Parkinson's disease revealed by solution NMR spectroscopy. *Biochemistry* **2010**, *49*, 862–871.

78. Maltsev, A.S.; Ying, J.; Bax, A. Impact of N-terminal acetylation of α-synuclein on its random coil and lipid binding properties. *Biochemistry* **2012**, *51*, 5004–5013.

79. Waudby, C.A.; Camilloni, C.; Fitzpatrick, A.W.; Cabrita, L.D.; Dobson, C.M.; Vendruscolo, M.; Christodoulou, J. In-cell NMR characterization of the secondary structure populations of a disordered conformation of α-synuclein within E. coli cells. *PloS one* **2013**, *8*, e72286.

80. Okazaki, H.; Ohori, Y.; Komoto, M.; Lee, Y.-H.; Goto, Y.; Tochio, N.; Nishimura, C. Remaining structures at the N-and C-terminal regions of alpha-synuclein accurately elucidated by amide-proton exchange NMR with fitting. *FEBS Lett.* **2013**, *587*, 3709–3714.

81. Mantsyzov, A.B.; Maltsev, A.S.; Ying, J.; Shen, Y.; Hummer, G.; Bax, A. A maximum entropy approach to the study of residue-specific backbone angle distributions in α-synuclein, an intrinsically disordered protein. *Protein Sci.* **2014**, *23*, 1275–1290.

82. Tavassoly, O.; Nokhrin, S.; Dmitriev, O.Y.; Lee, J.S. Cu (II) and dopamine bind to α-synuclein and cause large conformational changes. *FEBS J.* **2014**, *281*, 2738–2753. [PubMed]

83. Schwalbe, M.; Ozenne, V.; Bibow, S.; Jaremko, M.; Jaremko, L.; Gajda, M.; Jensen, M.R.; Biernat, J.; Becker, S.; Mandelkow, E. Predictive atomic resolution descriptions of intrinsically disordered hTau40 and α-synuclein in solution from NMR and small angle scattering. *Structure* **2014**, *22*, 238–249. [PubMed]

84. Porcari, R.; Proukakis, C.; Waudby, C.A.; Bolognesi, B.; Mangione, P.P.; Paton, J.F.; Mullin, S.; Cabrita, L.D.; Penco, A.; Relini, A. The H50Q mutation induces a 10-fold decrease in the solubility of alpha-synuclein. Journal of biological chemistry. *J. Biol. Chem.* **2015**, *290*, 2395–2404. [PubMed]

85. Lin, W.; Insley, T.; Tuttle, M.D.; Zhu, L.; Berthold, D.A.; Král, P.; Rienstra, C.M.; Murphy, C.J. Control of protein orientation on gold nanoparticles. *J. Phys. Chem. C* **2015**, *119*, 21035–21043.

86. Rezaeian, N.; Shirvanizadeh, N.; Mohammadi, S.; Nikkhah, M.; Arab, S.S. The inhibitory effects of biomimetically designed peptides on α-synuclein aggregation. *Arch. Biochem. Biophys.* **2017**, *634*, 96–106.

87. Rezaei-Ghaleh, N.; Parigi, G.; Soranno, A.; Holla, A.; Becker, S.; Schuler, B.; Luchinat, C.; Zweckstetter, M. Local and Global Dynamics in Intrinsically Disordered Synuclein. *Angew. Chem.* **2018**, *57*, 15262–15266.

88. Murrali, M.G.; Schiavina, M.; Sainati, V.; Bermel, W.; Pierattelli, R.; Felli, I.C. 13C APSY-NMR for sequential assignment of intrinsically disordered proteins. *J. Biomol. NMR* **2018**, *70*, 167–175.

89. Lee, J.H.; Li, F.; Grishaev, A.; Bax, A. Quantitative residue-specific protein backbone torsion angle dynamics from concerted measurement of 3J couplings. *J. Am. Chem. Soc.* **2015**, *137*, 1432–1435.

90. De Simone, A.; Cavalli, A.; Hsu, S.-T.D.; Vranken, W.; Vendruscolo, M. Accurate random coil chemical shifts from an analysis of loop regions in native states of proteins. *J. Am. Chem. Soc.* **2009**, *131*, 16332–16333.

91. Rance, M.; Sørensen, O.; Bodenhausen, G.; Wagner, G.; Ernst, R.; Wüthrich, K. Improved spectral resolution in COSY 1H NMR spectra of proteins via double quantum filtering. *Biochem. Biophys. Res. Commun.* **1983**, *117*, 479–485.

92. Rance, M.; Wagner, G.; Sørensen, O.; Wüthrich, K.; Ernst, R. Application of ω1-decoupled 2D correlation spectra to the study of proteins. *J. Magn. Reson.* **1984**, *59*, 250–261.

93. Wagner, G.; Braun, W.; Havel, T.F.; Schaumann, T.; Gō, N.; Wüthrich, K. Protein structures in solution by nuclear magnetic resonance and distance geometry: the polypeptide fold of the basic pancreatic trypsin inhibitor determined using two different algorithms, DISGEO and DISMAN. *J. Mol. Biol.* **1987**, *196*, 611–639. [PubMed]

94. Carr, C.M.; Kim, P.S. A spring-loaded mechanism for the conformational change of influenza hemagglutinin. *Cell* **1993**, *73*, 823–832.

95. Jiang, Z.; de Messieres, M.; Lee, J.C. Membrane remodeling by α-synuclein and effects on amyloid formation. *J. Am. Chem. Soc.* **2013**, *135*, 15970–15973. [PubMed]

96. Wang, L.; Das, U.; Scott, D.A.; Tang, Y.; McLean, P.J.; Roy, S. α-synuclein multimers cluster synaptic vesicles and attenuate recycling. *Curr. Biol.* **2014**, *24*, 2319–2326. [PubMed]

97. Murphy, D.D.; Rueter, S.M.; Trojanowski, J.Q.; Lee, V.M.-Y. Synucleins are developmentally expressed, and α-synuclein regulates the size of the presynaptic vesicular pool in primary hippocampal neurons. *J. Neurosci.* **2000**, *20*, 3214–3220.

98. Burré, J.; Sharma, M.; Tsetsenis, T.; Buchman, V.; Etherton, M.R.; Südhof, T.C. α-Synuclein promotes SNARE-complex assembly in vivo and in vitro. *Science* **2010**, *329*, 1663–1667.

99. Zhu, M.; Fink, A.L. Lipid binding inhibits α-synuclein fibril formation. *J. Biol. Chem.* **2003**, *278*, 16873–16877.

100. Galvagnion, C.; Brown, J.W.; Ouberai, M.M.; Flagmeier, P.; Vendruscolo, M.; Buell, A.K.; Sparr, E.; Dobson, C.M. Chemical properties of lipids strongly affect the kinetics of the membrane-induced aggregation of α-synuclein. *Proc. Natl. Acad. Sci. USA* **2016**, *113*, 7065–7070.

101. Davidson, W.S.; Jonas, A.; Clayton, D.F.; George, J.M. Stabilization of α-synuclein secondary structure upon binding to synthetic membranes. *J. Biol. Chem.* **1998**, *273*, 9443–9449.

102. Rhoades, E.; Ramlall, T.F.; Webb, W.W.; Eliezer, D. Quantification of α-synuclein binding to lipid vesicles using fluorescence correlation spectroscopy. *Biophys. J.* **2006**, *90*, 4692–4700. [PubMed]

103. Ulmer, T.S.; Bax, A.; Cole, N.B.; Nussbaum, R.L. Structure and dynamics of micelle-bound human α-synuclein. *J. Biol. Chem.* **2005**, *280*, 9595–9603. [PubMed]

104. Lee, J.H.; Ying, J.; Bax, A. Nuclear magnetic resonance observation of α-synuclein membrane interaction by monitoring the acetylation reactivity of its lysine side chains. *Biochemistry* **2016**, *55*, 4949–4959. [PubMed]

105. Radhakrishnan, I.; Pérez-Alvarado, G.C.; Parker, D.; Dyson, H.J.; Montminy, M.R.; Wright, P.E. Solution structure of the KIX domain of CBP bound to the transactivation domain of CREB: a model for activator: coactivator interactions. *Cell* **1997**, *91*, 741–752.

106. Marnett, L.J. Decoding endocannabinoid signaling. *Nat. Chem. Biol.* **2009**, *5*, 8–9.

107. Russo, L.; Giller, K.; Pfitzner, E.; Griesinger, C.; Becker, S. Insight into the molecular recognition mechanism of the coactivator NCoA1 by STAT6. *Sci. Rep.* **2017**, *7*, 1–12.

108. Ulrich, E.L.; Akutsu, H.; Doreleijers, J.F.; Harano, Y.; Ioannidis, Y.E.; Lin, J.; Livny, M.; Mading, S.; Maziuk, D.; Miller, Z. BioMagResBank. *Nucleic Acids Res.* **2007**, *36*, D402–D408.

109. Selenko, P. Quo vadis biomolecular NMR spectroscopy? *Int J. Mol. Sci.* **2019**, *20*, 1278.

 *biomolecules*

Review

The Disordered Cellular Multi-Tasker WIP and Its Protein–Protein Interactions: A Structural View

Chana G. Sokolik, Nasrin Qassem and Jordan H. Chill *

Department of Chemistry, Bar Ilan University, Ramat Gan 52900, Israel; chanasokolik@gmail.com (C.G.S.); nasrin_q86@hotmail.com (N.Q.)
* Correspondence: Jordan.Chill@biu.ac.il

Received: 18 June 2020; Accepted: 18 July 2020; Published: 21 July 2020

Abstract: WASp-interacting protein (WIP), a regulator of actin cytoskeleton assembly and remodeling, is a cellular multi-tasker and a key member of a network of protein–protein interactions, with significant impact on health and disease. Here, we attempt to complement the well-established understanding of WIP function from cell biology studies, summarized in several reviews, with a structural description of WIP interactions, highlighting works that present a molecular view of WIP's protein–protein interactions. This provides a deeper understanding of the mechanisms by which WIP mediates its biological functions. The fully disordered WIP also serves as an intriguing example of how intrinsically disordered proteins (IDPs) exert their function. WIP consists of consecutive small functional domains and motifs that interact with a host of cellular partners, with a striking preponderance of proline-rich motif capable of interactions with several well-recognized binding partners; indeed, over 30% of the WIP primary structure are proline residues. We focus on the binding motifs and binding interfaces of three important WIP segments, the actin-binding N-terminal domain, the central domain that binds SH3 domains of various interaction partners, and the WASp-binding C-terminal domain. Beyond the obvious importance of a more fundamental understanding of the biology of this central cellular player, this approach carries an immediate and highly beneficial effect on drug-design efforts targeting WIP and its binding partners. These factors make the value of such structural studies, challenging as they are, readily apparent.

Keywords: WASp interacting protein; protein–protein interactions; intrinsically disordered proteins; actin; cytoskeleton remodeling; SH3 domain; proline-rich motif

1. Introduction

1.1. Scope

Modern biochemical research emphasizes the importance of complementing the biological and functional description of cellular events with a structural understanding of these on the molecular level. Such a combined structure-function view of biology—and the biomacromolecules that power it—has been repeatedly established as a prerequisite for studying biological pathways, analyzing signaling and regulation cascades, efficient drug design and optimization, and other investigation avenues that focus the majority of research efforts today. Fortunately, this state of affairs has motivated the development and advancement of experimental techniques capable of addressing this need, the main ones being X-ray crystallography, nuclear magnetic resonance (NMR), cryo-electron microscopy (cryo-EM), mass-spectrometry (MS), fluorescence-based spectroscopy, and a variety of scattering methods. All these bear the potential to provide a detailed mechanistic view of key cellular processes, as well as how they interface with each other. A case in point is WASp-interacting protein (WIP), a ubiquitous central participant in remodeling of the actin cytoskeleton, and therefore involved in regulation of activation and proliferation of cells [1,2]. While several excellent reviews have focused on the biology and protein-interaction networks of this cellular multi-tasker [3–7], less attention has been given to these interactions on the molecular level. To some extent, this is due to the disordered nature of WIP, which does not exhibit a stable three-dimensional structure, and adopts a more rigid conformation only upon interacting with its various binding partners. The recognized importance of intrinsically disordered proteins (IDPs) in biology is constantly increasing [8–10]. In this review, we hope to bring forth and summarize our structural knowledge of WIP and its main biological interactions. After a brief description of WIP and its relevant protein–protein interaction map, we will devote a section to each of the main interactions, highlighting structural information that has become known over years of research. Finally, we will attempt to outline prospects for future structural study of this important system.

1.2. WIP—Biology and Cellular Roles

WIP, a member of the verprolin family of actin-binding proteins [11], is a versatile and significant player in a number of biological processes. Originally discovered as a binding partner of WASp (Wiskott–Aldrich syndrome protein) [1], WIP has come into the spotlight in its own right. WIP shows highest expression in hematopoietic cells, and its most prominent function is regulation of the assembly of cytoskeletal actin filaments [3]. These form actin-rich structures, membrane protrusions and projections differing according to cell type in form and function, such as podosomes, filopodia, dorsal ruffles, stress fibers, lamellipodia and invadopodia, all involved in cell motility and migration, cell invasion through matrix degradation, cell adhesion, formation of synapses, or endo/exocytosis.

WIP's function is crucial in immune cells, providing these cells with adhesive and migratory properties and an intact cortical actin cytoskeleton. WIP deficiency in B cells leads to distorted cortical actin and impaired signaling [12]. It is proposed that the actin cytoskeleton affects receptor diffusion and B cell surface receptor organization tuning receptor activation [12,13], which may be a mechanism relevant for other immune receptors as well [13]. Receptor-ligation in T cells and mast cells induces WIP-dependent actin polymerization and cytoskeletal rearrangement as a prerequisite for cell activation and proliferation [14–16]. WIP regulates the activation of both WASp, found in hematopoietic cells, and its ubiquitously expressed homolog N-WASp, nucleation-promoting factors (NPFs) that stimulate the molecular apparatus actin-related protein 2/3 (Arp2/3) complex to assemble filamentous actin. WIP also acts as a chaperone of WASp, protecting it from degradation and shuttling it to the sites of actin polymerization [15,17–20]. Research of the endocytosis mechanism in a yeast system suggests that threshold levels of WIP and WASp are needed to initiate actin assembly in the presence of a network of adaptor proteins, underscoring the central role of WIP and WASp in actin-nucleation scaffolds [21]. The pivotal role of the WIP-WASp complex in actin polymerization signaling is also exemplified by the fact that vaccinia virus and *Shigella* bacteria mimic regulators of the WIP-N-WASp complex (such as the adaptor Nck) to hijack the host's actin machinery [2]. However, WIP has important WASp-independent functions as well, since cells containing WIP capable of binding WASp yet lacking the actin-binding domain showed decreased F-actin content and defects in T cell [22] and B cell [23] function, in agreement with the finding that the presence of WIP stabilizes F-actin and inhibits its depolymerization [3,18]. In addition, the central proline-rich domain of WIP serves as a scaffold for indispensable interactions with adaptor proteins, linking it to up-stream and down-stream regulators, as detailed in Section 3.2. Finally, WIP's regulation of actin polymerization also affects maturation of neuronal cells and their synaptic activity [24].

WIP's activity in promoting actin-rich structures also implicates it in many pathologies and makes its binding interfaces potential drug targets. Actin-rich membrane protrusions of cancer cells known as invadopodia degrade the extracellular matrix which allows cancer cells to migrate and form metastases, high WIP levels correlating with high invasiveness in breast cancer cells [25]. In addition, bacterial and viral pathogens, such as *Shigella flexneri* and Vaccinia virus, can recruit the host's WIP-N-WASp complex to form actin-tails, propelling them and spreading infection [2,26]. As a regulatory protein, WIP impacts gene transcription and cell phenotype transitions. High WIP levels have been linked to enhanced stability of Yes associated protein (YAP) and transcriptional coactivator with PDZ-binding motif (TAZ) and oncogenic transformations [27–29]. WIP also controls through the G-actin/F-actin ratio the nuclear translocation of myocardin-related transcription factors (MRTFs), which in turn regulates the expression level of genes involved in focal adhesion as well as cancer cell migration and invasion [30]. Finally, the WIP-WASp complex affects T cell growth factor IL-2 gene transcription in T-cells through activation of the transcription factor NFAT which is needed for T cell proliferation [31].

1.3. Functional Domains and Sequences of WIP

Figure 1 schematically describes functional sequences along the WIP polypeptide, with a major division into three regions, (i) the N-terminal actin-binding domain, (ii) the central proline-rich domain, and (iii) the C-terminal WASp-binding domain. The first (residues 1–120) is homologous to verprolin, a yeast protein involved in cytoskeletal organization, and includes two WASp homology 2 (WH2) domains (residues 32–59 and 96–118) [1] with G-actin binding sequences LKKT (residues 46–49) and LRST (111–114) separated by a highly flexible glycine-rich stretch. The second region (residues 121–440) contains proline-rich motifs that bind Src homology 3 (SH3) domains of NPFs, such as cortactin and its hematopoietic homologue HLCS1 and various adaptor proteins (details in Section 3.2.). In addition, the SH3 domain of the Src family tyrosine kinase Hck is known to interact with WIP directly in vitro, yet its binding motif/segment, assumed to be in region two has not been specified [32]. In addition, several SH3 domains of Pombe Cdc 15 homology (PCH) family proteins from T cells have been found to precipitate WIP through interaction with proline-rich motifs presumably in the second region [33]. The third (residues 441–503) binds to the Ena/VASP homology 1 (EVH1) domain of (N-)WASp [1,34] and contains a consensus protein kinase C θ (PKCθ) recognition site for phosphorylation on S488 [15]. The consensus motif for binding to profilin, an actin-regulating protein (xPPPPP, x = A/S/L/G), appears three times and is assumed to be an actin-based motility homology-2 (ABM-2) motif [3].

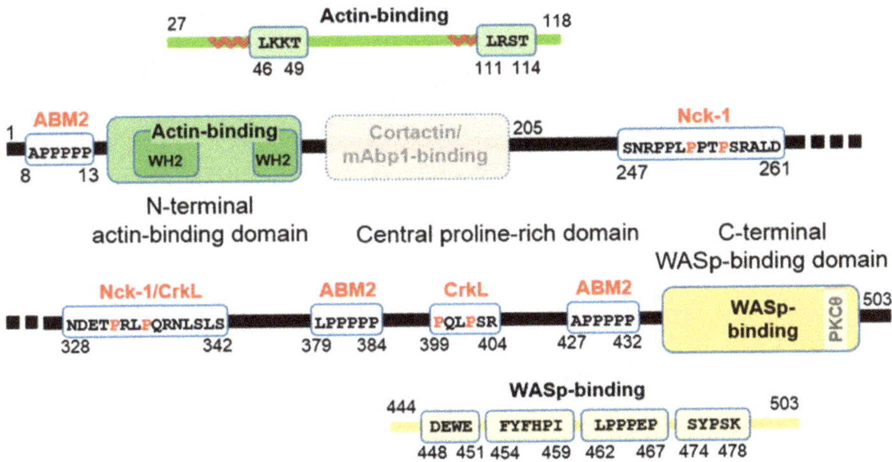

Figure 1. Functional domains of WIP. Schematic description of WIP (1–503) highlighting binding partners and motifs. Actin-binding, cortactin-binding, and WASp-binding regions are shown in green, pink (faded, indicating a putative binding domain), and orange, respectively. Polyproline motifs are shown in light blue (and extended in scale for clarity) with names of binding partners above (red). Sequence numbers are shown for motifs and domains. The actin-binding and WASp-binding regions are magnified (above and below, respectively) to highlight specific sequence features and epitopes. In the former, a red sawtooth pattern indicates the WH2 domain amphiphilic helix.

1.4. WIP Is a Disordered Polypeptide

WIP belongs to a class of proteins known as intrinsically disordered proteins (IDPs), defined as polypeptides lacking a well-defined secondary and tertiary structure under biologically native conditions [35–37]. This is a consequence of the WIP amino acid distribution, containing a low number (95 of 503, 19%) of hydrophobic residues and an excess of charged and polar residues (266 including Gly, 53%). In addition, WIP is rich in the disorder-promoting residue proline (142, 28%) that adopts locally rigid but globally flexible structures. As in other IDPs, the relatively small enthalpic gain of burying the few WIP hydrophobic residues that would normally drive the folding process is insufficient to compensate for the concomitant loss of entropy [38]. Although this lack of structure contradicts the structure-function paradigm that has motivated decades of protein investigations, IDPs have recently re-ignited the interest of the structural biology community as the idea of function without structure gains acceptance. It is now undisputed that IDPs are intimately involved in all central cellular processes, including gene expression, cell-cycle control and malignancy, signal transduction, protein aggregation and degradation, and are also disproportionally involved in human disease [39–41].

As is clear from the previous sections, WIP could be considered archetypical of this class of proteins. Characteristically, WIP can be described as an array of short interaction domains, each possessing independent binding capabilities, beaded together on a 'necklace' formed by connecting non-functional segments. However, contrary to multi-domain structured proteins, each of these 'beads' is actually an ensemble of rapidly interchanging unfolded and partially-folded states which on aggregate account for overall behavior in solution [42,43]. Accordingly, the energetic conformational landscape of such domains is a multi-minima surface lacking a distinct low-energy state. This description is consistent with the role of WIP as a multi-tasking interaction hub, with an ability to recruit proteins and elicit specific functionalities. As will be demonstrated below, it is clear that unstructured domains of WIP are induced to fold to specific structures upon binding of interaction partners. Typically for an IDP, the unfolded state of WIP is conducive to post-translational modifications (PTMs) [44], the main one being phosphorylation, and the coupling of multiple protein–protein interactions with their PTM-based modulation results in the potent regulatory network for which WIP is well-known. With the importance of IDPs on the rise in recent years, structural methods have evolved to address this intriguing class of proteins [45–52].

Beyond the phenomenological observation of the biological importance of IDPs, there remains the mechanistic question of how they exert their biological function in the absence of structure. As do the majority of IDPs, the encounter between a globular interaction partner and an unstructured functional domain of WIP induces the folding of the latter into a specific structure, with the binding protein serving as an 'external' hydrophobic core. Given that the typical WIP interaction domain is actually an ensemble of conformations, the binding protein could 'select' an appropriately quasi-folded conformation, or conversely induce a compaction of an unstructured conformation upon contact between binding surfaces. Determining the relative contributions of these two mechanisms is a fundamental question of IDP biology [53,54]. Hybrid mechanisms, in which residual disorder exists even in contact with the binding partner, have been described in some IDPs and are known as 'fuzzy complexes' [55–57]. Not surprisingly, the entropic penalty of a collapse of several possible states of WIP into a single bound state leads to complexes of varying affinity levels, and dissociation constants in the 0.1–100 µM range are known. This also highlights the challenging nature of WIP structural biology, since weak complexes often defy structural study by static approaches (crystallography, cryo-EM) and require methods that preserve molecular dynamics such as NMR, fluorescence, or scattering-based methods.

1.5. Rationale and Structure of Review—List of WIP Interaction Domains

The previous sections emphasize the central role played by WIP in a variety of cellular processes, and as a consequence the importance of understanding the molecular mechanisms underlying its interactions with its multiple binding partners and activity. In light of this, and the aforementioned relative paucity of such data, herein we aim to curate the available structural information on the cellular interactions of WIP. To place this molecular-level view in its biological context, Figure 2 illustrates the wingspan of WIP in terms of the proteins it engages in various stages of cellular homeostasis and highlights interactions for which structural information is available. This serves as a graphic illustration of the 'interaction hub' role assumed by WIP, while emphasizing sobering gaps of information (to be addressed by future investigations) separating the few interaction systems that have been well characterized. By nature, these interactions will form the focus of the following sections.

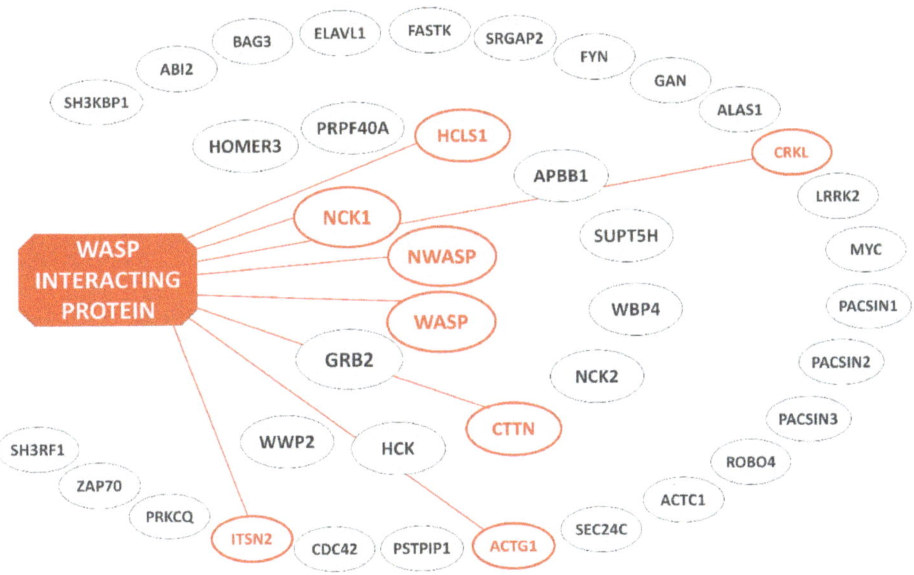

Figure 2. Interaction partners of WIP. The Human Integrated Protein-Protein Interaction rEference (HIPPIE) database [58] indicates proteins with a good probability ($p \geq 0.5$) of interacting with WIP. Full protein names can be found in Table S1. Inner circle—$0.94 \leq p \leq 0.99$, middle circle—$0.68 \leq p \leq 0.86$, outer circle—$0.52 \leq p \leq 0.63$. WIP binding partners mentioned in the structural context of this review are highlighted in red.

2. The Actin-Binding Region

2.1. Actin—A Cytoskeleton Protein—and Actin-Binding Domains

Actin is a cytoskeletal protein found in most eukaryotic cells. It participates in many crucial cellular processes, including muscle contraction, cell motility and migration, division and signaling, immune surveillance, angiogenesis, tissue repair, phagocytosis, and cell regeneration [59,60]. The constant and rapid reorganization of the actin microfilament system accompanying these depends on nucleation, elongation and depolymerization of actin filaments, and therefore cellular reorganization of actin is highly regulated [60]. Actin exists in two different forms in equilibrium, monomeric (globular, or G-) and polymeric (filamentous, or F-) form. The dynamic equilibrium between G- and F-actin is central to cellular behavior and is regulated by extracellular stimulation [61]. Monomeric G-actin, the basic unit for actin filaments, contains four subdomains: subdomains 1 (residues 1–32, 70–144, 338–374) and 2 (33–69) of the small main domain, and subdomains 3 (145–180, 270–337) and 4 (181–269) of the large main domain [61,62]. As shown in Figure 3, together these four subdomains create two structural clefts, a large nucleotide-binding cleft between subdomains 2 and 4 and a hydrophobic target-binding groove between subdomains 1 and 3 [62,63]. The former is the center of the enzymatic catalysis site where hydrolysis of ATP and binding of divalent cations (Mg^{2+} or Ca^{2+}) takes place, mediated by residues 11–18 and 154 [61]. The latter modulates the binding affinities of actin-binding modules (ABMs), leading to changes in the stability of the actin filament.

ABMs are actin-binding entities that control the formation of the actin cytoskeleton by regulating the transition between G- and F-actin in cells [64,65]. G-actin bound to ABMs or proteins of the profilin family is the major source of actin monomers for filament nucleation and elongation [66], and other roles of ABMs include disengaging, capping, and monomer sequestration. ABMs share a conserved motif that competes with actin for a common binding site. A main contributor to this essential site is a hydrophobic pocket that mediates significant interaction of actin complexes [62]. The hallmark of ABMs is a 9–10 residue segment that upon binding to the barbed end of actin forms a helical region followed by a conserved LKK(T/V) motif (with some variations). The five residues that follow this sequence play a key role in determining how the extended chain interacts with actin. Generally accepted is the subdivision of ABMs into WH2 domains, characterized by longer conserved regions preceding the amphiphilic helix, and β-thymosins, identified by a conserved linker connecting the helix and the LKK(T/V) motif and a second C-terminal helix following these that interacts with the pointed face of actin [67].

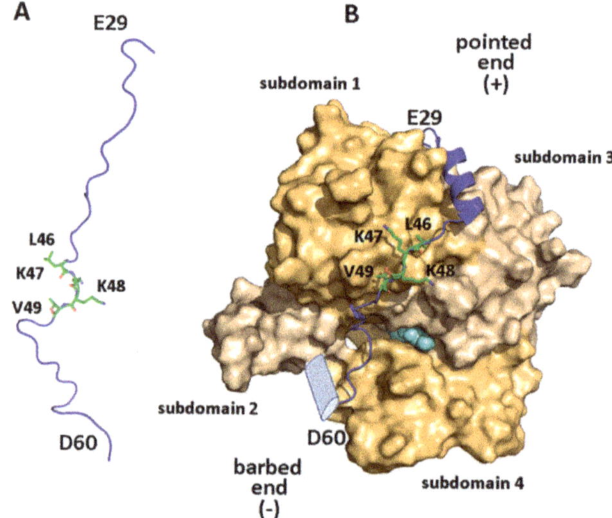

Figure 3. Structural view of the WIP-N/actin interaction. (**A**) A plausible representation of residues 28–61 of WIP (blue) and sidechains of residues 46–49 in their free form, (**B**) bound WIP(28–61) in complex with actin (pale/dark orange, PDB ID: 2A41 [68]), showing actin subdomains. Residues of the actin-binding LKKV motif are highlighted in green. The putative C-terminal helix (absent in WIP but present in other actin-binders) is portrayed as a light-blue cylinder, and the bound nucleotide (between subdomains 3 and 4) is shown as cyan-colored spheres.

2.2. Structural Aspects of the WIP-Actin Interaction

WIP and its homologs CR6 and WICH/WICH contain N-terminal ABMs belonging to the WH2 family [59,62,68]. In WIP, these span residues 32–60 (a 'long' WH2 domain) and 96–118 (a 'short' WH2 domain), including the conserved sequences $L^{46}KKT^{49}$ and $L^{111}RST^{114}$, respectively [11]. The crystal structure of the first of these ABMs in complex with actin (PDB ID: 2A41 [68]) revealed the structural details of this interaction. Residues 33–42 form a three-turn amphiphilic helix that embeds its hydrophobic face, including residues L36, L37, and I40, in a cleft at the barbed end of actin, and basic residues K47/K48 of the conserved motif are positioned close to a negatively charged surface including actin residues D24/D25, E99/E100 (subdomain 1), and E334 (subdomain 3). Characteristically for 'long' WH2 domains, the following segment (residues 52–55) runs parallel to the actin subdomain 1 β-sheet [69]. This creates an extensive binding interface (absent in 'short' WH2 domains) that includes a salt bridge between R54 and actin residue E93, while the small side-chain of adjacent S55 allows deeper penetration into the nucleotide cleft of actin [62,68,70,71]. The affinity to actin of smaller WIP fragments consisting of residues 29–46 and 46–63 drops 10-fold and over 1000-fold, respectively, demonstrating the importance of the amphiphilic helix in binding actin [68].

Since only a minor fraction of cellular WIP is in the actin-bound state, the ensemble of conformations adopted by its ABM sequences in the intrinsically disordered free form is relevant to their cellular behavior. NMR-based measurements were employed to characterize the conformational ensembles of residues 2–65 of WIP containing the N-terminal ABM. Secondary backbone chemical shifts, temperature-induced chemical shift effects, backbone heteronuclear coupling constants, and analysis of residual dipolar couplings for this segment all concurred in identifying a helical propensity for residues 30–42 and partial extended β-strand character for residues 45–62. These propensities echo the ABM actin-bound structure, suggesting this pre-formed conformation may contribute to the actin binding mode [72,73]. As shown by changes in backbone J-couplings, this structural bias in the WIP conformational ensemble was obviated by exposure to denaturing conditions [73]. Interestingly, a lysate mimicking actin-deficient cellular crowding effects found a decrease in these structural tendencies, presumably due to non-specific protein–protein interactions offering higher stabilization to unfolded conformations of the ABM. Thus, in the cellular environment, the ABM may be less structured than in its purified form. Notably, a partially pre-formed β-strand similar in significance to residues 45–62 was observed connecting the profilin-binding and amphipathic helix sequences (residues 17–25), a region highly conserved in WIP and its homologs. This may indicate a potential role for this linker in mediating the binding of actin, possibly by interacting with a yet-unknown binding partner [73].

3. The Proline-Rich Intermediate Region

In vitro and in vivo biological studies have pinpointed the proline-rich domain as a frontier of high clinical relevance with interaction motifs that are either heavily implicated in cancer metastasis formation [74,75] or may be vital for proper immune system functioning [15,76]. Thus, it is surprising to find that this major WIP segment has not been structurally investigated, particularly when compared to the terminal domains described in other sections of this review. Possibly because SH3/polyproline complexes have been characterized back in the 1990s, they have been considered research targets with less potential of novelty. However, this may be a misconception, as many issues of binding specificity, molecular determinants of affinity, and effects of extended motifs are extremely important for inhibitor design and remain largely unresolved. Another potential barrier faced by structural studies is the moderate affinity of these interactions that hinder both crystallization efforts and solution NMR investigations. Indeed, there is a lack of biophysical characterizations of these complexes using methods such as isothermal calorimetry (ITC) and microscale thermoephoresis (MST) for quantification of affinity, NMR, X-ray crystallography, and small-angle X-ray scattering (SAXS) for structure determination of the complexes, or NMR and single-molecule fluorescence techniques to characterize their dynamic nature. We therefore focus on more qualitative biological studies of these interactions.

3.1. SH3 Domains and Their Ligands

Src Homology-3 (SH3) domains are small modules of protein–protein interactions found in signaling and regulatory proteins. They usually are composed of 55–70 residues [77] with 5–8 β-strands arranged as two anti-parallel β-sheets or a β-barrel, with three loop regions, the RT loop (between β_1-β_2), N-Src loop (β_2-β_3), and distal loop (β_3-β_4), and a short 3_{10}-helix (β_4-β_5) [78]. Two of the three ligand-binding grooves are formed by highly conserved (mostly) aromatic residues, including a tryptophan (often the first in a β_3-WW motif), two additional aromatic residues (tyrosine or phenylalanine) located in the RT-loop and the 3_{10} helix, and a proline residue at the end of β_4

(see Figure 4A) [79]. Their sidechains adopt an orientation essentially unchanged by ligand binding, suggesting a preformed template [77].

SH3 ligands are proline-rich motifs of disordered protein segments that form a left-handed polyproline (PPII) helix seen as arches that place the (i) and (i + 3) residues at the same height (Figure 4B), usually with a qPxqP sequence. One qP dipeptide binds to each hydrophobic groove, q being a hydrophobic residue [78,80]. Ligand motifs include flanking basic residues which interact with acidic RT-loop residues in a third pocket called the canonical specificity pocket (the acidic residues seen behind tryptophan and marked in Figure 4, A and B and detail in C) [79]. Consensus ligands are classified as class 1 (consensus motif RxLPPxP) or class 2 (xPPLPxR), characterized by basic residues at the N- and C-terminal side of the PxxP motif, respectively and bind in opposite orientations [78]. Although all SH3 domain structures are highly similar and consensus motifs show small variations, SH3 domains do recognize specific ligands and, conversely, ligands recognize specific SH3 domains, to a certain degree. In particular, ligand interactions with RT-loop, N-Src loop, and β4 residues have been implicated in mediating both affinity and specificity. On the ligand side, residues outside the core binding motif have been associated with affinity and specificity [79,81–85]. In addition, non-canonical binding with recognition of non-PxxP ligands is not uncommon for certain SH3 domains and its prevalence may be underestimated [78,79,82,85,86].

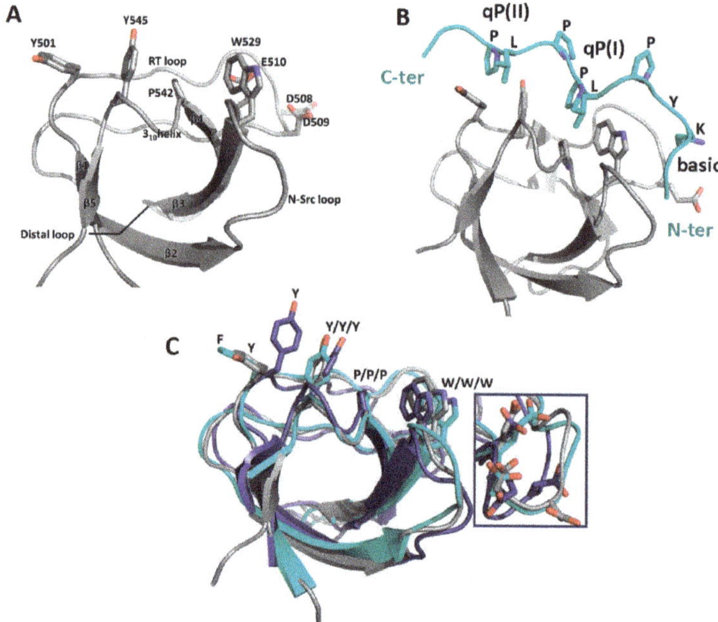

Figure 4. Common structural features of WIP-binding SH3 domains. (**A**) The cortactin SH3 domain showing the characteristic features with numbering according to human cortactin (PDB ID: 5NVJ) [87]. (**B**) Complex of a Hck SH3 domain (PDB ID: 2OJ2) [88] with a high-affinity class I peptide ligand KYPLPPLP showing the typical ligand PPII conformation and placement in binding grooves: The two LP dipeptides interact with the aromatic residues, while the N-terminal lysine of the ligand interacts with glutamate of the specificity pocket. (**C**) Overlay of the following SH3 domains, cortactin (grey, PDB ID: 5NVJ [87]), Nck SH3.2 (cyan, PDB ID: 2JS0 [89]), and N-terminal CrkL domain (blue, PDB ID: 2LQN [90]), demonstrating the high similarity of all SH3 domain structures. Key residues of the hydrophobic binding grooves are shown in stick representation. Inset shows an overlay of specificity pocket acidic residues that form salt bridges with the ligand.

3.2. Binding Partners and Binding Motifs

As mentioned above, information on WIP binding partners is often limited to identification of the interacting protein and, in some cases, the interacting segment or binding epitope sequence. This was generally obtained using biological methods, including yeast-two-hybrid assays, pull-down assays with immobilized SH3 domains followed by Western blot analysis of the binding partners, and further pull-down assays using purified WIP to verify direct binding. Deletion mutants were then used to identify WIP binding segments and/or assess the various affinities in cases of multiple SH3 domains. Alternatively, mutations of the critical SH3 tryptophan residue to lysine resulting in loss of affinity were employed to confirm SH3-mediated binding and ligand-SH3 domain pairings. Techniques used in cells were co-immunoprecipitation, fluorescence assays to verify co-localization and reveal cellular distribution, and assays to assess the cellular effects of binding. Table 1 lists the interaction partners discovered through these techniques.

Very rudimentary information is available for mammalian actin-binding protein 1 (mAbp1) and cortactin, two proteins with a similar domain organization including an N-terminal F-actin binding motif and a C-terminal SH3 domain. The high sequence identity of their SH3 domains (62% amino acid identity of mAbp1 and cortactin) suggests interaction with the same ligands [91]. A yeast-two-hybrid assay identified WIP residues 136–205 as a cortactin-binding segment while cortactin failed to interact with full-length WIP lacking residues 110–170 (Δ110–170) [92]. mAbp1, too, was found to bind WIP, and deletion of WIP residues 110–170 reduced the interaction by more than 70% [93]. The W→K mutation of the binding site tryptophan of both the cortactin and mAbp1 SH3 domains, W525K and W415K respectively, blocked binding of cortactin/mAbp1 to WIP, confirming SH3-mediated binding for cortactin and mAbp1 [92,93]. In addition, binding of WIP to the hematopoietic homologue of cortactin, hematopoietic lineage cell-specific protein 1 (HLCS1) was proven and W→Y mutation of the HLCS1 SH3 domain abolished binding [94]. Notably, the dissociation constant for the complex of full-length cortactin and WIP was estimated as 0.3 μM by densitometry (based on the correlation between the concentration of a complex and the intensity of a Western-visualized SDS-PAGE band), constituting a relatively high affinity for SH3-mediated interactions [92]. Similarly, the intersectin adaptor proteins intersectin-1 (ITSN1, the short variant ITSN1-S and the long variant ITSN1-L) and intersectin-2 (ITSN2) have been found to interact with the 318–450 and 13–450 segments of WIP respectively (overlapping the CrkL/Nck sites, see below), both omitting the N-WASp-binding segment to confirm that the interaction is not mediated by N-WASp [95], while in yeast-two-hybrid assays the 353–503 segment interacted with ITSN2, among others [96]. In vitro binding assays indicated that of their five different SH3 domains (labeled A–E), the interaction with WIP occurs via the A/C/E domains, whereas the B/D domains have no WIP affinity [95].

Specifically located binding motifs have been suggested for only two WIP binding partners. Yeast-two-hybrid assays mapped the Crk-like protein (CrkL) binding site in WIP to the 321–415 region, and established that WIP residues 321–376 and 377–503, but not 416–503, interact with CrkL, suggesting two binding sites in residues 321–376 and 377–415. An additional yeast-two-hybrid assay mapped the CrkL WIP binding site to the N-terminal SH3 domain (SH3.1), while the SH3.2 domain failed to interact with WIP. The 321–415 segment contains two copies of the Crk SH3.1 consensus binding motif PxLPx(K/R) [97], in residues 332–337 and 399–404, in complete agreement with the yeast-two-hybrid assay [15].

The most detailed information is available for the WIP–Nck interaction. The adaptor Nck is composed of three tandem SH3 domains (SH3.1, SH3.2, and SH3.3) followed by one SH2 domain. The latter interacts with phosphotyrosine residues in ligand-activated receptor tyrosine kinases (RTKs) and transmits the signals to effector molecules (such as WIP) interacting with its SH3 domains [98]. Affinity-precipitation of WIP with individual Nck SH3 domains demonstrated that WIP bound to SH3.2, but poorly to SH3.1 and SH3.3. Mapping of the Nck-binding site of WIP by yeast-two-hybrid system demonstrated binding to a region spanning residues 321–415 [98]. A peptide-array analysis consisting of WIP-derived 15-residue segments revealed that Nck-binding is mediated by class 2 peptide sequences SNRPPLPPTPSRALD (residues 247–261) and NDETPRLPQRNLSLS (residues 328–342), both sharing the PxxPxRxL motif, while the second one conforms also to the Nck SH3.2 consensus motif PxxPxRxxS [99]. Alanine substitution of the proline residues in the PxxPxR motif abolished Nck binding by the peptides in vitro. Alanine substitution in both motifs was needed in WIP mutants to eliminate Nck binding completely, indicating that each motif can bind Nck independently. Selective affinity of these peptides to SH3.2 was confirmed upon observing that the W143K mutation (in SH3.2), but not W38K (SH3.1) or W229K (SH3.3), was sufficient for eliminating Nck binding [26]. A summary of SH3-binding WIP epitopes appears in Table 1.

Table 1. Summary of SH3-binding WIP epitopes.

Partner	WIP Segment	Binding Motif	Effect	Ref.
Cortactin SH3 (NPF)	136–205	Not determined (ND)	Increases cortactin's activation of the Arp2/3 complex, cortactin recruits WIP in invadopodium formation	[75,92]
mAbp1 SH3 (adaptor)	110–170	ND	Regulates dorsal ruffle formation	[93]
ITSN1-S/ ITSN1-L 1st/3rd/5th of 5 SH3 domains (adaptor)	318–450	ND	enhances association of ITSN1 with N-WASp and β-actin, facilitates formation of filopodia-like protrusions, regulates intra-cellular vesicle trafficking	[95,100]
ITSN2 1st/3rd/5th of 5 (adaptor)	13–450	ND		[95,96]
CrkL 1st SH3 of 2 (adaptor)	321–415	^{332}PRLPQR^{337} (class 2) ^{399}PQLPSR^{404} (class 2) (comply with Crk SH3.1 consensus binding motif PxLPxK/R)	Presumably preformed CrkL-WIP-WASp complex associates with phos-ZAP70 after T cell receptor (TCR) ligation	[15]
Nck-1 2nd SH3 of 3 (adaptor)	247–261 328–342	^{247}SNRPPLPPTPSRALD^{261} ^{328}NDETPRLPQRNLSLS^{342} (both class 2, 328–342 complies with the consensus motif for Nck SH3.2 PxxPxRxxS)	Couples extracellular signals to cytoskeleton assembly system	[26,98,99]

4. The WIP-C/WASp Interface

4.1. Structure and Binding Epitopes in the WIP-C/N-WASp Complex

The interaction—for which WIP is named—between the C-terminal domain (last 50–60 residues) of WIP and the N-terminal EVH1 domain of WASp/N-WASp has been well characterized both biochemically and structurally. WIP was first identified by a yeast two-hybrid assay that linked it to WASp [1], and the interaction was pinpointed a few years later to the WASp EVH1 domain [2], consistent with the location of several WAS-causing mutations in this region [15,101–103]. NMR-based structure determination of complexes between short (residues 461–485) and extended (451–485) WIP-derived peptides fused to the EVH1 domain of N-WASp, an ubiquitously expressed homolog of WASp, revealed the molecular basis of this interaction in detail (Figure 5A) [34,104]. The most striking feature of the complex is the extensive interface involving multiple epitopes along the WIP sequence. Specifically, the canonical EVH1-binding polyproline motif (DLPPPEP, 461–467) nestles into a groove on the second EVH1 β-sheet formed by the characteristic tryptophan residue W64 and conserved residues Y54, F114, and T116 (all numbering based on the WASp sequence). However, this buried surface is flanked by two additional interaction regions, a hydrophobic motif (FYFHPIS, 454–460) identified in an earlier pull-down assay [105] interacting with 'bend' residues V50/V51 of the β-sandwich, and a helical motif (KSYPSK, 473–478) forming a salt bridge between K478 and residue E100 [34,104].

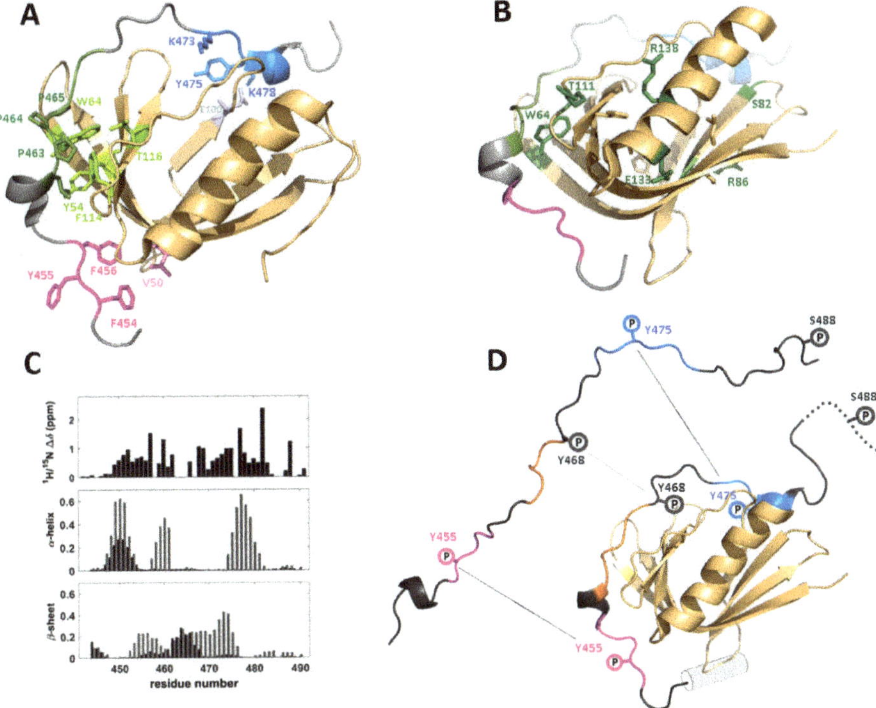

Figure 5. Structural view of the interaction between the C-terminal domain of WIP and the WASp EVH1 domain. Structures are based on the complex between WIP residues 451–485 tethered to residues 26–147 of rat N-WASp (PDB ID: 2IFS [104]). Residue numbers are based on the analogous WASp sequence. (**A**) Structure of the WIP-N-WASp complex (adapted from [104]). N-WASp is shown in light orange, and three WIP epitopes are shown in magenta, green, and blue. Sidechain atoms of these epitopes and key N-WASp residues forming the binding interface are shown as sticks with a similar coloring scheme. (**B**) Distribution of WAS-causing mutations; residues that when mutated result in severe WAS, are highlighted with sidechains in stick representation. Buried mutation hotspot residues are colored in light-orange, and surface-exposed hotspot residues are colored in green and labeled. T111 represents the location of analogous N-WASp residue R601. (**C**) Chemical shift data indicating a binding-induced conformational change in WIP, including (top) HSQC perturbations along the sequence, (middle) predicted helical content for free (black) and bound (gray) WIP, (bottom) same as previous but for β-strand content. (**D**) Model of binding induced changes in residues 442–492 of WIP showing the additional helical motif binding to WASp (gray cylinder) and phospho-sites along the sequence.

In its commonly found WASp-bound state, WIP adopts a tightly constrained conformation that positions the binding epitopes near their respective EVH1 interaction surfaces. Some secondary structural elements induced by EVH1-binding are also transiently present in free WIP, although this region is intrinsically disordered so that there is no prevalent conformation. An analysis of secondary chemical shifts and solvent exchange protection factors along the backbone of a C-terminal (residues 407–503) WIP domain revealed a structural propensity echoing the structure of EVH1-bound WIP residues 461–485, a lefthanded polyproline helix followed by a helical segment for residues 462–467, and 474–478, respectively. Thus, the complex may form by a conformational selectivity mechanism, accompanied by a tightening of the flexible linker (residues 469–472) between these two motifs. More importantly, the analysis also identified a previously undetected fourth segment (residues EDEWES, 447–452) with a strong helical tendency and high conservation level (DDFE, residues 417–420 in CR16, or 394–397 of WICH), suggesting a potential involvement in EVH1 binding (Figure 5C,D) [106]. Indeed, an NMR investigation of a complex of the T cell WASp EVH1 domain bound to an extended WIP polypeptide including this additional epitope (residues 442–492) showed the DEWE segment to adopt a turn conformation and interact with a helical segment (ENQRLFE, WASp residues 31–37) preceding the β-sandwich and overlooked in earlier structural studies [107]. Homologs of this additional helix appear in N-WASp (ENESLFT, residues 23–29) and in related pleckstrin homology (PH) domains [108,109], suggesting it should be included in the functional EVH1 domain.

4.2. Functional Implications of the WIP/WASp Interface

Over half of WAS-inducing mutations are missense mutants in the EVH1 domain (http://www.hgmd.cf.ac.uk, search term WAS, and [34]), emphasizing the importance of the WIP/WASp interaction. Structural data for the N-WASp/WIP and WASp/WIP complexes illuminate the mechanism by which these mutations exert their deleterious effect. Of 19 mutations identified as causing strong/severe WAS, ten (L35, C73, F74, V75, W97, H115, G125, L126, F128, A134, numbering based on the WASp sequence) are buried amino acids (defined as $f_{ASA} < 0.2$, where f_{ASA} is the side-chain fractional accessible solvent area), and likely to cause the disease by disrupting native WASp structure. Specifically, W97, H115, F128, A134 form a packing unit (together with Y107 that is hydrogen-bonded to H115) that directly affects the polyproline-binding groove, and other residues of this group are located in the hydrophobic core of the β-sandwich structure. The other surface-exposed nine mutations (S24, E31, L39, W64, S82, R86, T111, E133, R138) may directly impact WIP binding, as in the case of conserved polyproline-interacting residue W64, but also indirectly, as in the case of the helix-destabilizing mutation R138P that may affect the β-sandwich structure (Figure 5B) [110]. Particularly interesting is mutation hotspot R86, for which four different mutation phenotypes are known, and confirmed by yeast two-hybrid assay [103], located on the WASp face diametrically opposed to the polyproline binding site. Contrary to a previous hypothesis, the NMR analysis did not suggest a direct contact with WIP in this region, and it is also possible that its interaction with nearby negatively charged residues, including the critical E100 (homologous to N-WASp E90 forming an intermolecular salt-bridge) is the WAS-causing factor.

In cell imaging FRET techniques used to identify WASp/WIP dissociation in cells expressing various WIP mutants provided further insight into the positioning and contribution to WASp/WIP function of various WIP epitopes. While WIP mutated at the FYF (454–456) epitope lost the ability to bind WASp, loss of the polyproline and DEWE (448–451) epitopes incurred equally significant reductions in affinity to WASp. In addition, of all epitopes, the DEWE was shown to have the greatest effect upon ubiquitylation levels, indicating that this additional binding interface was important for protecting WASp from proteasomal degradation [107]. The mechanism by which this occurs in yet unclear, since confirmed WASp ubiquitylation sites K76 and K81 [111] are distant from DEWE interaction surface, and it is possible that this interaction interferes with another component of the ubiquitylation machinery.

4.3. Phosphorylation-Induced Dissociation of the WIP/WASp Complex

It is well established that phosphorylation-induced dissociation of the WIP C-terminal domain from WASp mediates both activation and eventual proteasomal degradation of WASp, but the molecular mechanism underlying this phosphorylation has been controversial. Soon after identification of the WIP/WASp interaction, it was shown that PKCθ-mediated phosphorylation occurring on S488 (in the sequence RSGSNR, residues 485–490) is correlated with dissociation in Jurkat cells, and that the S488D phospho-mimicking mutation in WIP abolished its affinity to WASp [15]. In contrast, a later study showed similar levels of WASp pulldown by WIP and its unphosphorylated (S488A) and phosphorylated (S488D) mimicking mutants [31]. However, a more recent in-cell molecular imaging approach attributed this to an independent actin-mediated interaction surface between the two proteins, and by following the movement of WIP-containing clusters in real time established a clear difference between unphosphorylated and phosphorylated mutants and directly implicated PKCθ phosphorylation at S488 as a mediator of complex dissociation [112]. Surprisingly, this proposed mechanism has found little structural support. Structures of single-chain tethered complexes (in which the WIP polypeptide was connected to the N-WASp N-terminal) were either missing the relevant residues [34,104] or unable to observe changes induced by phospho-mimicking mutants S488D/S488E [113]. The later NMR-based analysis of the WIP/WASp complex did not find secondary structure differences between free and bound WIP for residues 483–492, and WIP dissociation-inducing mutations (as shown in cells) caused little (if any) change in chemical shifts within this region [107]. Concomitantly, introduction of the phospho-mimicking S488E mutation had no effect on WASp resonance frequencies (Halle-Bikovski A, Baluom S, Chill JH, unpublished results). The lack of structural evidence from these systems supporting PKCθ-mediated dissociation hints at a possible indirect effect on the state of the WIP/WASp complex. Phosphorylation on tyrosine residues in the WASp-binding domain, specifically Y455, Y468, and Y475 by Bruton's tyrosine kinase (BtK) has been suggested as an alternative inducer of dissociation [114]. This appears to be in better agreement with available structural information, since these phospho-sites reside well within the proven WIP interaction surface (Figure 5D), but further studies would be required to provide experimental support of this notion.

5. Discussion and Summary

WIP is a multi-tasking protein forming a 'hub' of protein–protein interactions, and is involved in a variety of inter-connected and intricately regulated biochemical pathways. Although its first discovered role was in mediating the immune response, much research since has established important functions in cytoskeletal changes via its interaction with G- and F-actin under different conditions, regulation via interaction with several adaptor proteins, and maturation and synaptic activity of neuronal cells. Commensurately to its wide-ranging biological roles, WIP is involved in several pathological conditions and has become recognized as an important biomarker of aggressive cancer [28,74]. Many hypothesized WIP epitopes have been identified using only bioinformatics methods and lack experimental verification. Even so, important binding epitopes within the three major WIP sections have been investigated structurally, either by full structure determination or by structural biophysical approaches. In some cases, the average conformation adopted by such epitopes in their free form exhibits a structural propensity that is reminiscent of the epitope in the bound structure, hinting at a plausible conformational selectivity mechanism of binding. However, this effect may be sequence dependent. Some epitopes will be more 'pre-formed' due to local conformational constraints (i.e., in the case of a polyproline motif), while other epitopes may adopt a conformation corresponding to an alternative local energetic minimum, only to be 're-configured' upon interaction with a binding partner.

These important structural studies have invariably required a reductionist approach, in which each interaction epitope in complex with its binding partner is treated independently. The simplification achieved in such studies must constantly be weighed against the potential loss of biological context, specifically the interdependence of such interaction pairs and/or the possibility of multi-protein interactions. This highlights the importance of complementary cellular and in vivo studies, based on fluorescence cellular imaging or in-cell NMR, in addressing this concern by offering a more holistic and potentially temporally resolved view of the network of WIP interactions that is central to its biological function. Ultimately, a multidisciplinary approach to WIP structure-function studies (and other disordered proteins), in which critical epitopes are first predicted and later investigated by qualitative and quantitative structural approaches, is a promising path towards a better understanding of key biological processes on the molecular level, with potential therapeutic implications.

Supplementary Materials: The following are available online at http://www.mdpi.com/2218-273X/10/7/1084/s1, Table S1: WASp interacting protein binding partners from HIPPIE database.

Funding: This research was funded by the Israel Science Foundation (grants 491/10 - Heritage Legacy Fund - and 964/19), as well as the Christians for Israel Chair for Medical Research. The APC was funded by Israel Science Foundation grant 964/19.

Acknowledgments: We are grateful to Hila Elazari-Shalom, Eva Rozentur-Shkop, Adi Halle-Bikovsky, Hadassa Shaked, and Noam Haba and Saja Baluom for work contributing to this review, Keren Keinan-Adamsky, Hugo Gottlieb and Michal Afri for spectrometer assistance, and Israel Tabakman for technical assistance. The important contribution of Mira Barda-Saad (Bar Ilan University) and her co-workers to related studies is acknowledged. Establishment of the 700 MHz spectrometer system was supported by Fundácion Adar and a Converging Technologies award (Israel Science Foundation).

Conflicts of Interest: The authors declare no conflicts of interest.

References

1. Ramesh, N.; Antón, I.M.; Hartwig, J.H.; Geha, R.S. WIP, a Protein Associated with Wiskott-Aldrich Syndrome Protein, Induces Actin Polymerization and Redistribution in Lymphoid Cells. *Proc. Natl. Acad. Sci. USA* **1997**, *94*, 14671–14676. [CrossRef]

2. Moreau, V.; Frischknecht, F.; Reckmann, I.; Vincentelli, R.; Rabut, G.; Stewart, D.; Way, M. A Complex of N-WASp and WIP Integrates Signalling Cascades That Lead to Actin Polymerization. *Nat. Cell Biol.* **2000**, *2*, 441–448. [CrossRef] [PubMed]

3. Antón, I.M.; Jones, G.E. WIP: A Multifunctional Protein Involved in Actin Cytoskeleton Regulation. *Eur. J. Cell Biol.* **2006**, *85*, 295–304. [CrossRef] [PubMed]

4. Antón, I.M.; Jones, G.E.; Wandosell, F.; Geha, R.; Ramesh, N. WASp-Interacting Protein (WIP): Working in Polymerisation and Much More. *Trends Cell Biol.* **2007**, *17*, 555–562.

5. Ramesh, N.; Geha, R. Recent Advances in the Biology of WASp and WIP. *Immunol. Res.* **2009**, *44*, 99–111. [CrossRef]

6. Noy, E.; Fried, S.; Matalon, O.; Barda-Saad, M. WIP Remodeling Actin behind the Scenes: How WIP Reshapes Immune and Other Functions. *Int. J. Mol. Sci.* **2012**, *13*, 7629–7647. [CrossRef]

7. Sasahara, Y. WASp-WIP Complex in the Molecular Pathogenesis of Wiskott-Aldrich Syndrome. *Pediatr. Int.* **2016**, *58*, 4–7. [CrossRef] [PubMed]

8. Tompa, P.; Schad, E.; Tantos, A.; Kalmar, L. Intrinsically Disordered Proteins: Emerging Interaction Specialists. *Curr. Opin. Struct. Biol.* **2015**, *35*, 49–59. [CrossRef]

9. Uversky, V.N. Intrinsic Disorder, Protein–Protein Interactions, and Disease. *Adv. Prot. Chem. Struct. Biol.* **2010**, *110*, 85–120.

10. Berlow, R.B.; Dyson, H.J.; Wright, P.E. Expanding the Paradigm: Intrinsically Disordered Proteins and Allosteric Regulation. *J. Mol. Biol.* **2018**, *430*, 2309–2320. [CrossRef]

11. Aspenström, P. The Verprolin Family of Proteins: Regulators of Cell Morphogenesis and Endocytosis. *FEBS Lett.* **2005**, *579*, 5253–5259. [CrossRef] [PubMed]

12. Keppler, S.J.; Gasparrini, F.; Burbage, M.; Aggarwal, S.; Frederico, B.; Geha, R.S.; Way, M.; Bruckbauer, A.; Batista, F.D. Wiskott-Aldrich Syndrome Interacting Protein Deficiency Uncovers the Role of the Co-Receptor CD19 as a Generic Hub for PI3 Kinase Signaling in B Cells. *Immunity* **2015**, *43*, 660–673. [CrossRef] [PubMed]

13. Mattila, P.K.; Batista, F.D.; Treanor, B. Dynamics of the Actin Cytoskeleton Mediates Receptor Cross Talk: An Emerging Concept in Tuning Receptor Signaling. *J. Cell Biol.* **2016**, *212*, 267–280. [CrossRef] [PubMed]

14. Kettner, A.; Kumar, L.; Antón, I.M.; Sasahara, Y.; De La Fuente, M.; Pivniouk, V.I.; Falet, H.; Hartwig, J.H.; Geha, R.S. WIP Regulates Signaling via the High Affinity Receptor for Immunoglobulin E in Mast Cells. *J. Exp. Med.* **2004**, *199*, 357–368. [CrossRef]

15. Sasahara, Y.; Rachid, R.; Byrne, M.J.; De La Fuente, M.A.; Abraham, R.T.; Ramesh, N.; Geha, R.S. Mechanism of Recruitment of WASp to the Immunological Synapse and of Its Activation Following TCR Ligation. *Mol. Cell.* **2002**, *10*, 1269–1281. [CrossRef]

16. Antón, I.M.; De la Fuente, M.A.; Sims, T.N.; Freeman, S.; Ramesh, N.; Hartwig, J.H.; Dustin, M.L.; Geha, R.S. WIP Deficiency Reveals a Differential Role for WIP and the Actin Cytoskeleton in T and B Cell Activation. *Immunity* **2002**, *16*, 193–204. [CrossRef]

17. Ho, H.Y.H.; Rohatgi, R.; Lebensohn, A.M.; Le, M.; Li, J.; Gygi, S.P.; Kirschner, M.W. Toca-1 Mediates Cdc42-Dependent Actin Nucleation by Activating the N-WASp-WIP Complex. *Cell* **2004**, *118*, 203–216. [CrossRef]

18. Martinez-Quiles, N.; Rohatgi, R.; Antón, I.M.; Medina, M.; Saville, S.P.; Miki, H.; Yamaguchi, H.; Takenawa, T.; Hartwig, J.H.; Geha, R.S.; et al. WIP Regulates N-WASp-Mediated Actin Polymerization and Filopodium Formation. *Nat. Cell Biol.* **2001**, *3*, 484–491. [CrossRef]

19. Calle, Y.; Antón, I.M.; Thrasher, A.J.; Jones, G.E. WASp and WIP Regulate Podosomes in Migrating Leukocytes. *J. Microsc.* **2008**, *231*, 494–505. [CrossRef]

20. De La Fuente, M.A.; Sasahara, Y.; Calamito, M.; Antón, I.M.; Elkhal, A.; Gallego, M.D.; Suresh, K.; Siminovitch, K.; Ochs, H.D.; Anderson, K.C.; et al. WIP Is a Chaperone for Wiskott-Aldrich Syndrome Protein. *Proc. Natl. Acad. Sci. USA* **2007**, *104*, 926–931. [CrossRef]

21. Sun, Y.; Leong, N.; Jiang, T.; Tangara, A.; Darzacq, X.; Drubin, D. Switch-like Arp2/3 Activation upon WASp and WIP Recruitment to an Apparent Threshold Level by Multivalent Linker Proteins in vivo. *Elife* **2017**, *6*, e29140. [CrossRef] [PubMed]

22. Massaad, M.J.; Oyoshi, M.K.; Kane, J.; Koduru, S.; Alcaide, P.; Nakamura, F.; Ramesh, N.; Luscinskas, F.W.; Hartwig, J.; Geha, R.S. Binding of WIP to Actin Is Essential for T Cell Actin Cytoskeleton Integrity and Tissue Homing. *Mol. Cell. Biol.* **2014**, *34*, 4343–4354. [CrossRef] [PubMed]

23. Keppler, S.J.; Burbage, M.; Gasparrini, F.; Hartjes, L.; Aggarwal, S.; Massaad, M.J.; Geha, R.S.; Bruckbauer, A.; Batista, F.D. The Lack of WIP Binding to Actin Results in Impaired B Cell Migration and Altered Humoral Immune Responses. *Cell Rep.* **2018**, *24*, 619–629. [CrossRef] [PubMed]

24. Franco, A.; Knafo, S.; Banon-Rodriguez, I.; Merino-Serrais, P.; Fernaud-Espinosa, I.; Nieto, M.; Garrido, J.J.; Esteban, J.A.; Wandosell, F.; Anton, I.M. WIP is a Negative Regulator of Neuronal Maturation and Synaptic Activity. *Cereb. Cortex* **2012**, *22*, 1191–1202. [CrossRef]

25. García, E.; Machesky, L.M.; Jones, G.E.; Antón, I.M. WIP Is Necessary for Matrix Invasion by Breast Cancer Cells. *Eur. J. Cell Biol.* **2014**, *93*, 413–423. [CrossRef] [PubMed]

26. Donnelly, S.K.; Weisswange, I.; Zettl, M.; Way, M. WIP Provides an Essential Link between Nck and N-WASp during Arp2/3-Dependent Actin Polymerization. *Curr. Biol.* **2013**, *23*, 999–1006. [CrossRef]

27. Escoll, M.; Gargini, R.; Cuadrado, A.; Antón, I.M.; Wandosell, F. Mutant P53 Oncogenic Functions in Cancer Stem Cells Are Regulated by WIP through YAP/TAZ. *Oncogene* **2017**, *36*, 3515–3527. [CrossRef]

28. Gargini, R.; Escoll, M.; García, E.; García-Escudero, R.; Wandosell, F.; Antón, I.M. WIP Drives Tumor Progression through YAP/TAZ-Dependent Autonomous Cell Growth. *Cell Rep.* **2016**, *17*, 1962–1977. [CrossRef]

29. Rivas, S.; Antón, I.M.; Wandosell, F. WIP-YAP/TAZ as a New pro-Oncogenic Pathway in Glioma. *Cancers (Basel)* **2018**, *10*, 191. [CrossRef]

30. Ramesh, N.; Massaad, M.J.; Kumar, L.; Koduru, S.; Sasahara, Y.; Anton, I.; Bhasin, M.; Libermann, T.; Geha, R. Binding of the WASp/N-WASp-Interacting Protein WIP to Actin Regulates Focal Adhesion Assembly and Adhesion. *Mol. Cell. Biol.* **2014**, *34*, 2600–2610. [CrossRef]

31. Dong, X.; Patino-Lopez, G.; Candotti, F.; Shaw, S. Structure-function analysis of WIP role in TCR-stimulated NFAT activation: Evidence that WIP/WASp dissociation is not required and that the WIP N-terminus is inhibitory. *J. Biol. Chem.* **2007**, *282*, 30303–30310. [CrossRef] [PubMed]

32. Scott, M.P.; Zappacosta, F.; Kim, E.Y.; Annan, R.S.; Miller, W.T. Identification of Novel SH3 Domain Ligands for the Src Family Kinase Hck: WASp, WIP, and ELMO1. *J. Biol. Chem.* **2002**, *277*, 28238–28246. [CrossRef]

33. Linkermann, A.; Gelhaus, C.; Lettau, M.; Qian, J.; Kabelitz, D.; Janssen, O. Identification of Interaction Partners for Individual SH3 Domains of Fas Ligand Associated Members of the PCH Protein Family in T Lymphocytes. *Biochim. Biophys. Acta—Proteins Proteom.* **2009**, *1794*, 168–176. [CrossRef]

34. Volkman, B.F.; Prehoda, K.E.; Scott, J.A.; Peterson, F.C.; Lim, W.A. Structure of the N-WASP EVH1 Domain-WIP Complex: Insight into the Molecular Basis of Wiskott-Aldrich Syndrome. *Cell* **2002**, *111*, 565–576. [CrossRef]

35. Eliezer, D. Biophysical Characterization of Intrinsically Disordered Proteins. *Curr. Opin. Struct. Biol.* **2009**, *19*, 23–30. [CrossRef] [PubMed]

36. Jensen, M.R.; Zweckstetter, M.; Huang, J.R.; Blackledge, M. Exploring Free-Energy Landscapes of Intrinsically Disordered Proteins at Atomic Resolution Using NMR Spectroscopy. *Chem. Rev.* **2014**, *114*, 6632–6660. [CrossRef]

37. Marsh, J.A.; Teichmann, S.A.; Forman-Kay, J.D. Probing the Diverse Landscape of Protein Flexibility and Binding. *Curr. Opin. Struct. Biol.* **2012**, *22*, 643–650. [CrossRef]

38. Uversky, V.N.; Dunker, A.K. Understanding Protein Non-Folding. *Biophys. Biochim. Act.* **2010**, *1804*, 1231–1264. [CrossRef]

39. Tompa, P. The Interplay between Structure and Function in Intrinsically Unstructured Proteins. *FEBS Lett.* **2005**, *579*, 3346–3354. [CrossRef]

40. Uversky, V.N. Intrinsic Disorder in Proteins Associated with Neurodegenerative Disease. *Front. Biosci.* **2009**, *14*, 5188–5238. [CrossRef]

41. Uyar, B.; Weatheritt, R.J.; Dinkel, H.; Davey, N.E.; Gibson, T.J. Proteome-Wide Analysis of Human Disease Mutations in Short Linear Motifs: Neglected Players in Cancer? *Mol. Biosyst.* **2014**, *10*, 2626–2642. [CrossRef]

42. Dunker, A.K.; Silman, I.; Uversky, V.N.; Sussman, J.L. Function and Structure of Inherently Disordered Proteins. *Curr. Opin. Struct. Biol.* **2008**, *18*, 756–764. [CrossRef]

43. Rezaei-Ghaleh, N.; Blackledge, M.; Zweckstetter, M. Intrinsically Disordered Proteins: From Sequence and Conformational Properties toward Drug Discovery. *ChemBioChem* **2012**, *13*, 930–950. [CrossRef] [PubMed]

44. Bah, A.; Forman-Kay, J.D. Modulation of Intrinsically Disordered Protein Function by Post-Translational Modifications. *J. Biol. Chem.* **2016**, *291*, 6696–6705. [CrossRef] [PubMed]

45. Schramm, A.; Bignon, C.; Brocca, S.; Grandori, R.; Santambrogio, C.; Longhi, S. An Arsenal of Methods for the Experimental Characterization of Intrinsically Disordered Proteins – How to Choose and Combine Them? *Arch. Biochem. Biophys.* **2019**, *676*, 108055. [CrossRef] [PubMed]

46. Selenko, P. Quo Vadis Biomolecular NMR Spectroscopy? *Int. J. Mol. Sci.* **2019**, *20*, 1278. [CrossRef]

47. Milles, S.; Salvi, N.; Blackledge, M.; Jensen, M.R. Characterization of Intrinsically Disordered Proteins and Their Dynamic Complexes: From in Vitro to Cell-like Environments. *Prog. Nucl. Magn. Reson. Spectrosc.* **2018**, *109*, 79–100. [CrossRef] [PubMed]

48. Gomes, G.N.; Gradinaru, C.C. Insights into the Conformations and Dynamics of Intrinsically Disordered Proteins Using Single-Molecule Fluorescence. *Biochim. Biophys. Acta—Proteins Proteom.* **2017**, *1865*, 1696–1706. [CrossRef] [PubMed]

49. Best, R.B. Computational and Theoretical Advances in Studies of Intrinsically Disordered Proteins. *Curr. Opin. Struct. Biol.* **2017**, *42*, 147–154. [CrossRef] [PubMed]

50. Kachala, M.; Valentini, E.; Svergun, D.I. Application of SAXS for the Structural Characterization of IDPs. *Adv. Exp. Med. Biol.* **2015**, *870*, 261–289.

51. Schuler, B.; Soranno, A.; Hofmann, H.; Nettels, D. Single-Molecule FRET Spectroscopy and the Polymer Physics of Unfolded and Intrinsically Disordered Proteins. *Annu. Rev. Biophys.* **2016**, *45*, 207–231. [CrossRef] [PubMed]

52. Stuchfield, D.; Barran, P. Unique Insights to Intrinsically Disordered Proteins Provided by Ion Mobility Mass Spectrometry. *Curr. Opin. Chem. Biol.* **2018**, *42*, 177–185. [CrossRef]

53. Ma, B.; Kumar, S.; Tsai, C.; Nussinov, R. Folding Funnels and Binding Mechanisms. *Protein Eng.* **1999**, *12*, 713–720. [CrossRef]

54. Koshland, D.E. Application of a Theory of Enzyme Specificity to Protein Synthesis. *Proc. Natl. Acad. Sci. USA* **1958**, *44*, 98–104. [CrossRef]

55. Fuxreiter, M.; Tompa, P. Fuzzy complexes: A more stochastic view of protein function. *Adv. Exp. Med. Biol.* **2012**, *725*, 1–14.

56. Csermely, P.; Palotai, R.; Nussinov, R. Induced Fit, Conformational Selection and Independent Dynamic Segments: An Extended View of Binding Events. *Trends Biochem. Sci.* **2010**, *35*, 539–546. [CrossRef]

57. Olsen, J.G.; Teilum, K.; Kragelund, B.B. Behaviour of Intrinsically Disordered Proteins in Protein–Protein Complexes with an Emphasis on Fuzziness. *Cell. Mol. Life Sci.* **2017**, *74*, 3175–3183. [CrossRef]

58. Alanis-Lobato, G.; Andrade-Navarro, M.A.; Schaefer, M.H. HIPPIE v2.0: Enhancing Meaningfulness and Reliability of Protein-Protein Interaction Networks. *Nucleic Acids Res.* **2017**, *45*, D408–D414. [CrossRef] [PubMed]

59. Roberto Dominguez and Kenneth, C. Holmes. Actin Structure & Function. *Annu Rev. Biophys.* **2011**, *45*, 169–186.

60. Lassing, I.; Schmitzberger, F.; Björnstedt, M.; Holmgren, A.; Nordlund, P.; Schutt, C.E.; Lindberg, U. Molecular and Structural Basis for Redox Regulation of β-Actin. *J. Mol. Biol.* **2007**, *370*, 331–348. [CrossRef] [PubMed]

61. Oda, T.; Iwasa, M.; Aihara, T.; Maéda, Y.; Narita, A. The Nature of the Globular- to Fibrous-Actin Transition. *Nature* **2009**, *457*, 441–445. [CrossRef] [PubMed]

62. Dominguez, R. Actin-Binding Proteins—A Unifying Hypothesis. *Trends Biochem. Sci.* **2004**, *29*, 572–578. [CrossRef]

63. Otterbein, L.R.; Graceffa, P.; Dominguez, R. The Crystal Structure of Uncomplexed Actin in the ADP State. *Science* **2001**, *293*, 708–711. [CrossRef]

64. Durer, Z.A.O.; Kudryashov, D.S.; Sawaya, M.R.; Altenbach, C.; Hubbell, W.; Reisler, E. Structural States and Dynamics of the D-Loop in Actin. *Biophys. J.* **2012**, *103*, 930–939. [CrossRef] [PubMed]

65. Fixe, P. Actin-Binding Proteins (ABPs) review. 2014. Available online: https://www.tebu-bio.com/blog/2014/06/12/actin-binding-proteins-abps-review/ (accessed on 17 July 2020).

66. Paavilainen, V.O.; Bertling, E.; Falck, S.; Lappalainen, P. Regulation of Cytoskeletal Dynamics by Actin-Monomer-Binding Proteins. *Trends Cell Biol.* **2004**, *14*, 386–394. [CrossRef] [PubMed]

67. Edwards, J.; Lappalainen, P.; Mattila, P. Are β-Thymosins WH2 Domains? *FEBS Lett.* **2004**, *573*, 231–232. [CrossRef]

68. Chereau, D.; Kerff, F.; Graceffa, P.; Grabarek, Z.; Langsetmo, K.; Dominguez, R. Actin-Bound Structures of Wiskott-Aldrich Syndrome Protein (WASp)-Homology Domain 2 and the Implications for Filament Assembly. *Proc. Natl. Acad. Sci. USA* **2005**, *102*, 16644–16649. [CrossRef]

69. Aguda, A.H.; Xue, B.; Irobi, E.; Préat, T.; Robinson, R.C. The Structural Basis of Actin Interaction with Multiple WH2/β-Thymosin Motif-Containing Proteins. *Structure* **2006**, *14*, 469–476. [CrossRef]

70. Paunola, E.; Mattila, P.K.; Lappalainen, P. WH2 Domain: A Small, Versatile Adapter for Actin Monomers. *FEBS Lett.* **2002**, *513*, 92–97. [CrossRef]

71. Husson, C.; Cantrelle, F.X.; Roblin, P.; Didry, D.; Le, K.H.D.; Perez, J.; Guittet, E.; Van Heijenoort, C.; Renault, L.; Carlier, M.F. Multifunctionality of the β-Thymosin/WH2 Module: G-Actin Sequestration, Actin Filament Growth, Nucleation, and Severing. *Ann. N. Y. Acad. Sci.* **2010**, *1194*, 44–52. [CrossRef]

72. Elazari-Shalom, H.; Shaked, H.; Esteban-Martin, S.; Salvatella, X.; Barda-Saad, M.; Chill, J.H. New Insights into the Role of the Disordered WIP N-Terminal Domain Revealed by NMR Structural Characterization. *FEBS J.* **2015**, *282*, 700–714. [CrossRef] [PubMed]

73. Rozentur-Shkop, E.; Goobes, G.; Chill, J.H. A J-Modulated Protonless NMR Experiment Characterizes the Conformational Ensemble of the Intrinsically Disordered Protein WIP. *J. Biomol. NMR* **2016**, *66*, 243–257. [CrossRef] [PubMed]

74. García, E.; Ragazzini, C.; Yu, X.; Cuesta-García, E.; Bernardino De La Serna, J.; Zech, T.; Sarrió, D.; Machesky, L.M.; Antón, I.M. WIP and WICH/WIRE Co-Ordinately Control Invadopodium Formation and Maturation in Human Breast Cancer Cell Invasion. *Sci. Rep.* **2016**, *6*, 23590. [CrossRef] [PubMed]

75. Van Audenhove, I.; Boucherie, C.; Pieters, L.; Zwaenepoel, O.; Vanloo, B.; Martens, E.; Verbrugge, C.; Hassanzadeh-Ghassabeh, G.; Vandekerckhove, J.; Cornelissen, M.; et al. Stratifying Fascin and Cortactin Function in Invadopodium Formation Using Inhibitory Nanobodies and Targeted Subcellular Delocalization. *FASEB J.* **2014**, *28*, 1805–1818. [CrossRef]

76. Bañón-Rodríguez, I.; Monypenny, J.; Ragazzini, C.; Franco, A.; Calle, Y.; Jones, G.E.; Antón, I.M. The Cortactin-Binding Domain of WIP Is Essential for Podosome Formation and Extracellular Matrix Degradation by Murine Dendritic Cells. *Eur. J. Cell Biol.* **2011**, *90*, 213–223. [CrossRef] [PubMed]

77. Yu, H.; Chen, J.K.; Feng, S.; Dalgarno, D.C.; Brauer, A.W.; Schreiber, S.L. Structural Basis for the Binding of Proline-Rich Peptides to SH3 Domains. *Cell* **1994**, *76*, 933–945. [CrossRef]

78. Kurochkina, N.; Guha, U. SH3 Domains: Modules of Protein-Protein Interactions. *Biophys. Rev.* **2013**, *5*, 29–39. [CrossRef] [PubMed]

79. Saksela, K.; Permi, P. SH3 Domain Ligand Binding: What's the Consensus and Where's the Specificity? *FEBS Lett.* **2012**, *586*, 2609–2614. [CrossRef]

80. Kay, B.K.; Williamson, M.P.; Sudol, M. The Importance of Being Proline: The Interaction of Proline-rich Motifs in Signaling Proteins with Their Cognate Domains. *FASEB J.* **2000**, *14*, 231–241. [CrossRef]

81. Lee, C.H.; Leung, B.; Lemmon, M.A.; Zheng, J.; Cowburn, D.; Kuriyan, J.; Saksela, K. A Single Amino Acid in the SH3 Domain of Hck Determines Its High Affinity and Specificity in Binding to HIV-1 Nef Protein. *EMBO J.* **1995**, *14*, 5006–5015. [CrossRef] [PubMed]

82. Li, S.S.C. Specificity and Versatility of SH3 and Other Proline-Recognition Domains: Structural Basis and Implications for Cellular Signal Transduction. *Biochem. J.* **2005**, *390*, 641–653. [CrossRef] [PubMed]

83. Dalgarno, D.C.; Botfield, M.C.; Rickles, R.J. SH3 Domains and Drug Design: Ligands, Structure, and Biological Function. *Biopolymers* **1997**, *43*, 383–400. [CrossRef]

84. Mayer, B.J. SH3 Domains: Complexity in Moderation. *J. Cell. Sci.* **2001**, *114*, 1253–1263. [PubMed]

85. Zarrinpar, A.; Bhattacharyya, R.P.; Lim, W.A. The Structure and Function of Proline Recognition Domains. *Sci. STKE* **2003**, *2003*, 1–10. [CrossRef]

86. Teyra, J.; Huang, H.; Jain, S.; Guan, X.; Dong, A.; Liu, Y.; Tempel, W.; Min, J.; Tong, Y.; Kim, P.M.; et al. Comprehensive Analysis of the Human SH3 Domain Family Reveals a Wide Variety of Non-Canonical Specificities. *Structure* **2017**, *25*, 1598–1610. [CrossRef]

87. Twafra, S.; Gil-Henn, H.; Dessau, M. PDB ID: 5NVJ. Available online: https://pdbj.org/emnavi/quick.php?id=pdb-5nvj (accessed on 21 July 2020).

88. Schmidt, H.; Hoffmann, S.; Tran, T.; Stoldt, M.; Stangler, T.; Wiesehan, K.; Willbold, D. Solution Structure of a Hck SH3 Domain Ligand Complex Reveals Novel Interaction Modes. *J. Mol. Biol.* **2007**, *365*, 1517–1532. [CrossRef]

89. Hake, M.J.; Choowongkomon, K.; Kostenko, O.; Carlin, C.R.; Sönnichsen, F.D. Specificity Determinants of a Novel Nck Interaction with the Juxtamembrane Domain of the Epidermal Growth Factor Receptor. *Biochemistry* **2008**, *47*, 3096–3108. [CrossRef]

90. Jankowski, W.; Saleh, T.; Pai, M.T.; Sriram, G.; Birge, R.B.; Kalodimos, C.G. Domain Organization Differences Explain Bcr-Abl's Preference for CrkL over CrkII. *Nat. Chem. Biol.* **2012**, *8*, 590–596. [CrossRef]

91. Kessels, M.M.; Engqvist-Goldstein, Å.E.Y.; Drubin, D.G. Association of Mouse Actin-Binding Protein 1 (MAbp1/SH3P7), an Src Kinase Target, with Dynamic Regions of the Cortical Actin Cytoskeleton in Response to Rac1 Activation. *Mol. Biol. Cell* **2000**, *11*, 393–412. [CrossRef]

92. Kinley, A.W.; Weed, S.A.; Weaver, A.M.; Karginov, A.V.; Bissonette, E.; Cooper, J.A.; Parsons, J.T. Cortactin Interacts with WIP in Regulating Arp2/3 Activation and Membrane Protrusion. *Curr. Biol.* **2003**, *13*, 384–393. [CrossRef]

93. Cortesio, C.L.; Perrin, B.J.; Bennin, D.A.; Huttenlocher, A. Actin-Binding Protein-1 Interacts with WASp-Interacting Protein to Regulate Growth Factor-Induced Dorsal Ruffle Formation. *Mol. Biol. Cell* **2010**, *21*, 186–197. [CrossRef] [PubMed]

94. Klos Dehring, D.A.; Clarke, F.; Ricart, B.G.; Huang, Y.; Gomez, T.S.; Williamson, E.K.; Hammer, D.A.; Billadeau, D.D.; Argon, Y.; Burkhardt, J.K. Hematopoietic Lineage Cell-Specific Protein 1 Functions in Concert with the Wiskott–Aldrich Syndrome Protein To Promote Podosome Array Organization and Chemotaxis in Dendritic Cells. *J. Immunol.* **2011**, *186*, 4805–4818. [CrossRef] [PubMed]

95. Gryaznova, T.; Kropyvko, S.; Burdyniuk, M.; Gubar, O.; Kryklyva, V.; Tsyba, L.; Rynditch, A. Intersectin Adaptor Proteins Are Associated with Actin-Regulating Protein WIP in Invadopodia. *Cell. Signal.* **2015**, *27*, 1499–1508. [CrossRef] [PubMed]

96. Wong, K.A.; Wilson, J.; Russo, A.; Wang, L.; Okur, M.N.; Wang, X.; Martin, N.P.; Scappini, E.; Carnegie, G.K.; O'Bryan, J.P. Intersectin (ITSN) Family of Scaffolds Function as Molecular Hubs in Protein Interaction Networks. *PLoS ONE* **2012**, *7*, 1–10. [CrossRef] [PubMed]

97. Sattler, M.; Salgia, R. Hematopoietic and BCR/ABL-Transformed Cells. *Leukemia* **1998**, *12*, 637–644. [CrossRef] [PubMed]

98. Antón, I.M.; Lu, W.; Mayer, B.J.; Ramesh, N.; Geha, R.S. The Wiskott-Aldrich Syndrome Protein-Interacting Protein (WIP) Binds to the Adaptor Protein Nck. *J. Biol. Chem.* **1998**, *273*, 20992–20995. [CrossRef]

99. Zhao, Z.-S.; Manser, E.; Lim, L. Interaction between PAK and Nck: A Template for Nck Targets and Role of PAK Autophosphorylation. *Mol. Cell. Biol.* **2000**, *20*, 3906–3917. [CrossRef]

100. Gryaznova, T.; Gubar, O.; Burdyniuk, M.; Kropyvko, S.; Rynditch, A. WIP/ITSN1 Complex Is Involved in Cellular Vesicle Trafficking and Formation of Filopodia-like Protrusions. *Gene* **2018**, *674*, 49–56. [CrossRef]

101. Derry, J.M.; Ochs, H.D.; Francke, U. Isolation of a Novel Gene Mutated in Wiskott-Aldrich Syndrome. *Cell* **1994**, *79*, 635–644. [CrossRef]

102. Schindelhauer, D.; Weiss, M.; Hellebrand, H.; Golla, A.; Hergersberg, M.; Seger, R.; Belohradsky, B.H.; Meindl, A. Wiskott-Aldrich Syndrome: No Strict Genotype-Phenotype Correlations but Clustering of Missense Mutations in the Amino-Terminal Part of the WASP Gene Product. *Hum. Genet.* **1996**, *98*, 68–76. [CrossRef]

103. Stewart, D.M.; Tian, L.; Nelson, D.L. Mutations That Cause the Wiskott-Aldrich Syndrome Impair the Interaction of Wiskott-Aldrich Syndrome Protein (WASp) with WASp Interacting Protein. *J. Immunol.* **1999**, *162*, 5019–5024. [PubMed]

104. Peterson, F.C.; Deng, Q.; Zettl, M.; Prehoda, K.E.; Lim, W.A.; Way, M.; Volkman, B.F. Multiple WASp-Interacting Protein Recognition Motifs Are Required for a Functional Interaction with N-WASp. *J. Biol. Chem.* **2007**, *282*, 8446–8453. [CrossRef] [PubMed]

105. Zettl, M.; Way, M. The WH1 and EVH1 Domains of WASp and Ena/VASP Family Members Bind Distinct Sequence Motifs. *Curr. Biol.* **2002**, *12*, 1617–1622. [CrossRef]

106. Haba, N.Y.; Gross, R.; Novacek, J.; Shaked, H.; Zidek, L.; Barda-Saad, M.; Chill, J.H. NMR Determines Transient Structure and Dynamics in the Disordered C-Terminal Domain of WASp Interacting Protein. *Biophys. J.* **2013**, *105*, 481–493. [CrossRef] [PubMed]

107. Halle-Bikovski, A.; Fried, S.; Rozentur-Shkop, E.; Biber, G.; Shaked, H.; Joseph, N.; Barda-Saad, M.; Chill, J.H. New Structural Insights into Formation of the Key Actin Regulating WIP-WASp Complex Determined by NMR and Molecular Imaging. *ACS Chem. Biol.* **2018**, *13*, 100–109. [CrossRef]

108. Ferguson, K.M.; Lemmon, M.A.; Schlessinger, J.; Sigler, P.B. Structure of the High Affinity Complex of Inositol Trisphosphate with a Phospholipase C Pleckstrin Homology Domain. *Cell* **1995**, *83*, 1037–1046. [CrossRef]

109. Li, S.C.; Zwahlen, C.; Vincent, S.J.F.; Jane McGlade, C.; Kay, L.E.; Pawson, T.; Forman-Kay, J.D. Structure of a Numb PTB Domain-Peptide Complex Suggests a Basis for Diverse Binding Specificity. *Nat. Struct. Biol.* **1998**, *5*, 1075–1083. [CrossRef]

110. Luthi, J.N.; Gandhi, M.J.; Drachman, J.G. X-Linked Thrombocytopenia Caused by a Mutation in the Wiskott-Aldrich Syndrome (WAS) Gene That Disrupts Interaction with the WAS Protein (WASp)-Interacting Protein (WIP). *Exp. Hematol.* **2003**, *31*, 150–158. [CrossRef]

111. Reicher, B.; Joseph, N.; David, A.; Pauker, M.H.; Perl, O.; Barda-Saad, M. Ubiquitylation-Dependent Negative Regulation of WASp Is Essential for Actin Cytoskeleton Dynamics. *Mol. Cell. Biol.* **2012**, *32*, 3153–3163. [CrossRef]

112. Fried, S.; Reicher, B.; Pauker, M.H.; Eliyahu, S.; Matalon, O.; Noy, E.; Chill, J.; Barda-Saad, M. Triple-Color FRET Analysis Reveals Conformational Changes in the WIP-WASp Actin-Regulating Complex. *Sci. Signal.* **2014**, *7*, ra60. [CrossRef]

113. Peterson, F.C.; Volkman, B.F. Diversity of Polyproline Recognition by EVH1 Domains. *Front. Biosci.* **2009**, *14*, 833–846. [CrossRef] [PubMed]

114. Vijayakumar, V.; Monypenny, J.; Chen, X.J.; Machesky, L.M.; Lilla, S.; Thrasher, A.J.; Antón, I.M.; Calle, Y.; Jones, G.E. Tyrosine Phosphorylation of WIP Releases Bound WASp and Impairs Podosome Assembly in Macrophages. *J. Cell Sci.* **2015**, *128*, 251–265. [CrossRef] [PubMed]

MDPI

St. Alban-Anlage 66

4052 Basel

Switzerland

Tel. +41 61 683 77 34

Fax +41 61 302 89 18

www.mdpi.com

Biomolecules Editorial Office

E-mail: biomolecules@mdpi.com

www.mdpi.com/journal/biomolecules